Regional Climate Change and Application of Remote Sensing

Regional Climate Change and Application of Remote Sensing

Jun Qin
Hou Jiang

MDPI

Basel • Beijing • Wuhan • Barcelona • Belgrade • Novi Sad • Cluj • Manchester

Editors

Jun Qin
State Key Laboratory of
Resources and Environmental
Information System
Chinese Academy of Sciences
Beijing
China

Hou Jiang
State Key Laboratory of
Resources and Environmental
Information System
Chinese Academy of Sciences
Beijing
China

Editorial Office
MDPI AG
Grosspeteranlage 5
4052 Basel, Switzerland

This is a reprint of articles from the Special Issue published online in the open access journal *Sustainability* (ISSN 2071-1050) (available at: www.mdpi.com/journal/sustainability/special_issues/climate_remote_sus).

For citation purposes, cite each article independently as indicated on the article page online and as indicated below:

Lastname, A.A.; Lastname, B.B. Article Title. *Journal Name* **Year**, *Volume Number*, Page Range.

ISBN 978-3-7258-2610-0 (Hbk)
ISBN 978-3-7258-2609-4 (PDF)
doi.org/10.3390/books978-3-7258-2609-4

Contents

About the Editors

Jun Qin

Dr. Jun Qin, born in 1978 in Sichuan, China, is a distinguished professor and doctoral advisor at the Institute of Geographic Sciences and Natural Resources Research, Chinese Academy of Sciences. His research primarily focuses on data assimilation, quantitative remote sensing, and climate change. Over his career, Dr. Qin has published more than 70 SCI-indexed papers, which have received over 6,000 citations, with his research being referenced by leading journals such as *Science* and *Nature*. He holds one authorized invention patent.

Dr. Qin's research has led to several groundbreaking achievements. He independently developed a particle-filter-based satellite data assimilation algorithm, enhancing soil moisture estimation accuracy. He has also contributed to the creation of an operational variational land data assimilation system, which has produced a decade-long soil moisture dataset for the Tibetan Plateau. Additionally, Dr. Qin played a key role in establishing a multi-scale soil temperature and moisture observation network in Naqu, Tibet, facilitating soil moisture validation for satellite missions through a novel spatial upscaling algorithm. Furthermore, he has advanced the understanding of climate variability on the Tibetan Plateau by integrating high-resolution satellite and station data to analyze temperature and atmospheric water content trends across varying altitudes, addressing data limitations at elevations above 5,000 meters.

Hou Jiang

Dr. Hou Jiang, born in 1993 in Enshi, Hubei, is an Associate Research Fellow at the State Key Laboratory of Resources and Environmental Information System, Institute of Geographic Sciences and Natural Resources Research, Chinese Academy of Sciences. He holds a Ph.D. in Cartography and Geographic Information Systems from the University of Chinese Academy of Sciences. His research interests encompass remote sensing information analysis and applications, renewable energy, deep learning, and sustainable development.

Dr. Jiang has made significant contributions to the convergence of renewable energy research and remote sensing applications. Over the past five years, he has published 35 papers, including 28 SCI-indexed papers, with 15 as the first author in top journals such as *Renewable & Sustainable Energy Reviews* and *Remote Sensing of Environment*. His work has garnered an H-index of 15 with over 600 citations.

He has led several major projects, including those funded by the National Natural Science Foundation of China, the China Postdoctoral Science Foundation, and the National Key Laboratory of Remote Sensing Science. Dr. Jiang has also co-authored two books and has been recognized with multiple awards, including the Special and First-Class Awards for Scientific and Technological Progress in Geographic Information, the Li Xiaowen Young Scientist Award in Remote Sensing, and the Chinese Academy of Sciences President's Award.

Preface

This Reprint focuses on leveraging remote sensing technology to address regional challenges associated with climate change from a sustainable development perspective. Climate change is reshaping ecosystems, economies, and social systems worldwide, prompting the urgent need for innovative solutions that can mitigate its adverse impacts. Remote sensing, with its ability to provide detailed, scalable, and continuous monitoring of the Earth's surface, has proven indispensable in advancing our understanding of climate phenomena and supporting regionally targeted adaptation and mitigation efforts.

The scope of this Reprint spans a diverse array of topics, including satellite-based monitoring of extreme weather events, the analysis of biological responses to climate shifts, assessments of climate-induced losses, and the evaluation of nature's and humanity's adaptive capacity to changing conditions. By offering a comprehensive look at the various applications of remote sensing in climate-related issues, this Reprint aims to serve as a resource for researchers, policymakers, and practitioners looking to gain insights into regional climate adaptation strategies and sustainable practices.

This Reprint was driven by the need to consolidate pioneering studies and technological advancements that demonstrate remote sensing's role in understanding and addressing climate change. Our motivation stemmed from the recognition that although climate impacts are global, their manifestations and effective countermeasures often vary significantly at regional levels. Thus, highlighting innovative remote sensing applications tailored to specific regional contexts can foster more effective, localized strategies to counter climate change.

We hope that this Reprint will be valuable to the scientific community, including researchers, students, and industry professionals involved in remote sensing, climate science, and sustainable development. Our intended audience also includes government and policy stakeholders who can benefit from the insights provided by these studies in making data-driven decisions.

This work represents the collaborative effort of numerous authors whose dedication and expertise have been instrumental in compiling this collection. We extend our sincere gratitude to each contributing author for their invaluable work and insights. Additionally, we acknowledge the support of the reviewers, technical editors, and others who provided constructive feedback throughout the editorial process, ensuring that this Reprint is of a high quality and relevant.

Our deepest thanks go to all those who have contributed to this project and to the institutions that supported our authors in their research efforts. This Reprint is a testament to the collective efforts within the scientific community to advancing knowledge and finding solutions in the face of one of the most pressing challenges of our time—climate change.

Jun Qin and Hou Jiang
Editors

sustainability

Article

Assessing Carbon Reduction Potential of Rooftop PV in China through Remote Sensing Data-Driven Simulations

Hou Jiang [1], Ning Lu [1,*] and Xuecheng Wang [2]

1 State Key Laboratory of Resources and Environmental Information System, Institute of Geographic Sciences and Natural Resources Research, Chinese Academy of Sciences, Beijing 100101, China
2 School of Geography and Planning, Nanning Normal University, Nanning 530001, China
* Correspondence: lvn@lreis.ac.cn; Tel.: +86-10-6488-9981

Abstract: Developing rooftop photovoltaic (PV) has become an important initiative for achieving carbon neutrality in China, but the carbon reduction potential assessment has not properly considered the spatial and temporal variability of PV generation and the curtailment in electricity dispatch. In this study, we propose a technical framework to fill the gap in assessing carbon reduction potential through remote sensing data-driven simulations. The spatio-temporal variations in rooftop PV generations were simulated on an hourly basis, and a dispatch analysis was then performed in combination with hourly load profiles to quantify the PV curtailment in different scenarios. Our results showed that the total rooftop PV potential in China reached 6.5 PWh yr^{-1}, mainly concentrated in the eastern region where PV generation showed high variability. The carbon reduction from 100% flexible grids with 12 h of storage capacity is close to the theoretical maximum, while without storage, the potential may be halved. To maximize the carbon reduction potential, rooftop PV development should consider grid characteristics and regional differences. This study has important implications for the development of rooftop PV and the design of carbon-neutral pathways based on it.

Keywords: rooftop PV; carbon emission reduction; dispatch modeling; remote sensing

Citation: Jiang, H.; Lu, N.; Wang, X. Assessing Carbon Reduction Potential of Rooftop PV in China through Remote Sensing Data-Driven Simulations. *Sustainability* **2023**, *15*, 3380. https://doi.org/10.3390/su15043380

Academic Editor: Grigorios L. Kyriakopoulos

Received: 9 January 2023
Revised: 8 February 2023
Accepted: 9 February 2023
Published: 13 February 2023

1. Introduction

Solar photovoltaics (PV) has become an important pathway for achieving carbon emission reduction around the world [1,2]. Globally installed PV capacity has grown more than eightfold in the last 10 years, providing about 3.6% of the world's total electricity consumption in 2021 [3]. Various forms of PVs are proposed for extensive and widespread development, such as floating PV [4], agricultural PV [5], building-integrated PV [6], rooftop PV [7], etc. Compared to ground-mounted PVs, rooftop PV has unique advantages, such as proximity to consumers [8] and no need for additional land [9], making it highly favored in recent years. In China, the government is implementing a county-based strategy to promote rooftop PV development to reduce carbon emissions [10].

Several studies have proposed methods to assess rooftop PV potential, which can be broadly classified into geographic information system (GIS)-based methods and remote sensing (RS)-based methods [11]. GIS-based methods are used for energy planning at city scales that involve building structures (e.g., roof slope and façade orientation) [12] and building interrelationships (e.g., shading and lighting) [13]. They are typically based on a series of spatial or non-spatial data on available solar radiation and building features, and combine GIS technology with machine learning, physical modeling, geostatistics, and sampling methods for analysis [14–16]. GIS technology plays the role in capturing, storing, manipulating, analyzing, managing, and presenting all types of data. For example, Bergamasco and Asinari [17] proposed a method integrating GIS and solar radiation maps to estimate the power generation of rooftop PV in Piedmont, northwestern Italy. Assouline et al. [18] combined support vector regression and GIS to estimate the electricity generation potential of rooftop PV in Switzerland. GIS-based methods are usually suitable for fine-scale spatio-temporal assessments, and the results can be employed to design effective policies for rooftop PV development in built environments [11]. However, the intensive computational demand is a main obstacle to their application on a large scale [13,16].

The typical RS-based approach integrates multi-source remote sensing data to assess the regional potential of rooftop PV, and the main work usually consists of two aspects, namely, building footprint extraction and solar resource estimation [7,19]. Mapping the building footprints from very high-resolution images has attracted a lot of attention because high-resolution images contain more spatial-detailed contexts on ground objects [20]. For example, Guo et al. [21] proposed a coarse-to-fine boundary refinement network to extract building footprints from aerial imagery and addressed the challenge of extracting sharp building boundaries caused by obstructions from nearby shadows or trees, diversity of roof shapes, and variation in building scales. Guo et al. [22] further developed a model for automatic building footprint updates using bi-temporal remote sensing images. Regarding solar resource estimation, geostationary meteorological satellites are widely used to retrieve the total solar radiation and the direct/diffuse fraction at high spatial and temporal resolutions [23–25]. For example, Jiang et al. [24] introduced a convolutional neural network to extract spatial patterns from satellite imagery to deal with the spatial proximity effects in solar radiation inversion. Li et al. [26] applied transfer learning for global estimates of surface solar radiation, which combines the advantage of radiative transfer simulations and ground measurements. Compared to GIS-based methods, remote sensing makes large-scale assessment a reality, and the integration of deep learning significantly improves computational efficiency [7,11]. Therefore, RS-based methods are typically applied to large-scale resource estimation and spatial planning but are not applicable to the design and integration of individual rooftop PV systems [19,27].

In parallel, light detection and ranging (LiDAR) technology has contributed to the accurate simulation of PV electricity generations at urban scales [28,29]. The LiDAR-based method allows for consideration of the shading effects of local topography on PV generation. LiDAR can be combined with GIS tools to enrich the services, such as optimizing the PV installations, balance-of-system costs, return on investment, payback time, and potential carbon reductions [12,14]. For example, Jacques et al. [30] presented a methodology that combines roof segmentation algorithms with LiDAR data to estimate potential PV capacity for buildings in Leeds, UK. Gagnon et al. [31] integrated GIS with a regression statistical approach and LiDAR dataset to determine the potential electricity generation of rooftop PV across the United States. This kind of solution possesses the advantages of both RS-based and GIS-based ones, while the high cost of LiDAR is the main reason why they are not yet widely used [28,31].

In contrast to the refined assessment of PV power generation potential, the estimation of PV carbon reduction capacity is relatively crude [32]. Typically, the estimated potential is regarded as the activity level and multiplied by the grid emission factors published by government departments to calculate carbon reduction potential [32,33]. There are two main problems associated with such an approach. First, the fine-scale spatial and temporal variability of PV power generation is not considered. The variability leads to a mismatch between PV generation and user-side demand, and thus, a portion of PV electricity is to be curtailed during dispatching; that is, not all PV electricity can be delivered to the grid and then consumed by end users [34,35]. Second, the impact of the grid's own characteristics is ignored. It is known that the grid's ability to absorb variable generations varies with different system flexibility and energy storage capacity [36,37]. In addition, since clean energy already exists in the grid [2,38], it is unlikely that one unit of rooftop PV electricity will replace an equivalent amount of power in the current grid.

In this study, we attempted to solve the above problems by simulating fine-scale variability in rooftop PV power generation using multi-source remote sensing data and performing dispatch analysis by combining hourly PV generation with the hourly customer-side load. In this way, the portion of rooftop PV generation that was actually consumed was calculated to correct for the overestimation of carbon reduction. The main contributions of this study embody three aspects:

(1) The high-resolution mapping of the distribution of China's rooftop PV potential. An empirical relationship was established to estimate the rooftop area from the settlement area. Multi-source remote sensing data were integrated to simulate the spatio-temporal variation in rooftop PV electricity generation.

(2) The technical framework for calibrating the overestimation of carbon emission reduction. Rooftop PV generation curves and customer-side load profiles were combined to obtain the PV curtailment rates during electricity dispatch. Current PV penetration rates were used to calibrate the grid emission factors for PV-specific ones.

(3) The quantification of carbon reductions in China's rooftop PV. We designed twelve scenarios with 80%, 90%, and 100% flexibility and 0, 4, 8, and 12 h of storage capacity to reflect the differences in the grid's ability to absorb intermittent PV electricity.

The paper is organized as follows. Section 2 describes the remote sensing data, including settlement footprints, building footprints, solar radiation, and air temperature, as well as the models used for PV generation simulation and dispatch analysis. Section 3 shows the results for rooftop PV potential and carbon reduction potential. Further discussion is presented in Section 4, and conclusions are drawn in Section 5.

2. Materials and Methods

The workflow for assessing the carbon reduction potential of rooftop PV through remote sensing data-driven simulations is shown in Figure 1.

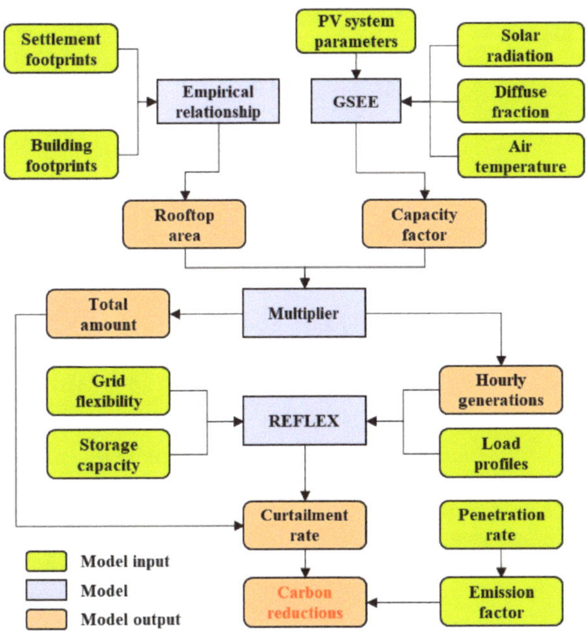

Figure 1. Workflow for assessing the carbon reduction potential of rooftop PV.

First, an empirical relationship between settlement footprints and building footprints was built to map the national rooftop area and its distribution. Second, remote sensing data on total solar radiation, diffuse fraction, and air temperature, as well as PV system parameters, were fed into the Global Solar Energy Estimator (GSEE) [39] to simulate the PV system's electricity generation efficiency, which is measured by the capacity factor (CF) defined as the ratio of a PV system's actual output over a given period to the maximum output under standard test conditions over that period. Third, the Renewable Energy Flexibility (REFLEX) [40] model was used for dispatch modeling based on hourly PV generations and load profiles, given grid flexibility, storage capacity, and other settings. Finally, potential carbon reductions were calculated according to the simulated curtailment rate, the total amount of rooftop PV generation and known penetration rates, and emission factors.

2.1. Estimation of Rooftop Resources

Accurate surveys of rooftop area rely on very high resolution (e.g., WorldView, GeoEye and Pleiades satellites) and LiDAR remote sensing [7,20,41]. However, this approach faces challenges of high cost and low efficiency when applied over large areas. Therefore, it is impractical to conduct rooftop surveys over the 9.6 million square kilometers of China's territory. In contrast, extracting settlement footprints from moderate- or high-resolution remote sensing is more cost-effective, and settlement footprint products with global coverage are already available [42,43]. Figure 2a shows the settlement area in China at 500 m resolution, which is aggregated from the world settlement footprint products [43]. These settlement footprints indicate the extent to which the buildings locate and can be used as a reference for rooftop area estimation.

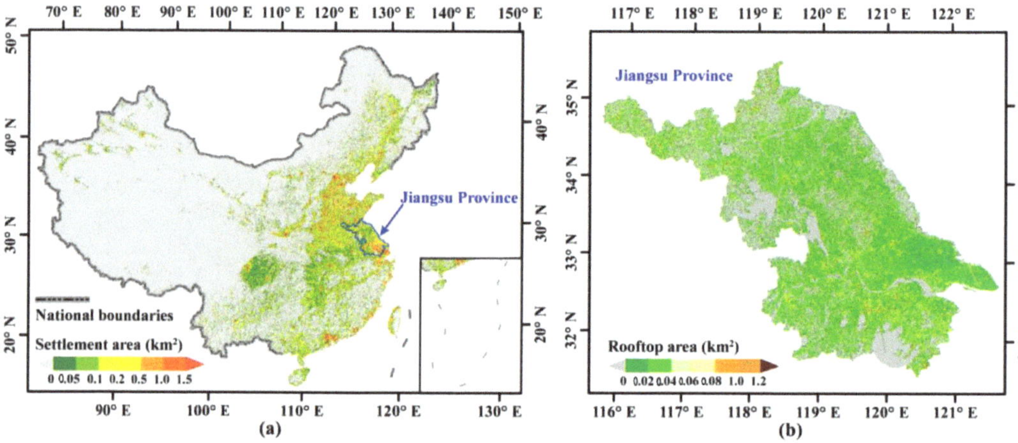

Figure 2. Settlement and building footprints. (**a**) Settlement area in China at a spatial resolution of 500 m, aggregated from the world settlement footprint products [43]; (**b**) building rooftop area at a spatial resolution of 500 m, which was calculated based on sub-meter building footprints in Jiangsu Province, China [7].

Regional building footprint extraction has been conducted in many regions [44,45]. Previously, we conducted a rooftop footprint survey in Jiangsu Province by fusing multi-source remote sensing images [7]. Figure 2b shows the rooftop area distribution based on the sub-meter building footprints. We counted the settlement area and rooftop area of each town in Jiangsu Province and found that these two areas had a significant linear correlation with a coefficient of determination of 0.9247 at a 95% confidence level (Figure 3a). Such a correlation was also observed at the global scale [27]. We further validated this correlation at the county level and observed a coefficient of determination of 0.9495 at a 95% confidence level (Figure 3b), implying the stability of the relationship across different scales. Here we extended this relationship to the entire country to map the rooftop area based on the settlement area at 500 m resolution. In addition, according to the experiences in Europe and China, only about 60% of the rooftop is suitable for PV installations [7,8]. Therefore the effective rooftop area (A_r) was calculated from the settlement area (A_s) as:

$$A_r = 0.6 \times 0.3633 \times A_s \tag{1}$$

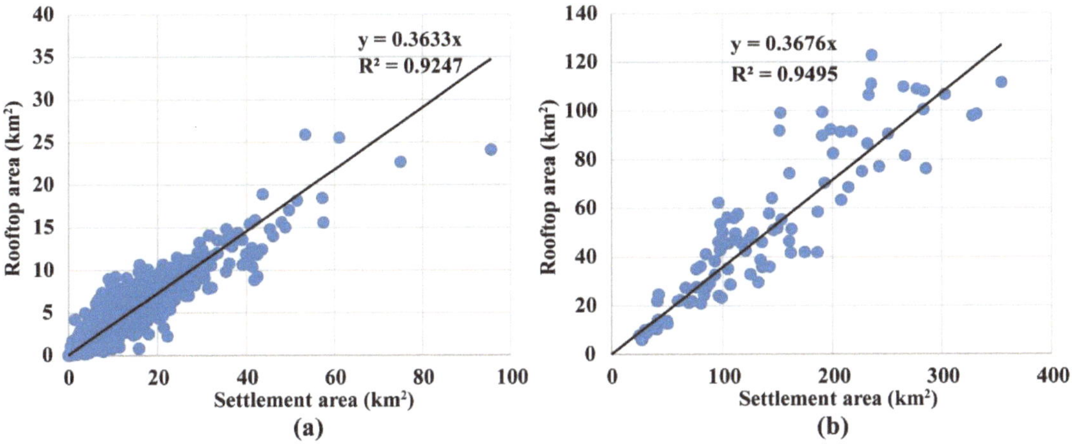

Figure 3. Scatterplot of rooftop area and settlement area. (**a**) Areas are counted by towns in Jiangsu Province; (**b**) areas are counted by counties in Jiangsu Province.

2.2. Simulation of PV Generation

In this study, we used GSEE [39] to simulate the PV system's performance, pixel by pixel, at a spatial resolution of 500 m, and on an hourly basis. Its inputs consisted of climate data from remote sensing inversion (mainly including solar radiation and air temperature) and PV system parameters (including installed capacity, panel orientation, panel tilt angle, panel material, and inverter efficiency). Hourly solar radiation data were from geostationary meteorological satellite images. Jiang et al. [46] produced hourly solar radiation products in China using an outstanding deep learning algorithm, and the accuracy was proven to be superior over other products [47]. Here, we collected the hourly data from 2007 to 2018 and used multi-year averages for simulation. The distribution of annual total solar radiation and the fraction of diffuse solar radiation are illustrated in Figure 4a,b, respectively. We assumed that all rooftop PV systems were south-facing and tilted at an optimal angle (θ_T) that varied with the latitude (φ) and diffuse fraction (R_d), as [48]:

$$\theta_T = 4.521 + 0.430 \times \varphi + 0.006 \times \varphi^2 + 54.504 \times R_d - 80.712 \times {R_d}^2 \quad (2)$$

This empirical relationship was validated at 98 radiation stations in China [48]. Figure 4c shows the calculated distribution of θ_T in China according to the empirical relationship and spatial estimates of the diffuse fraction. Hourly air temperatures (Figure 4d) were retrieved from the ERA5-Land reanalysis data [49]. We simulated the performance of PV modules composed of crystalline silicon material, assuming an inverter loss of 10%. The installed capacity was set to 1 kW; hence, the output of GSEE was equal to CF.

When the CF is known, the PV electricity generation (E_{PV}) can be calculated as:

$$E_{PV} = A_r \times D_P \times CF, \quad (3)$$

with D_P denoting the PV installation density (here, a density of 74 W/m^2 was adopted according to the practices in Jiangsu Province, China [7]). In addition, the coefficient of variation (CV) in daily averaged CF was calculated to provide a comparable understanding of the variability in rooftop PV generations [50]:

$$CV = \frac{\delta}{\mu}, \quad (4)$$

where δ and μ denote the standard deviation and mean of CF, respectively.

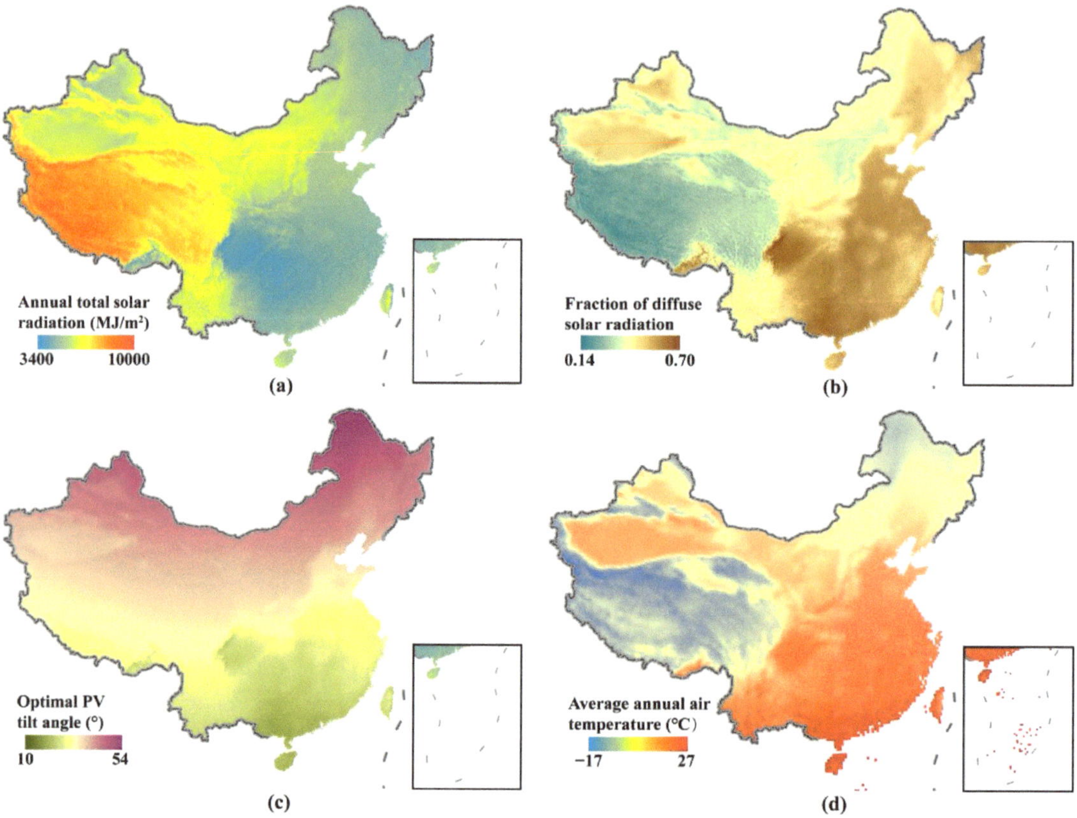

Figure 4. Spatial distribution of key parameters for simulating the capacity factor. (**a**) Total solar radiation; (**b**) fraction of diffuse solar radiation; (**c**) optimal PV tilt angle; (**d**) air temperature.

2.3. Calculation of Carbon Reductions

If the electricity generated by rooftop PV is delivered to the electrical grid to replace the power from other sources, rooftop PV contributes to carbon emission reductions. We calculated CO_2 equivalent emission reductions (C_r) based on the approach provided by the Intergovernmental Panel on Climate Change (IPCC) [51]:

$$C_r = (1 - R_c) \times AD \times \frac{EF}{(1 - R_p)}, \quad (5)$$

where AD represents activity data, equaling the amount of PV electricity consumption (E_{PV}) in this study; EF represents the emission factor that is associated with each unit of electricity supplied by a grid (tCO$_2$e MWh^{-1}); and R_c and R_p denote the curtailment rate and the penetration rate, respectively.

PV generation varies depending on local weather conditions, so they do not always correlate well with customer-side demand. If PV generation exceeds the net demand the excess generation is at risk of being curtailed. In Equation (5), we introduce $(1 - R_c)$ to correct for the decline in AD due to the curtailment; that is, this item indicates the proportion of rooftop PV generation that is delivered to the grid. We used the REFLEX model to simulate the electricity dispatch of each regional grid (Figure 5a) to obtain the R_c value when their respective rooftop PV potential was fully released. The REFLEX model compares the generations from rooftop PV and other plants with the net system load, hour by hour, to calculate the fraction of load met by PV generation. When PV generation exceeds the net load, the excess portion is curtailed or placed into storage, if available. The amount of storage in a grid is characterized by the average hourly load, for example, if the average hourly load is 30 GW, 2 h storage represents 60 GWh of energy capacity. The REFLEX model can evaluate the ability of an entire grid system to accommodate variable generations (e.g., solar PV and wind) according to the minimum generation level across the system. The minimum generation level represents the limit of both baseload generators, as well as generators that must remain online to reliably meet the variability and uncertainty of the net load and can be more generally expressed as the system flexibility, defined as the fraction of the must-run generators below the annual peak [36]. In this study, we designed a set of scenarios with different flexibility (80%, 90%, and 100%) and storage capacity (4 h, 8 h, and 12 h) levels. The typical load profiles of each grid were extracted from the report issued by the National Energy Administration [52]. The example of the Beijing grid is displayed in Figure 5b,c. The hourly loads throughout the year were calculated as [52]:

$$L_{i,j} = L_{peak,i} - \frac{H_{max} - H_j}{H_{max} - H_{min}}\left(L_{peak,i} - L_{base,i}\right), \tag{6}$$

where H_j denotes the load at hour j, H_{max} is the maximum load, and H_{min} is the minimum load, whose values vary depending on whether the day belongs to weekdays (blue line in Figure 5b) or weekends (brown line in Figure 5b). $L_{i,j}$ denotes the load at hour j of day i, $L_{peak,i}$ is the peak load of day i (red line in Figure 5c), and $L_{base,i}$ is the base load of day i (green line in Figure 5c).

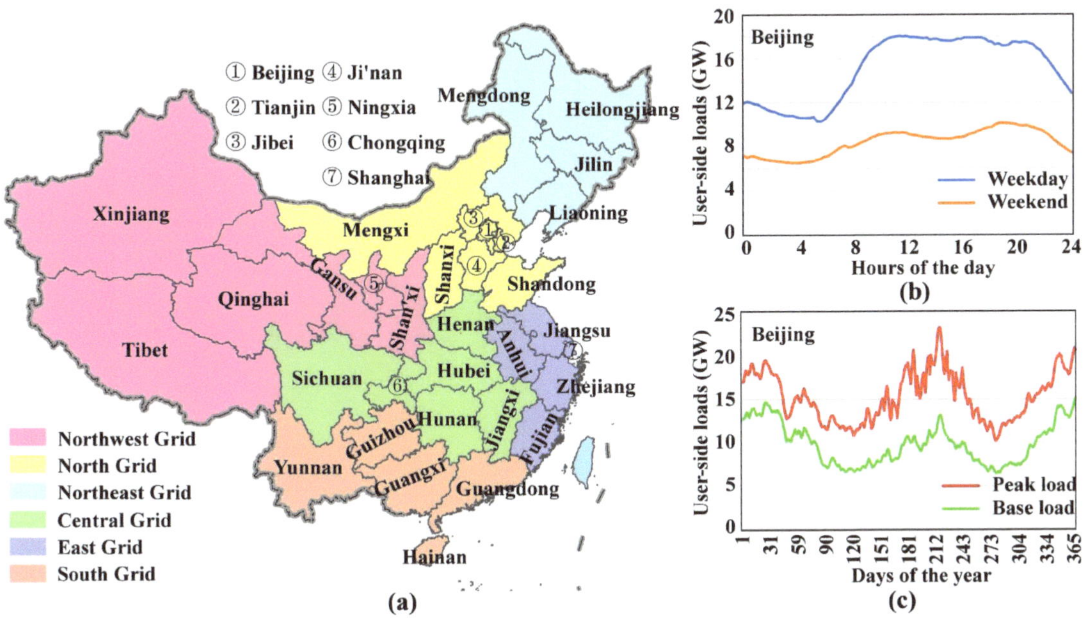

Figure 5. (**a**) Spatial division of China's regional grids [53]; (**b**) typical daily load profiles of the Beijing grid; (**c**) annual variations in base and peak loads of the Beijing grid.

Emission factors are sourced from China's Regional Grid Emissions Factors 2019 [53]. We used the operating margins (OMs) that represented emission factors from existing power plants in the electric grid, and the values for the northwest, north, northeast, central, east, and south grids were 0.8922, 0.9419, 1.0826, 0.8587, 0.7921, and 0.8042 tCO_2/MWh, respectively. Since these values were calculated for a generation mix that already included zero-emission PV electricity and rooftop PV did not displace existing clean energy on the grid, the emission reductions calculated by directly applying these factors were underestimated. Here, we used the current renewable energy penetration rate (Figure 6a sourced from https://www.bjx.com.cn/, accessed on 30 December 2022) to correct for this underestimation, as $\frac{EF}{(1-R_p)}$.

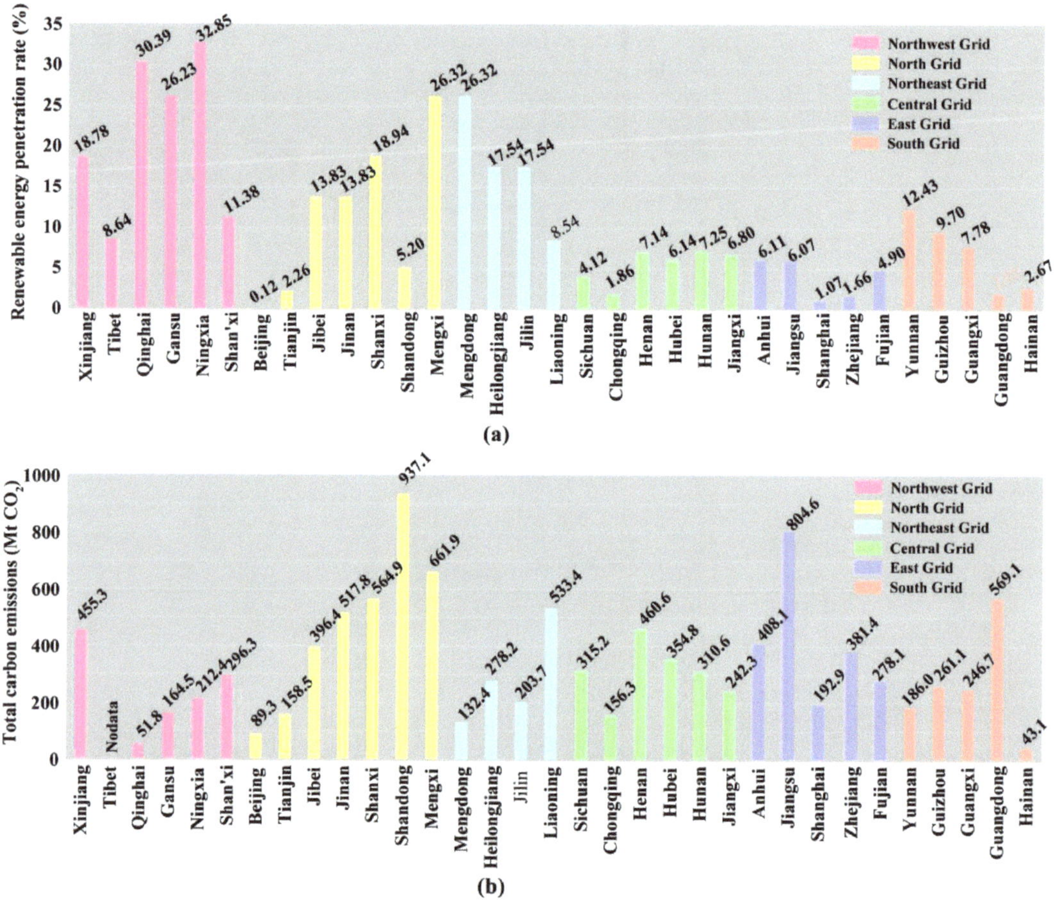

Figure 6. (**a**) The renewable energy penetration rate of each regional grid (numbers were sourced from https://www.bjx.com.cn/, accessed on 30 December 2022); (**b**) total carbon emissions of the corresponding regions in each regional grid in 2019 [54].

For clear recognition of the magnitude of emission reductions from rooftop PV, we defined a metric called the carbon offset rate (R_o):

$$R_o = \frac{C_r}{C_a} \quad (7)$$

This metric compared the emission reductions (C_r) to the total CO_2 emissions in 2019 (C_a). The emission inventories for each regional grid (Figure 6b) were collected from the China Emission Accounts and Datasets (https://www.ceads.net.cn/, accessed on 30 December 2022) [54]. $R_o \geq 1$ indicated that carbon neutrality could be achieved within the grid by developing rooftop PV.

3. Results

3.1. Electricity Generation Potential

Figure 7 shows the spatial distribution of rooftop PV electricity generation potential, as well as statistics by regional grid. In total, the theoretical maximum potential reached 6.5 PWh yr^{-1}, of which, more than 80% was concentrated in the eastern part of China. The North China Plain, Yangtze River Delta, and Pearl River Delta represented the most notable hotspots. The potential per unit area (500 m × 500 m) in these regions could exceed 5 PWh yr^{-1}. However, the western parts with high-quality solar energy resources accounted for less than 20% of the potential. As for the 33 regional grids, Shandong (681.9 TWh yr^{-1}), Henan (521.0 TWh yr^{-1}), Jiangsu (489.3 TWh yr^{-1}), Jinan (400.3 TWh yr^{-1}), and Guangdong (381.7 TWh yr^{-1}) grids had the greatest potential. According to energy statistics, these grids were in the front ranks of electricity consumption (Figure 7b). Such a coincidence reflects the advantages of rooftop PVs; that is, their electricity generation is close to the energy demand.

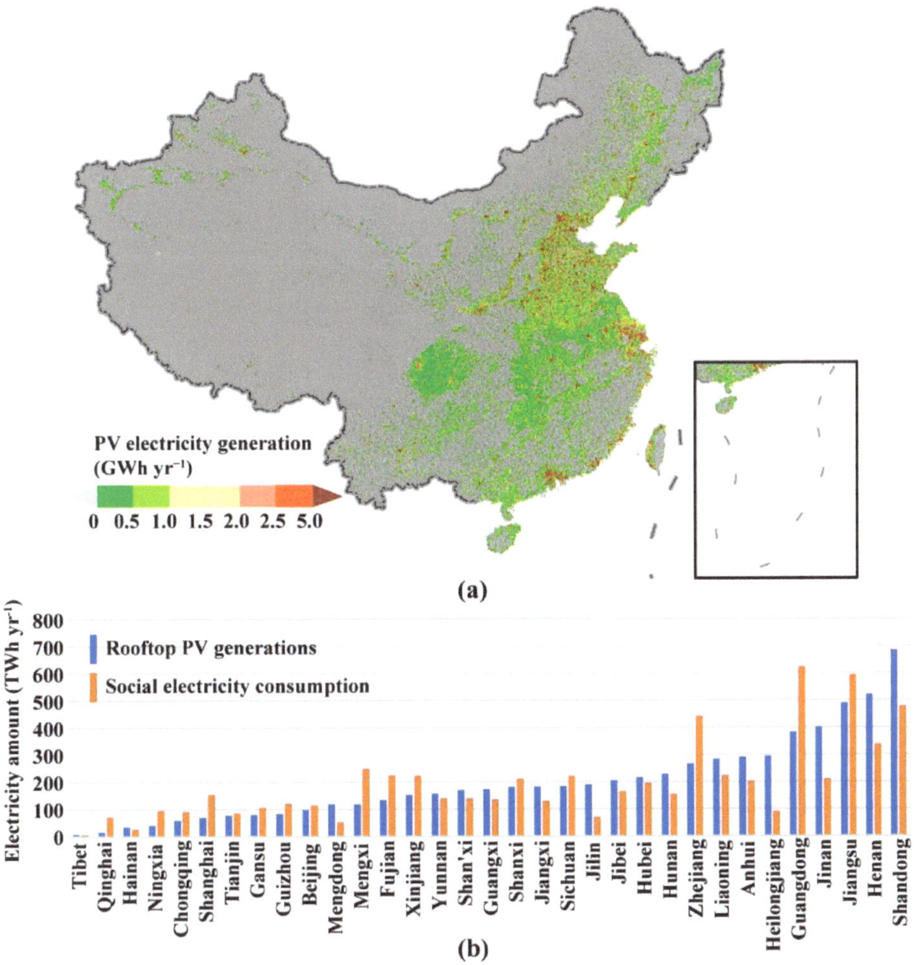

Figure 7. Rooftop PV electricity generation potential in China. (**a**) Spatial distribution at a spatial resolution of 500 m; (**b**) statistics by regional grid, where rooftop PV generations are compared to social electricity consumption.

Figure 8a illustrates the variability of rooftop PV electricity generation, expressed as the CV of the daily averaged CF. In general, the variability was small in western China while large in eastern China (especially in the south). The variability was extremely high (CV > 0.6) in the Sichuan basin and the middle and lower reaches of the Yangtze River, due to frequently cloudy and rainy weather. The spatial pattern of CV implied that most rooftop PV generations were characterized by significant volatility and uncertainty. Spatial aggregation was an effective way to reduce volatility. As demonstrated by the case of the Beijing grid (Figure 8b), the variability remained high after aggregation. Moreover, the fluctuations were highly stochastic and irregular in nature, making them difficult to forecast [55]. This issue was compounded by the shortened forecasting horizon because the fluctuations were more frequent and drastic at finer time scales (cf. the black and blue lines in Figure 8b). When comparing the hourly rooftop PV generations with the load profiles of the Beijing grid (Figures 5c and 8b), the mismatch between the two was clear. This mismatch magnified the difficulty of electricity dispatch [34]. Meanwhile, the increased variability imposed a more cyclic operating profile on dispatchable generation, with considerable cost implications [56].

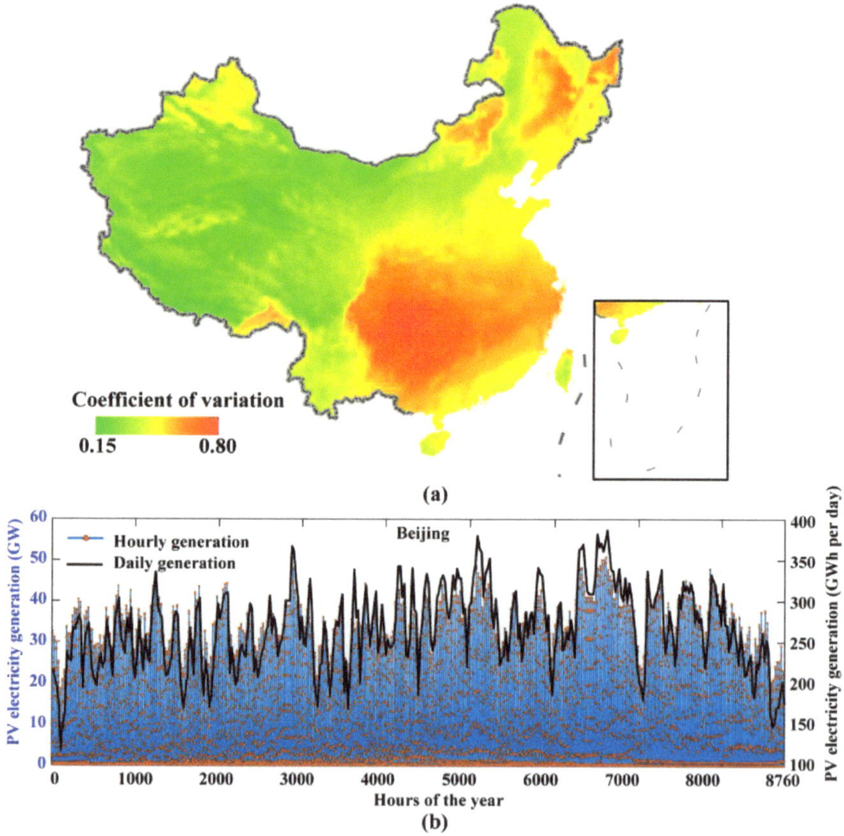

Figure 8. Variability in rooftop PV generation. (**a**) Coefficient of variation for the capacity factor (**b**) variation in rooftop PV generation for the Beijing grid.

3.2. Carbon Reduction Potential

We calculated the ratio of the minimum base load relative to the maximum peak load as a lower bound for grid flexibility in our scenario design. Based on the calculation results we simulated the potential carbon reductions in the 33-region grid under 80%, 90%, and

100% flexibility and 0, 4, 8, and 13 h of storage capacity (Table 1). As shown in Figure 9a, the carbon reduction potential declined with decreasing flexibility. A less flexible grid had more electricity from must-run units, so the net load to be met by variable PV generations was lower, putting rooftop PV at greater risk of curtailment [36]. The adoption of energy storage technology enhanced the capability of reducing carbon emissions (Figure 9a). Rooftop PV systems generate electricity during the daytime and store the excess to meet the load during the nighttime, ensuring that more electricity can eventually be consumed. Given 100% grid flexibility, 4 h storage capacity increased carbon reductions by 30% compared to the no-storage scenario. In contrast, at the same 100% flexibility, 12 h storage capacity increased the carbon reduction potential by only 9% compared to the 8 h storage case, suggesting that the role of storage diminished with expanding capacity. Because of the diminishing effect, 8–12 h storage capacity hardly worked at 80% flexibility. In the scenario with 90% grid flexibility and 8 h storage capacity, the total carbon reductions reached 4471.2 $MtCO_2$, consisting of 457.3 $MtCO_2$ from the Northwest grid, 1226.8 $MtCO_2$ from the North grid, 469.0 $MtCO_2$ from the Northeast grid, 824.9 $MtCO_2$ from the Central grid, 886.7 $MtCO_2$ from the East grid, and 606.6 $MtCO_2$ from the South grid (Figure 9b).

Table 1. Potential carbon reductions (unit: $MtCO_2$) in each regional grid under different scenarios.

	100% Flexibility				**90% Flexibility**				**80% Flexibility**			
	0 h	**4 h**	**8 h**	**12 h**	**0 h**	**4 h**	**8 h**	**12 h**	**0 h**	**4 h**	**8 h**	**12 h**
Xinjiang	111.5	153.1	166.5	166.8	101.9	145.3	165.8	166.8	91.6	136.1	163.8	166.4
Tibet	3.6	4.8	5.9	6.6	3.0	4.3	5.3	5.6	2.5	3.7	4.4	4.5
Qinghai	20.4	20.4	20.4	20.4	20.4	20.4	20.4	20.4	20.3	20.4	20.4	20.4
Gansu	59.8	82.6	95.6	96.6	54.2	77.5	93.9	96.5	48.2	71.8	90.8	94.5
Ningxia	47.2	54.6	54.6	54.6	43.9	54.5	54.6	54.6	39.9	54.3	54.6	54.6
Shan'xi	74.6	100.6	126.0	144.5	65.6	91.6	117.2	128.5	56.2	82.2	102.8	107.7
Beijing	54.4	73.4	86.8	89.6	47.5	67.0	80.5	83.2	39.9	59.2	68.8	70.3
Tianjin	41.5	56.4	68.0	72.3	36.7	51.8	64.6	68.5	31.6	46.8	58.0	59.7
Jibei	92.4	125.5	158.2	186.6	82.2	115.4	148.3	168.9	71.6	104.7	136.9	144.9
Jinan	129.8	172.1	214.3	245.4	113.9	156.2	197.3	213.0	97.5	139.7	171.4	176.1
Shanxi	117.9	162.2	196.5	208.3	105.7	150.7	188.8	204.2	92.7	138.0	178.6	188.9
Shandong	253.3	341.1	428.3	504.1	221.9	309.7	396.0	442.8	189.2	277.0	354.5	367.4
Mengxi	124.9	151.2	151.3	151.3	115.8	150.7	151.3	151.3	105.3	148.4	151.3	151.3
Mengdong	43.3	57.6	71.9	84.0	37.9	52.2	66.5	72.7	32.4	46.7	59.1	60.3
Heilongjiang	70.9	92.8	114.7	127.0	62.7	84.6	104.7	110.9	54.4	76.3	90.6	92.8
Jilin	54.4	71.6	88.8	98.9	48.1	65.3	80.8	86.6	41.7	58.9	70.2	72.9
Liaoning	134.3	183.1	230.3	271.1	120.4	169.3	217.0	247.1	106.0	154.8	202.3	216.3
Sichuan	100.0	135.2	158.6	163.5	88.2	124.3	152.2	158.1	75.4	112.0	136.9	139.5
Chongqing	37.6	48.8	51.5	51.5	32.8	45.8	50.8	51.2	27.2	41.1	44.9	45.1
Henan	168.2	225.7	283.2	330.4	143.9	201.5	258.5	282.1	118.7	176.3	218.9	225.9
Hubei	93.2	126.2	156.9	176.8	80.8	113.9	145.0	156.9	67.6	100.7	124.6	129.7
Hunan	78.7	105.0	131.1	149.4	67.9	94.2	118.7	128.3	56.6	82.9	99.8	103.7
Jiangxi	64.3	86.3	108.1	125.3	56.4	78.3	99.7	110.0	48.0	70.0	87.5	91.1
Anhui	89.4	120.7	151.9	177.6	76.4	107.8	138.5	152.7	62.9	94.2	117.6	123.0
Jiangsu	242.2	330.3	389.8	406.6	214.6	305.0	373.3	394.4	184.8	276.0	346.4	358.3
Shanghai	48.5	55.2	55.4	55.4	44.5	54.5	55.4	55.4	39.1	53.0	54.9	55.2
Zhejiang	162.3	205.7	212.0	212.5	146.8	197.9	209.1	210.0	129.0	184.4	203.4	204.4
Fujian	84.2	107.8	110.7	110.8	76.8	104.5	110.4	110.6	68.4	99.1	109.0	109.5
Yunnan	67.7	91.5	114.4	130.7	60.8	84.6	107.9	118.9	53.7	77.4	99.1	102.9
Guizhou	51.2	68.3	73.7	73.9	45.4	63.9	72.9	73.5	38.9	58.2	67.4	68.1
Guangxi	64.4	86.2	107.9	124.9	56.5	78.4	99.9	109.3	48.2	70.1	87.7	89.7
Guangdong	230.1	299.2	310.8	311.2	207.2	286.2	306.4	307.5	180.9	265.2	297.4	299.8
Hainan	12.5	16.8	21.1	24.6	11.0	15.3	19.6	21.8	9.5	13.8	17.6	18.2
Total	**3028.8**	**4011.9**	**4715.0**	**5153.1**	**2691.9**	**3722.3**	**4471.2**	**4762.1**	**2329.9**	**3393.5**	**4091.4**	**4212.7**

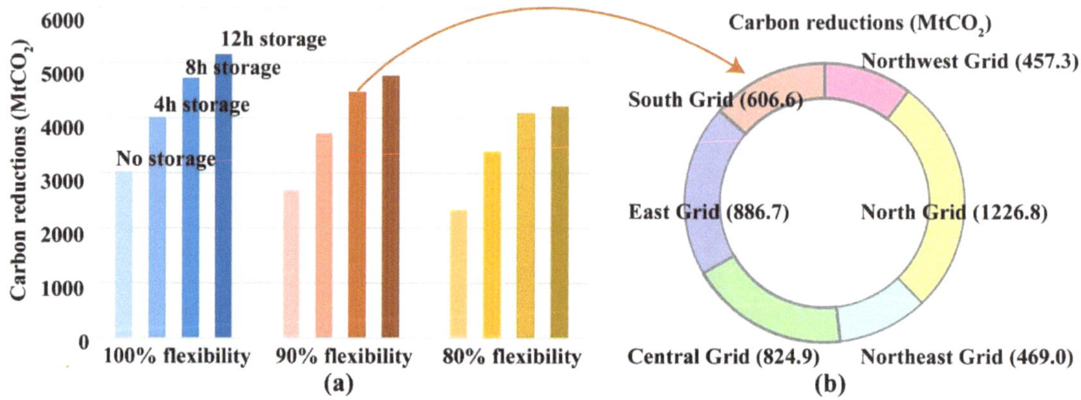

Figure 9. Potential carbon reductions from rooftop PV. (**a**) Total carbon reductions in China under different scenarios; (**b**) distribution of potential in the six primary regional grids.

Using the total CO_2 emissions in 2019 as a reference, we calculated the minimum and maximum offset rates for each grid among the twelve scenarios simulated, as shown in Figure 10a,b, respectively. Overall, the North and Central grids were relatively poor, with most grids having minimum offsets of less than 20%; while the East and South grids fared well, with most grids having maximum offsets exceeding 50%. Except for the Qinghai grid, all other grids witnessed varying degrees of improvement in the offset, owing to the increased flexibility and storage capacity. The Qinghai grid itself had little rooftop PV potential, so curtailment rarely occurred, even when the peak output of the rooftop PV came across the lowest base load. Energy storage did not make sense for the Qinghai grid (Table 1). By comparison, the Guizhou, Mengxi, Ningxia, and Chongqing grids were the most worrisome, while the Guangdong, Zhejiang, and Beijing grids were the most exciting in terms of both minimum and maximum offset rates. In addition, we can conclude that carbon neutrality cannot be achieved in China by relying on rooftop PV alone, given that the offset rates of all grids were below 100% in all scenarios.

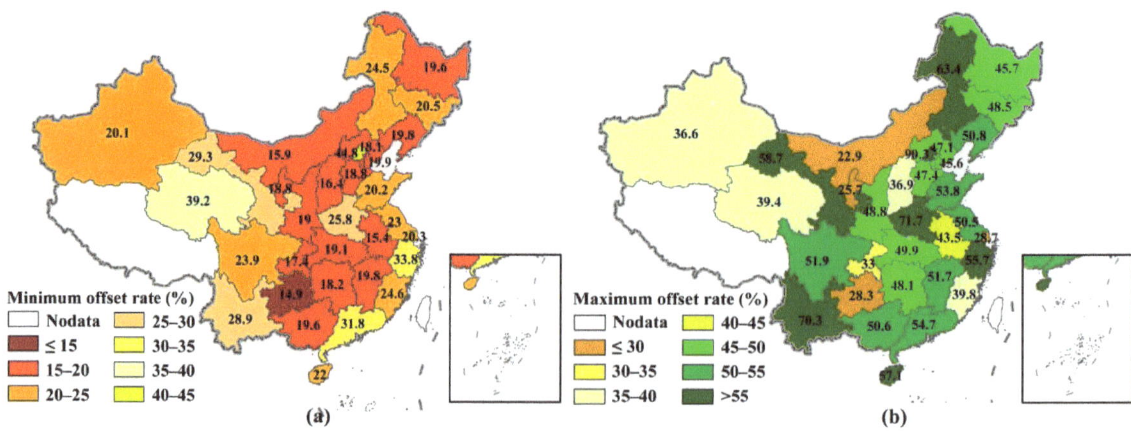

Figure 10. Carbon offset rate of rooftop PV for each grid. (**a**) Minimum offset rate; (**b**) Maximum offset rate.

4. Discussion

Our evaluations revealed that China has a considerable amount of rooftop PV electricity generation. Although rooftop PV potential showed great spatial heterogeneity, it exhibited a good match with energy demand (Figure 7). On the one hand, the major energy-consuming provinces, such as Guangdong, Shandong, Jiangsu, and Zhejiang, were at the forefront in terms of the total potential; on the other hand, the potential was mainly concentrated in urban areas with high intensities of energy consumption. This is precisely the advantage of rooftop PV over large ground-based PV plants: helping not only to reduce energy losses in transmission but also to save on electricity transmission costs [37,57]. In addition, rooftop PV offers the option of bypassing land competition. If the same amounts of rooftop PV electricity were obtained by installing ground-based PVs, a minimum net area of approximately 30,196 square kilometers would be required, which would be equivalent to six Shanghai cities. This is almost unacceptable for the densely populated eastern region, where land resources are inherently competitive [9].

In addition to being directly related to the number of electricity generations, carbon emission reduction is also affected by market consumption [58]. In this study, we quantified this effect through electricity dispatch based on hourly generation simulations and load profiles. The results revealed that fully releasing the potential of rooftop PV could reduce CO_2 equivalent emissions by 2.3–5.2 Gt, lower than those (~5.9 Gt) emitted when the generated electricity is completely consumed without curtailment. Nevertheless, the potential reduction is equal to 21%–47% of China's total emissions in 2019 [54]. Such a scale is sufficient to make a significant contribution to China's carbon neutrality [59,60]. Our study highlights the importance of increasing grid flexibility and preparing energy storage to obtain a greater amount of carbon reduction. When grid flexibility increased from 80% to 100%, carbon reductions increased by about 25%, and an additional increase of about 60% was attainable if 8–12 h of energy storage were available. It is worth noting that the effect of increased flexibility and energy storage varied from grid to grid (Table 1), suggesting that rooftop PV development planning should be tailored to local conditions.

The amount of both roof resources and energy consumption reflects, to some extent, the level of regional development, and thus, the two coincide spatially. In contrast, carbon emission intensity was higher in the developed eastern regions than in the central and western regions, so the total emissions were not consistent with the rooftop PV potential in space. This leads to significant spatial differences in the process of carbon neutrality that depend on rooftop PV (Figure 10). Based on the twelve scenarios simulated, we presented a plausible range of offset rates. It should be noted that the offset rates can be further enhanced by using larger energy storage or other means, such as load shifting, and may be decreased if the flexibility is lowered. However, the uncertainty will not be great, as 12 h of energy storage and 80% flexibility is close to the limit. In addition, increased flexibility and greater storage capacity are not necessarily better because their effectiveness also depends on the characteristics of customer-side loads and PV generations in the grid. This, once again, emphasizes the importance of differentiated layout and planning when developing rooftop PV, which may also be true for other PV development.

5. Conclusions

In this study, we designed a technical framework for integrating multi-source remote sensing data to assess the carbon reduction potential of rooftop PV. The key point was to obtain the PV curtailment rate through a dispatch model based on the spatio-temporal simulations of hourly PV generation and the load profiles of each grid. The main conclusions include:

(1) The maximum electricity generation of rooftop PV in China reached 6.5 PWh yr^{-1}, of which more than 80% was concentrated in densely populated areas in the east and characterized by high variability.

(2) Unlocking China's full rooftop PV potential could reduce CO_2 equivalent emissions by 2.3–5.2 Gt, depending on the grid flexibility and storage capacity.

(3) The potential carbon reductions could offset 21%–47% of China's total emissions using the data in 2019 as a reference; thus, it could make a significant contribution to carbon neutrality.
(4) Both carbon reductions and their offset rates vary greatly from grid to grid, highlighting the need for rooftop PV development plans tailored to local conditions.

Author Contributions: Conceptualization, H.J. and N.L.; methodology, H.J.; software, N.L.; validation, X.W.; formal analysis, H.J.; investigation, N.L.; resources, H.J. and X.W.; data curation X.W.; writing—original draft preparation, H.J.; writing—review and editing, N.L.; visualization, H.J. supervision, N.L.; project administration, N.L.; funding acquisition, H.J. All authors have read and agreed to the published version of the manuscript.

Funding: This research was funded by the Open Fund of the State Key Laboratory of Remote Sensing Science (Grant No. OFSLRSS202204), the China Postdoctoral Science Foundation (Grant No 2021M703176), and the National Natural Science Foundation of China (Grant No. 42201382).

Institutional Review Board Statement: Not applicable.

Informed Consent Statement: Not applicable.

Data Availability Statement: The settlement footprint datasets are available at https://download geoservice.dlr.de/WSF2019/files/ (accessed on 30 December 2022). The solar radiation data are freely accessible from the Pangaea at https://doi.org/10.1594/PANGAEA.904136 (accessed on 30 December 2022). The load profiles for the 33 regional grids are from the report at https://www.ndrc gov.cn/xwdt/tzgg/202012/P020201202546044875868.pdf (accessed on 30 December 2022). Other datasets are available from the authors on reasonable request.

Conflicts of Interest: The authors declare no conflict of interest.

References

1. Chen, X.; Liu, Y.; Wang, Q.; Lv, J.; Wen, J.; Chen, X.; Kang, C.; Cheng, S.; McElroy, M.B. Pathway toward carbon-neutral electrical systems in China by mid-century with negative CO_2 abatement costs informed by high-resolution modeling. *Joule* **2021** *5*, 2715–2741. [CrossRef]
2. Lu, X.; Chen, S.; Nielsen Chris, P.; Zhang, C.; Li, J.; Xu, H.; Wu, Y.; Wang, S.; Song, F.; Wei, C.; et al. Combined solar power and storage as cost-competitive and grid-compatible supply for China's future carbon-neutral electricity system. *Proc. Natl. Acad. Sci USA* **2021**, *118*, e2103471118. [CrossRef] [PubMed]
3. IEA. *Solar PV*; International Energy Agency (IEA): Paris, France, 2022.
4. Liu, L.; Sun, Q.; Li, H.; Yin, H.; Ren, X.; Wennersten, R. Evaluating the benefits of integrating floating photovoltaic and pumped storage power system. *Energy Convers. Manag.* **2019**, *194*, 173–185. [CrossRef]
5. Xue, J. Photovoltaic agriculture—New opportunity for photovoltaic applications in China. *Renew. Sustain. Energy Rev.* **2017** *73*, 1–9. [CrossRef]
6. Cheng, C.L.; Sanchez Jimenez, C.S.; Lee, M.-C. Research of BIPV optimal tilted angle, use of latitude concept for south orientated plans. *Renew. Energy* **2009**, *34*, 1644–1650. [CrossRef]
7. Jiang, H.; Yao, L.; Lu, N.; Qin, J.; Liu, T.; Liu, Y.; Zhou, C. Geospatial assessment of rooftop solar photovoltaic potential using multi-source remote sensing data. *Energy AI* **2022**, *10*, 100185. [CrossRef]
8. Bódis, K.; Kougias, I.; Jäger-Waldau, A.; Taylor, N.; Szabó, S. A high-resolution geospatial assessment of the rooftop solar photovoltaic potential in the European Union. *Renew. Sustain. Energy Rev.* **2019**, *114*, 109309. [CrossRef]
9. Sacchelli, S.; Garegnani, G.; Geri, F.; Grilli, G.; Paletto, A.; Zambelli, P.; Ciolli, M.; Vettorato, D. Trade-off between photovoltaic systems installation and agricultural practices on arable lands: An environmental and socio-economic impact analysis for Italy *Land Use Policy* **2016**, *56*, 90–99. [CrossRef]
10. NEA. Notice of Publishing the List of Pilot Projects for Distributed Rooftop Photovoltaic Development in the Whole County (City or District). Available online: http://zfxxgk.nea.gov.cn/2021-09/08/c_1310186582.htm (accessed on 30 December 2022).
11. Gassar, A.A.A.; Cha, S.H. Review of geographic information systems-based rooftop solar photovoltaic potential estimation approaches at urban scales. *Appl. Energy* **2021**, *291*, 116817. [CrossRef]
12. Mohajeri, N.; Assouline, D.; Guiboud, B.; Bill, A.; Gudmundsson, A.; Scartezzini, J.-L. A city-scale roof shape classification using machine learning for solar energy applications. *Renew. Energy* **2018**, *121*, 81–93. [CrossRef]
13. Machete, R.; Falcão, A.P.; Gomes, M.G.; Moret Rodrigues, A. The use of 3d GIS to analyse the influence of urban context on buildings' solar energy potential. *Energy Build.* **2018**, *177*, 290–302. [CrossRef]
14. Wu, A.N.; Biljecki, F. Roofpedia: Automatic mapping of green and solar roofs for an open roofscape registry and evaluation of urban sustainability. *Landsc. Urban Plan.* **2021**, *214*, 104167. [CrossRef]

15. Debbarma, M.; Sudhakar, K.; Baredar, P. Thermal modeling, exergy analysis, performance of BIPV and BIPVT: A review. *Renew. Sustain. Energy Rev.* **2017**, *73*, 1276–1288. [CrossRef]
16. Hong, T.; Koo, C.; Park, J.; Park, H.S. A GIS (geographic information system)-based optimization model for estimating the electricity generation of the rooftop PV (photovoltaic) system. *Energy* **2014**, *65*, 190–199. [CrossRef]
17. Bergamasco, L.; Asinari, P. Scalable methodology for the photovoltaic solar energy potential assessment based on available roof surface area: Application to piedmont region (Italy). *Sol. Energy* **2011**, *85*, 1041–1055. [CrossRef]
18. Assouline, D.; Mohajeri, N.; Scartezzini, J.-L. Quantifying rooftop photovoltaic solar energy potential: A machine learning approach. *Sol. Energy* **2017**, *141*, 278–296. [CrossRef]
19. Zhong, T.; Zhang, Z.; Chen, M.; Zhang, K.; Zhou, Z.; Zhu, R.; Wang, Y.; Lü, G.; Yan, J. A city-scale estimation of rooftop solar photovoltaic potential based on deep learning. *Appl. Energy* **2021**, *298*, 117132. [CrossRef]
20. Huang, J.; Zhang, X.; Xin, Q.; Sun, Y.; Zhang, P. Automatic building extraction from high-resolution aerial images and LiDAR data using gated residual refinement network. *ISPRS J. Photogramm. Remote Sens.* **2019**, *151*, 91–105. [CrossRef]
21. Guo, H.; Du, B.; Zhang, L.; Su, X. A coarse-to-fine boundary refinement network for building footprint extraction from remote sensing imagery. *ISPRS J. Photogramm. Remote Sens.* **2022**, *183*, 240–252. [CrossRef]
22. Guo, H.; Shi, Q.; Marinoni, A.; Du, B.; Zhang, L. Deep building footprint update network: A semi-supervised method for updating existing building footprint from bi-temporal remote sensing images. *Remote Sens. Env.* **2021**, *264*, 112589. [CrossRef]
23. Peng, Z.; Letu, H.; Wang, T.; Shi, C.; Zhao, C.; Tana, G.; Zhao, N.; Dai, T.; Tang, R.; Shang, H.; et al. Estimation of shortwave solar radiation using the artificial neural network from Himawari-8 satellite imagery over China. *J. Quant. Spectrosc. Radiat. Transf.* **2020**, *240*, 106672. [CrossRef]
24. Jiang, H.; Lu, N.; Qin, J.; Tang, W.J.; Yao, L. A deep learning algorithm to estimate hourly global solar radiation from geostationary satellite data. *Renew. Sustain. Energy Rev.* **2019**, *114*, 109327. [CrossRef]
25. Huang, G.H.; Li, Z.Q.; Li, X.; Liang, S.L.; Yang, K.; Wang, D.D.; Zhang, Y. Estimating surface solar irradiance from satellites: Past, present, and future perspectives. *Remote Sens. Environ.* **2019**, *233*, 111371. [CrossRef]
26. Li, R.; Wang, D.; Liang, S.; Jia, A.; Wang, Z. Estimating global downward shortwave radiation from VIIRS data using a transfer-learning neural network. *Remote Sens. Env.* **2022**, *274*, 112999. [CrossRef]
27. Joshi, S.; Mittal, S.; Holloway, P.; Shukla, P.R.; Ó Gallachóir, B.; Glynn, J. High resolution global spatiotemporal assessment of rooftop solar photovoltaics potential for renewable electricity generation. *Nat. Commun.* **2021**, *12*, 5738. [CrossRef]
28. Abd Latif, Z.; Mohd Zaki, N.; Salleh, S. GIS-based estimation of rooftop solar photovoltaic potential using LiDAR. In Proceedings of the 2012 IEEE 8th International Colloquium on Signal Processing and Its Applications, Mahkota, Malaysia, 23–25 March 2012; pp. 388–392.
29. Mesude Bayrakci, B.; Kirby, C.; Jeffrey, R.S.B. An automated model for rooftop PV systems assessment in ArcGIS using LiDAR. *AIMS Energy* **2015**, *3*, 401–420. [CrossRef]
30. Jacques, D.A.; Gooding, J.; Giesekam, J.J.; Tomlin, A.S.; Crook, R. Methodology for the assessment of PV capacity over a city region using low-resolution LiDAR data and application to the city of Leeds (UK). *Appl. Energy* **2014**, *124*, 28–34. [CrossRef]
31. Gagnon, P.; Margolis, R.; Melius, J.; Phillips, C.; Elmore, R. Estimating rooftop solar technical potential across the us using a combination of GIS-based methods, lidar data, and statistical modeling. *Environ. Res. Lett.* **2018**, *13*, 024027. [CrossRef]
32. Wang, M.; Mao, X.; Gao, Y.; He, F. Potential of carbon emission reduction and financial feasibility of urban rooftop photovoltaic power generation in Beijing. *J. Clean. Prod.* **2018**, *203*, 1119–1131. [CrossRef]
33. Taborianski, V.M.; Pacca, S.A. Carbon dioxide emission reduction potential for low income housing units based on photovoltaic systems in distinct climatic regions. *Renew. Energy* **2022**, *198*, 1440–1447. [CrossRef]
34. Knuepfer, K.; Rogalski, N.; Knuepfer, A.; Esteban, M.; Shibayama, T. A reliable energy system for Japan with merit order dispatch, high variable renewable share and no nuclear power. *Appl. Energy* **2022**, *328*, 119840. [CrossRef]
35. Zhang, X.; Wang, J.-X.; Cao, Z.; Shen, S.; Meng, S.; Fan, J.-L. What is driving the remarkable decline of wind and solar power curtailment in China? Evidence from China and four typical provinces. *Renew. Energy* **2021**, *174*, 31–42. [CrossRef]
36. Denholm, P.; Hand, M. Grid flexibility and storage required to achieve very high penetration of variable renewable electricity. *Energy Policy* **2011**, *39*, 1817–1830. [CrossRef]
37. Ziegler, M.S.; Mueller, J.M.; Pereira, G.D.; Song, J.; Ferrara, M.; Chiang, Y.-M.; Trancik, J.E. Storage requirements and costs of shaping renewable energy toward grid decarbonization. *Joule* **2019**, *3*, 2134–2153. [CrossRef]
38. Wang, B.; Ji, F.; Zheng, J.; Xie, K.; Feng, Z. Carbon emission reduction of coal-fired power supply chain enterprises under the revenue sharing contract: Perspective of coordination game. *Energy Econ.* **2021**, *102*, 105467. [CrossRef]
39. Pfenninger, S.; Staffell, I. Long-term patterns of European PV output using 30 years of validated hourly reanalysis and satellite data. *Energy* **2016**, *114*, 1251–1265. [CrossRef]
40. Hargreaves, J.; Hart, E.K.; Jones, R.; Olson, A. Reflex: An adapted production simulation methodology for flexible capacity planning. *IEEE Trans. Power Syst.* **2015**, *30*, 1306–1315. [CrossRef]
41. Zhou, K.; Lindenbergh, R.; Gorte, B.; Zlatanova, S. Lidar-guided dense matching for detecting changes and updating of buildings in airborne lidar data. *ISPRS J. Photogramm. Remote Sens.* **2020**, *162*, 200–213. [CrossRef]
42. Esch, T.; Brzoska, E.; Dech, S.; Leutner, B.; Palacios-Lopez, D.; Metz-Marconcini, A.; Marconcini, M.; Roth, A.; Zeidler, J. World settlement footprint 3d—A first three-dimensional survey of the global building stock. *Remote Sens. Env.* **2022**, *270*, 112877. [CrossRef]

43. Marconcini, M.; Metz-Marconcini, A.; Üreyen, S.; Palacios-Lopez, D.; Hanke, W.; Bachofer, F.; Zeidler, J.; Esch, T.; Gorelick, N.; Kakarla, A.; et al. Outlining where humans live, the world settlement footprint 2015. *Sci. Data* **2020**, *7*, 242. [CrossRef]
44. Liu, P.; Liu, X.; Liu, M.; Shi, Q.; Yang, J.; Xu, X.; Zhang, Y. Building footprint extraction from high-resolution images via spatial residual inception convolutional neural network. *Remote Sens.* **2019**, *11*, 830. [CrossRef]
45. Liu, T.; Yao, L.; Qin, J.; Lu, N.; Jiang, H.; Zhang, F.; Zhou, C. Multi-scale attention integrated hierarchical networks for high-resolution building footprint extraction. *Int. J. Appl. Earth Obs. Geoinf.* **2022**, *109*, 102768. [CrossRef]
46. Jiang, H.; Lu, N.; Qin, J.; Yao, L. Hourly 5-km surface total and diffuse solar radiation in China, 2007–2018. *Sci. Data* **2020**, *7*, 311. [CrossRef]
47. Jiang, H.; Yang, Y.; Wang, H.; Bai, Y.; Bai, Y. Surface diffuse solar radiation determined by reanalysis and satellite over east Asia: Evaluation and comparison. *Remote Sens.* **2020**, *12*, 1387. [CrossRef]
48. Liu, Y.; Yao, L.; Jiang, H.; Lu, N.; Qin, J.; Liu, T.; Zhou, C. Spatial estimation of the optimum PV tilt angles in China by incorporating ground with satellite data. *Renew. Energy* **2022**, *189*, 1249–1258. [CrossRef]
49. Munoz-Sabater, J.; Dutra, E.; Agusti-Panareda, A.; Albergel, C.; Arduini, G.; Balsamo, G.; Boussetta, S.; Choulga, M.; Harrigan, S.; Hersbach, H.; et al. Era5-land: A state-of-the-art global reanalysis dataset for land applications. *Earth Syst. Sci. Data* **2021**, *13*, 4349–4383. [CrossRef]
50. Wu, C.; Zhang, X.-P.; Sterling, M. Solar power generation intermittency and aggregation. *Sci. Rep.* **2022**, *12*, 1363. [CrossRef]
51. IPCC. Chapter 2: Stationary combustion. In *2006 IPCC Guidelines for National Greenhouse Gas Inventories Volume 2: Energy*; Intergovernmental Panel on Climate Change; IPCC: Geneva, Switzerland, 2006.
52. Typical Load Profiles of Provincial Power Grid. Available online: https://www.ndrc.gov.cn/xwdt/tzgg/202012/P020201202546044875868.pdf (accessed on 1 August 2022).
53. China's Regional Grid Emissions Factors. 2019. Available online: https://www.mee.gov.cn/ywgz/ydqhbh/wsqtkz/202012/t20201229_815386.shtml (accessed on 1 August 2022).
54. Guan, Y.; Shan, Y.; Huang, Q.; Chen, H.; Wang, D.; Hubacek, K. Assessment to China's recent emission pattern shifts. *Earth's Future* **2021**, *9*, e2021EF002241. [CrossRef]
55. Qin, J.; Jiang, H.; Lu, N.; Yao, L.; Zhou, C. Enhancing solar PV output forecast by integrating ground and satellite observations with deep learning. *Renew. Sustain. Energy Rev.* **2022**, *167*, 112680. [CrossRef]
56. Bravo, R.; Ortiz, C.; Chacartegui, R.; Friedrich, D. Multi-objective optimisation and guidelines for the design of dispatchable hybrid solar power plants with thermochemical energy storage. *Appl. Energy* **2021**, *282*, 116257. [CrossRef]
57. Asadi Majd, A.; Farjah, E.; Rastegar, M.; Bacha, S. Generation and transmission expansion planning for bulk renewable energy export considering transmission service cost allocation. *Electr. Power Syst. Res.* **2021**, *196*, 107197. [CrossRef]
58. Sánchez de la Nieta, A.A.; Paterakis, N.G.; Gibescu, M. Participation of photovoltaic power producers in short-term electricity markets based on rescheduling and risk-hedging mapping. *Appl. Energy* **2020**, *266*, 114741. [CrossRef]
59. Mallapaty, S. How China could be carbon neutral by mid-century. *Nature* **2020**, *586*, 482–483. [CrossRef] [PubMed]
60. Liu, L.; Wang, Y.; Wang, Z.; Li, S.; Li, J.; He, G.; Li, Y.; Liu, Y.; Piao, S.; Gao, Z.; et al. Potential contributions of wind and solar power to China's carbon neutrality. *Resour. Conserv. Recycl.* **2022**, *180*, 106155. [CrossRef]

sustainability

MDPI

Article

Vegetation Analysis in the Archaeological Area of Pasargadae WHS (Iran) Enhancing the Naturalistic Value of the Site within the Occurring Environmental Changes

Giulio Zangari, Zohreh Hosseini * and Giulia Caneva

Department of Science, University of Roma Tre, Viale Marconi 446, 00146 Rome, Italy; giulio.zangari@uniroma3.it (G.Z.); giulia.caneva@uniroma3.it (G.C.)

* Correspondence: seyedhzohreh.hosseini@uniroma3.it

Abstract: This study provides pioneering research on the vegetation of archaeological areas in Iran to enhance its naturalistic and bioindication values by selecting the Pasargadae World Heritage Site (WHS). Vegetation surveys were carried out in different homogeneous habitats, analyzing the plant communities through statistical elaboration, syntaxonomic role, mapping, and enhancement of plants with conservation interest. In an ecological approach, the study included an analysis of the recent climate changes and human interventions influencing the water resources. Results revealed seven main vegetation types reflecting ecological gradients shaped by environmental, edaphic, and anthropogenic factors. The syntaxonomic analysis showed a primary subdivision in semi-natural grasslands and synanthropic vegetation. Several key species were identified as bioindicators of multiple factors, such as: *Launaea acanthodes*, *Stipa barbata*, *Alhagi maurorum*, *Bellevalia saviczii*, *Glycyrrhiza glabra*, *Convolvulus arvensis*, and *Hordeum murinum*. The vegetation map showed how the hilly grassland communities hosted the highest number of species with conservation interest and their need to be better protected. Bio-climatic data, such as the construction of dams and the exploitation for irrigation purposes, pointed to the increasing xeric conditions, which make urging conservation efforts for the site's historical and naturalistic values. The study underscores the importance of preserving places with high plant diversity for effective site management, and enhances the intricate relationship between vegetation and natural features in the occurring environmental changes.

Keywords: archaeological site; biodiversity; climate change; syntaxonomy; vegetation mapping; nature conservation

Citation: Zangari, G.; Hosseini, Z.; Caneva, G. Vegetation Analysis in the Archaeological Area of Pasargadae WHS (Iran) Enhancing the Naturalistic Value of the Site within the Occurring Environmental Changes. *Sustainability* **2024**, *16*, 3784. https://doi.org/10.3390/su16093784

Academic Editors: Jun Qin and Hou Jiang

Received: 29 March 2024
Revised: 26 April 2024
Accepted: 29 April 2024
Published: 30 April 2024

1. Introduction

Archaeological sites display the interaction of the natural environment and human activities [1–6]. In this interaction, multiple dynamics are involved, in which vegetation can have both a negative and sometimes a positive action in the conservation of archaeological structures [7]. Indeed, the growth of plants in archaeological areas recurrently gives rise to biodeterioration phenomena, which is related to the development of their roots [8–17]. However, less consideration has been given to the benefits and values that the plants can provide, both in terms of the direct benefits they offer and the indirect advantages gained from understanding their role. In fact, the presence of plants in archaeological areas has a mitigating effect on the microclimate, thus ensuring higher stability and durability of the monuments and offering better comfort for those who visit these places [18,19]. Indeed, several studies demonstrated the mitigating action of plants by reducing solar radiation, maintaining a higher relative humidity, and reducing weathering, contaminant deposition, and wind erosion as well [20–23]. Moreover, some contributions [18,24–26] provided a methodological framework for evaluating the heritage value of vegetation. Furthermore, wild plants can also show indications of buried archaeological structures, thus adding information to the history of the site [27–31].

In addition, archaeological areas, being better protected from anthropogenic disturbance compared to other human-managed areas, have proven to be valuable refuges for biodiversity conservation [5,6,32,33]. The permanence of natural habitats and the floristic richness found in these areas is considerable, with the occurrence of a high number of species of conservation interest [6,32,34–37].

Analyzing plant communities in archaeological areas, such as their distribution and bioindication values, can greatly enhance the efficiency of site management and enrichment activities [18,20]. In fact, understanding how plant species are spatially and numerically distributed within the archaeological areas can facilitate management planning and activities, minimizing their negative impacts and promoting the benefits that plant communities can provide in these contexts [11,38,39]. Conversely, the vegetation growing on archaeological sites is influenced not only by human activities, but also by environmental and climatic conditions, as well as their ecological characteristics can serve as a bioindicator of these changes [40,41]. Due to severe climatic conditions, vegetation in arid or semi-arid environments shows a relative adaptive capacity, which becomes more significant considering current climate changes [42,43]. Vegetation maps of archaeological sites are useful tools for management planning, providing insight into the reading of the site characteristics, both for precise location of the various types of plant communities and contribution to the conservation of monuments [39].

In the Mediterranean areas, while several studies have addressed the vegetation of archaeological areas, research carried out on archaeological sites in arid or semi-arid environments is limited [6,32,44], despite their relevance and fragility caused by the occurring climatic changes. Notably, the UNESCO World Heritage Site (WHS) of Pasargadae, dating back to the 6th century BC [45,46], is of significant historical importance for its age, and its value since it is the place where the Persian Garden originated. The origin of the garden itself can be related to the favorable rainfall and hydrological conditions, since in the past the site had a high availability of water. However, the aridity has increased over the centuries, and this phenomenon has evidently intensified more recently. The present desiccation of the river was the consequence of both direct human interventions and the current climate change [47,48]. Moreover, Pasargadae is located in the border zone of the Zagros mountains, and the Irano-Turanian region, and it results in a rich ecotone from a biodiversity point of view, as well as a high naturalistic interest [6]. Following our recent floristic assessment [6] and the biodeterioration evaluation due to plant growth [17,49], we wish to deepen the knowledge of the site, by adding detailed insights into the vegetation's naturalistic values that characterize the various parts of the site and the emerging issues linked to the climatic changes. In particular, this study will analyze the different types of vegetation growing in the archaeological area to: (1) assess their ecological and syntaxonomic characteristics and enhance their bioindication values; (2) evaluate the naturalistic interest and the distribution in the area through their mapping; and (3) give a preliminary assessment of the effects of the increasing stresses induced by the warming and desiccation of the area.

The findings will be useful for both the enhancement of natural values and the protection of monuments on the site, thus ensuring a balance between the need to preserve natural and cultural values, which is an aspect that should be better considered in the management plan of archeological sites [5,10,24,39,49,50].

2. Materials and Methods

2.1. The Study Area: Pasargadae World Heritage Site

The Pasargadae Plain (Figure 1), also known as the Morghab Plain, is one of the vast alluvial sedimentary intermountain basins that is characteristic of the central and eastern part of the Zagros Mountains range [48] which form a plateau at 1400–1800 m above sea level, surrounded by a 2200–2500 m high mountain range [51]. It was formed during the Zagros orogeny, dating back to the Mesozoic era [52], and it is located in a unique position in High Zagros, where it experienced Wurm glaciation and later pluvial stages [51]

Polvar/Sivand River, the principal watercourse of Pasargadae, crosses the region from northeast to southwest, joining the plain of Persepolis downstream [48].

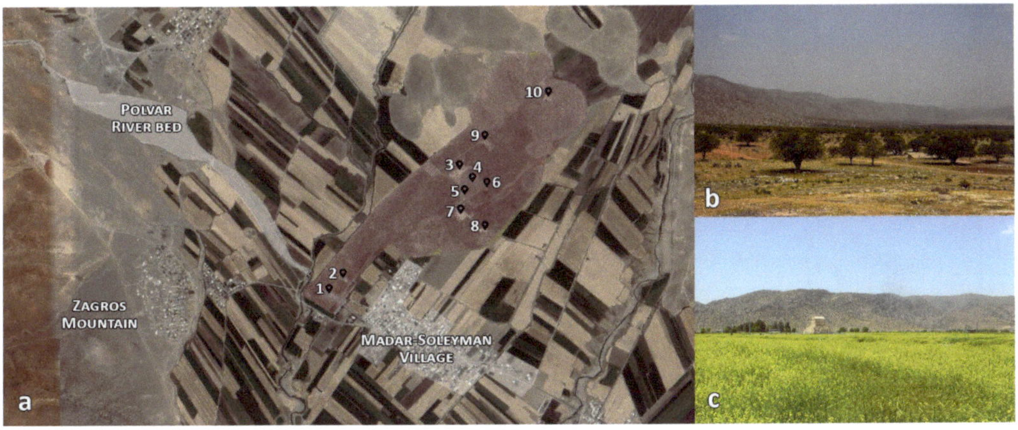

Figure 1. Views of the Pasargadae archaeological area: (**a**) the areal map of the site with the location of monuments in the site, riverbeds, Zagros Mountain, farmlands, and villages; 1. Cyrus the Great Tomb, 2. Caravanserai, 3. Private Palace, 4. Watercourses of Royal Garden, 5. Pavilion B, 6. Pavilion A, 7. Audience Hall, 8. Gate Palace, 9. Stone Tower, 10. Fortification terrace.; (**b**) the landscape of the surrounding area with trees of *Pistacia atlantica* (Babak Sedighi: archive of Pasargadae research center); and (**c**) the landscape of the Cyrus Tomb (Author, May 2019).

Pasargadae holds significant historical and geographical importance in the Fars Province of Iran since it was the location of the first capital of the Achaemenid Empire, the ancient Persian capital founded by Cyrus the Great around 546 BCE, due to the favorable orographic and hydrogeological conditions [45,46]. In fact, the plain is surrounded by a range of hills and mountains radiating from the Zagros Mountains chains, and the conditions of a well-watered basis and a wide area of arable land played an important role in the site's choice and provided a favorable environment for agricultural development [48,53]. The natural landscape of the area not only consists of Zagros and Irano-Turanian biodiversity but has also been impacted by human activities such as agriculture and pastoral activities over centuries. The plain features archaeological remains, notably the tomb of Cyrus the Great, along with other structures showcasing ancient Persian architecture and cultural achievements.

The climate of the Pasargadae area is classified within the Mediterranean xeric continental bioclimate, characteristic of southern Iran and northern Fars [54], which exhibits semi-arid features, including warm and dry summers and relatively cold winters.

Previous elaborations (2006–2021) [6] showed a certain variation in rainfall with average values of approximately only 222.8 mm, primarily concentrated from December to May, including occasional snow at higher altitudes, followed by a dry period from June to September. At the same time, the annual temperature typically ranged between 14.7–17.5 °C, with the lowest absolute temperature recorded at −10.6 °C in February and the highest at 44.4 °C in July. Relative humidity averaged around 39–41%, showcasing significant fluctuations throughout the day, with winter temperatures often falling below 0 °C [47].

2.2. Methodology

2.2.1. Vegetation Sampling

Following the phytosociological approach of the Zurich–Montpellier school [55], the vegetation survey was conducted (May 2022) in sampling areas selected on the basis of homogeneous edaphic and stational conditions in order to avoid ecotones and overlapping among different communities. In total, 33 plots of 10 m × 10 m (Figure 2) were carried out randomly in the different areas, such as highly disturbed areas near the monuments, dry riverbed areas, remnants of the Royal Garden watercourses, semi-natural grasslands, shrublands under several edaphic conditions, and stony and rocky hills. In each plot the vegetation survey was carried out visually estimating the plant coverage index of the Braun-Blanquet scale: + = <1%; 1 = 1–5%; 2 = 5–25%; 3 = 25–50%; 4 = 50–75%; 5 = 75–100%. In addition, the most relevant environmental variables and edaphic factors, such as slope, aspect, altitude, and soil characteristics, were collected.

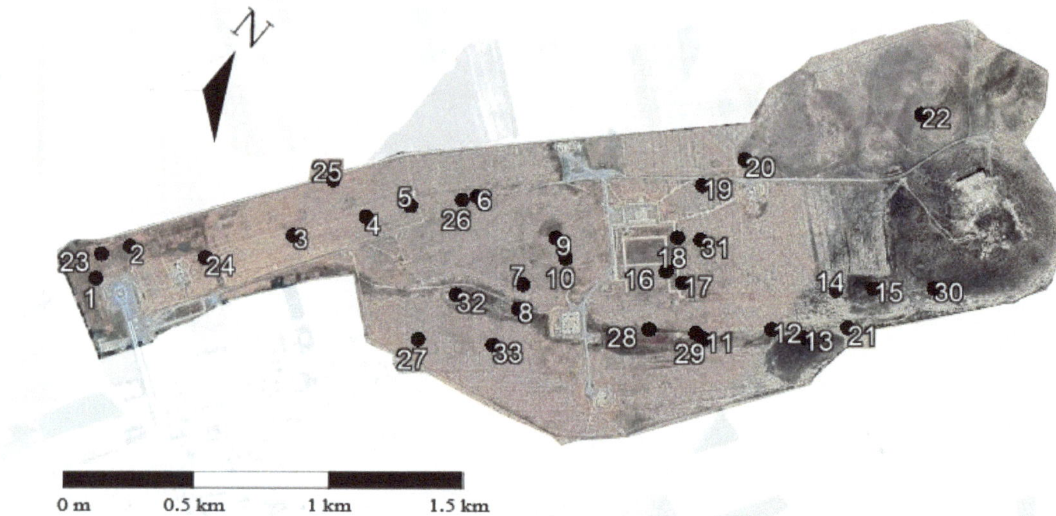

Figure 2. WHS of Pasargadae, Fars Province (Iran): map of the 33 surveys carried out within the site. The sampling locations covered: highly disturbed areas near the monuments (1–3,23,25), dry riverbed area (11,12,14,29,32), remnants of the Royal Garden watercourses (16–18), semi-natural grasslands and shrublands under several edaphic conditions (4–7,10,24,26,28,31,33), and stony and rocky hills (8,9,13,15,19–22,27,30).

For the vascular plant species identification, we used the Flora Iranica [56], comparing data with those obtained from the floristic study of Hosseini et al. [6] and with the herbarium specimens stored in the Herbarium of the University of Roma Tre. The nomenclature followed the "World of Flora Online" [57].

2.2.2. Statistical Elaborations and Syntaxonomic Analysis of Plant Communities

To analyze the different communities based on their plant composition, as well as the similarities and differences between them, a cluster analysis was performed to group plots into vegetation units based on a set of species and cover abundances. Data dissimilarity matrices were calculated using the Bray–Curtis dissimilarity index. A hierarchical cluster analysis was performed on this matrix using the mean agglomeration method (UPGMA), and the optimal number of clusters was determined using the Silhouette index [58]. A dendrogram was derived to illustrate the dissimilarities between samples and species, sorted according to the distance matrices [59]. Furthermore, to study the ecological gradients between the vegetation clusters, an ordination graph of sampling sites was created using the Non-Metric Multidimensional Scaling (NMDS) method. The latter is an unconstrained method which attempts to represent, as closely as possible, the pairwise dissimilarity between objects in a low-dimensional space, unlike maximizing the variance or correspondence between objects in an ordination, as other methods do [60]. We also passively projected the environmental variables measured in the field on the NMDS plots to highlight the ecological drivers between the vegetation clusters.

Furthermore, we conducted a syntaxonomic analysis of the plant communities based on the ecological interpretation, mainly following Zohary [61]. However, given that Zohary's work is somewhat dated and lacks detailed regional information, we compared our findings with the more recent analysis of vegetation types in Fars province [62]. To address this gap, species identified in our study but not listed in Zohary [61] were attributed to specific syntaxa based on the group divisions done by Khodagholi [62].

An investigation on bioindication values of the plant communities was carried out based on the scientific literature, considering the habitats that commonly host the most recurrent species within each community.

2.2.3. Evaluation of Plant and Communities of Conservation Interest

The natural and conservation status of the species found in the plots were cross-referenced with the Red Data Book of Iran [63], scientific literature [6,64–66], and the International Directives of CITES. New records and notes on species from the scientific literature were also evaluated.

2.2.4. Vegetation Mapping

The vegetation map was created through field surveys and photo interpretation. Indeed, for each vegetation type identified in the field, the extent and distribution within the perimeter of the archaeological site were estimated by recording the GPS coordinates of each area boundary The QGIS Software version 3.36 was utilized to map the distribution of the different vegetation units, providing a spatial representation of different plant communities. A photointerpretation of areas was also carried out, comparing collected data with orthophotos taken from Google Satellite.

Subsequently, we produced a map that illustrates the distribution of species with conservation interest, highlighting areas with the highest and lowest abundance among the different clusters.

2.2.5. Preliminary Evaluation of Recent Environmental Changes

The climate analysis involved meteorological data from Fars, sourced from nearby synoptic stations (Persepolis, Safashahr, and Arsenjan) through http://www.irimo.ir, (accessed on 11 July 2023). This analysis included diagrams covering the past 16 years (2006–2022), and we developed the occurring trends. We also conducted analytical diagrams focusing on the most recent 4 years to evaluate possible recent drastic changes.

We also evaluated the occurrence of further anthropic activities which could influence the climatic conditions of the site, analyzing the recent documentation referred to water management [67–71].

3. Results

3.1. Ecological and Syntaxonomic Characteristics of the Plant Communities and Their Bioindication Values

The resulting dendrogram of the 33 plots carried out in the site (Figure 3), based on similarities in species composition and cover abundances, highlighted the presence of seven main distinct clusters, which correspond to: 1. hilly grasslands; 2. grasslands dominated by *Stipa barbata* Desf.; 3. shrublands dominated by *Alhagi maurorum* Medik; 4. Grasslands dominated by *Bellevalia saviczii* Woronow; 5. grasslands dominated by *Glycyrrhiza glabra* L.; 6. ruderal vegetation of the Royal Garden watercourses; and 7. grasslands dominated by *Hordeum murinum* L. Three of them differ the most from all the others, in particular, in order of relevance, the groups 1, 2, and 3. The remaining four groups are divided into two subgroups (groups 4–5 and 6–7), which show a certain internal similarity:

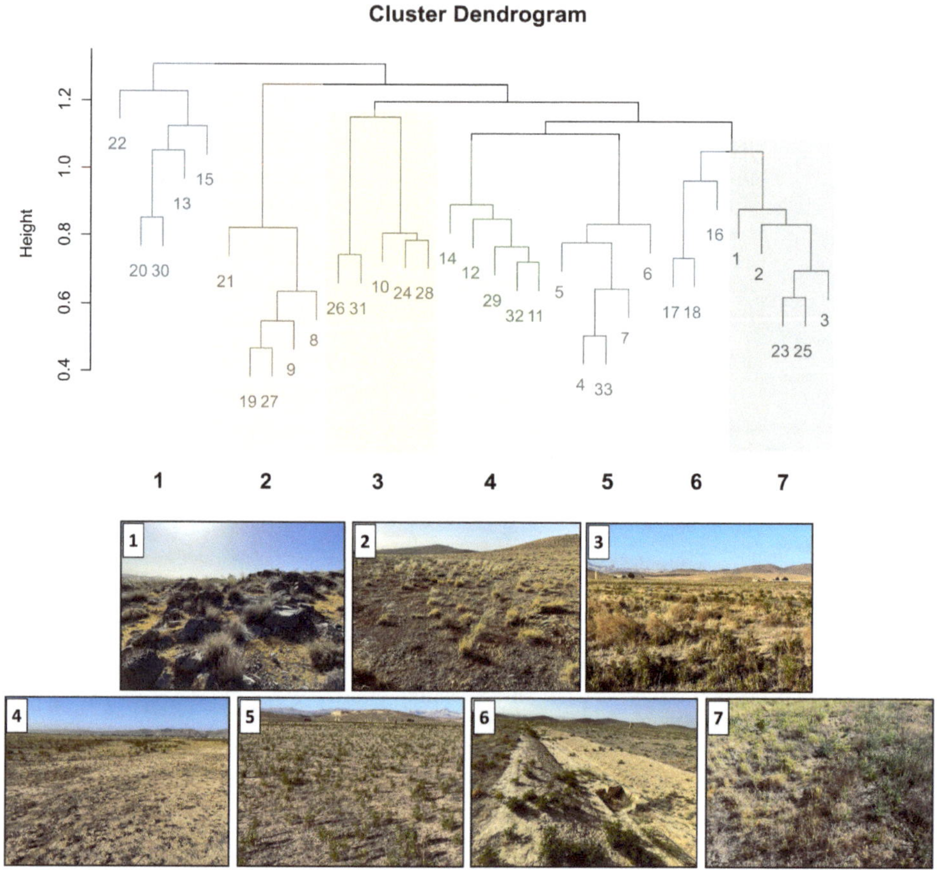

Figure 3. Dendrogram of vegetation trough cover data and corresponding pictures of vegetation types growing in the Pasargadae (May 2022). Plant communities (clusters): 1. hilly grasslands; 2. grasslands dominated by *Stipa barbata* Desf.; 3. shrublands dominated by *Alhagi maurorum* Medik.; 4. Grasslands dominated by *Bellevalia saviczii* Woronow; 5. grasslands dominated by *Glycyrrhiza glabra* L.; 6. ruderal vegetation of the Royal Garden watercourses; and 7. grasslands dominated by *Hordeum murinum* L. The numbers not in bold from 1 to 33 indicate the different sampling areas.

The ordinations with the cluster arrangements in an ecological space resulting from the NMDS are displayed in Figure 4. This ordination confirms the cluster analysis results, regarding the organization of the groups. Two ecological gradients are evident: one along the *x*-axis from cluster 1 to cluster 5, and another along the *y*-axis from clusters 1 and 2 to cluster 7. The *y*-axis variability appears to be influenced by edaphic and geomorphological factors, since the observed hilly grasslands and grasslands dominated by *Stipa barbata* (clusters 1 and 2) grow in rockier and more sloped soils or with the presence of humps, and the other grasslands grow mainly in clay and loamy soils. The variability observed along the *x*-axis seems to be linked mainly to the anthropogenic disturbance, which has the highest values in clusters 7 and 3, characterized also by higher plant covers, and the lowest in clusters 1 and 2.

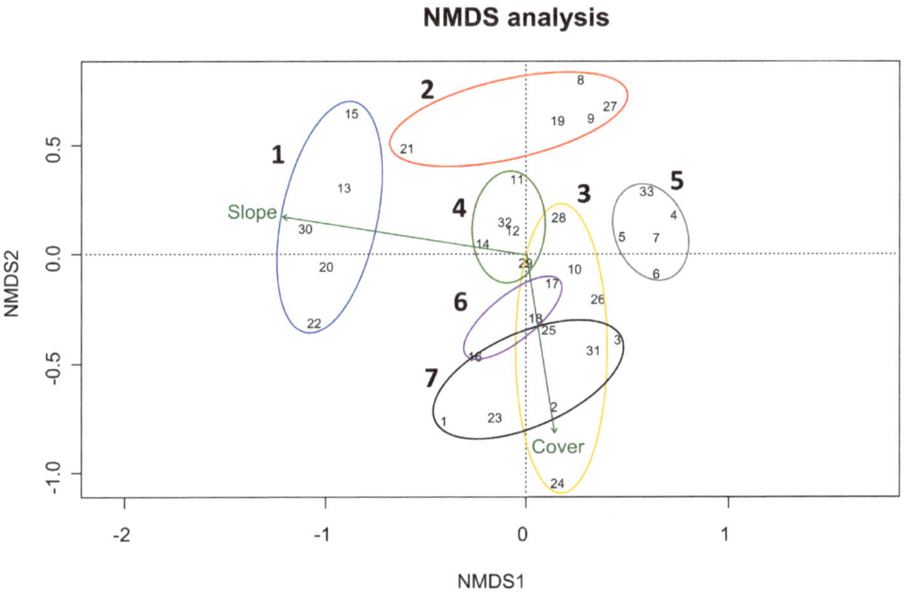

Figure 4. NMDS ordination of the vegetation samples carried out in the Pasargadae WHS, Iran (NMDS stress = 0.14; Shepard plot non-metric fit R^2 = 0.95 and linear fit R^2 = 0.78). Clusters are circled and over-imposed on NMDS plots. 1. Hilly grasslands; 2. Grasslands dominated by *Stipa barbata* Desf.; 3. Shrublands dominated by *Alhagi maurorum* Medik; 4. Grasslands dominated by *Bellevalia saviczii* Woronow; 5. Grasslands dominated by *Glycyrrhiza glabra* L.; 6. Ruderal vegetation of the Royal Garden watercourses; and 7. Grasslands dominated by *Hordeum murinum* L. The numbers not in bold from 1 to 33 indicate the different sampling areas.

The syntaxonomic analysis of the vegetation showed the presence of different alliances and orders of plant communities, which can be subdivided into the main categories of synanthropic vegetation and semi-natural grasslands (Table 1). The classes attributed to each category resulted in *Chenopodietea* Oberd. 1957 and *Secalinetea Orientalia* Zohary 1973, for the synanthropic vegetation, and *Artemisietea herbae-albae iranica* Zohary 1973 and *Astragaletea iranica* Zohary 1973 for the semi-natural grasslands (Tables 1, 2 and S1–S5) [61,62].

Table 1. Syntaxonomic scheme of the main resulting vegetation types.

1. SYNANTHROPIC VEGETATION
1.1 RUDERAL AND SEGETAL VEGETATION OF MAN-MADE HABITATS
CHENOPODIETEA Oberd. 1957
SECALINETEA ORIENTALIA Zohary 1973
Tricetalia 7apsica77a Zohary 1950
Prosopidion farctae segetale Zohary 1973
Triticetalia iranica Zohary 1973
Secalion cereale segetale Zohary 1973
Hulthemion persicae segetale Zohary 1973
2. SEMI-NATURAL GRASSLANDS
2.1 SEMI-DESERTS AND STEPPES VEGETATION
ARTEMISIETEA HERBAE-ALBAE IRANICA Zohary 1973
Artemisietalia iranica typica Zohary 1973
Artemisietalia iranica tragacantha Zohary 1973
ASTRAGALETEA IRANICA Zohary 1973

Table 2. Synoptic Table showing the classes of frequency (from I to V) of the species occurring among the different clusters. Clusters: 1. Hilly grasslands; 2. Grasslands dominated by *Stipa barbata* Desf.; 3. Shrublands dominated by *Alhagi maurorum* Medik; 4. Grasslands dominated by *Bellevalia saviczii* Woronow; 5. Grasslands dominated by *Glycyrrhiza glabra* L.; 6. Ruderal vegetation of the Royal Garden watercourses; and 7. Grasslands dominated by *Hordeum murinum* L. Life forms: Ch = chamaephyte, G = geophyte, H = hemicryptophyte, T = therophyte.

		Synoptic Table						
Clusters	**Life Form**	**1**	**2**	**3**	**4**	**5**	**6**	**7**
***Artemisietalia iranica tragacantha* Zohary 1973**								
Stipa barbata Desf.	H	III	V	I	I		I	I
Launaea acanthodes (Boiss.) Kuntze	H	IV	II	II	V		III	III
Astragalus cancellatus Bunge	H		III	III	III		III	II
Medicago sativa L.	H		II	II	III		I	IV
Helichrysum leucocephalum Boiss.	H	IV						
Salvia macrosiphon Boiss.	Ch	II	I		I	III		II
Noaea mucronata (Forssk.) Asch. & Schweinf.	Ch	III	I		II			II
Centaurea balsamita subsp. *kermanensis* (Bornm.) Wagenitz	T	I	II	II				I
Centaurea bruguierana subsp. *belangeriana* (DC.) Bornm.	T	I			IV			I
Lomelosia olivieri (Coult.) Greuter & Burdet	T	III	I		II			
Senecio glaucus L.	T		I	I				IV
Astragalus cemerinus Beck	Ch	II	I					
Astragalus fasciculifolius Boiss.	H	I						
Hertia angustifolia (DC.) Kuntze	Ch		II					
Picris strigosa M.Bieb.	H	I			II		I	
Astragalus borraginaceus Rech.f.	H		I	II				
Phlomis persica Boiss.	H	I					II	
Zosima absinthifolia Link	H							II
Cousinia gracilis Boiss.	H				II			
Peganum harmala L.	Ch		I	I				
Phlomis aucheri Boiss.	H		II					
Phlomis orientalis Mill.	H	I			I			
Pimpinella aurea DC.	H	I						I
Reseda alba L.	H					I		I
Acantholimon serotinum Rech.f. & Schiman-Czeika	H	I						
Centaurea calcitrapa L.	H				I			
Cousinia leptomera Rech.f.	H					I		
Cousinia nekarmanica Rech.f.	H	I						
Polygonum hyrcanicum Rech.f.	H							I
Silene sisianica Boiss. & Buhse	T	I						
Stachys inflata Benth.	H		I					
***Artemisietalia iranica typica* Zohary 1973**								
Bellevalia saviczii Woronow	G		IV	II	V	IV	I	II
Boissieria squarrosa (Banks & Sol.) Nevski	T	IV	I	IV	IV		III	III
Taeniatherum caput-medusae (L.) Nevski	T	IV	II	II	IV	II		I
Euphorbia dracunculoides Lam.	T	I	III	III	I		II	
Dianthus crinitus subsp. *kermanensis* Rech.fil	H	IV						
Aegilops tauschii Coss.	T			I		I		
Aegilops crassa Boiss.	T				I	I	I	
Lactuca orientalis (Boiss.) Boiss.	Ch		I					I
Stipa hohenackeriana Trin. & Rupr.	H	I	II					
Bellevalia glauca (Lindl.) Kunth	G				II			
Crupina vulgaris Pers. ex Cass.	T	I						I
Diarthron lessertii (Wikstr.) Kit Tan	Ch	I						
Eryngium billardieri Delile	H			I				
Euphorbia sororia Schrenk	T						I	
Stipa lessingiana Trin. & Rupr.	H						I	

Table 2. *Cont.*

Clusters	Life Form	Synoptic Table						
		1	2	3	4	5	6	7
***Astragaletea iranica* Zohary 1973**								
Achillea vermicularis Trin.	H		IV	III	V	V	I	II
Allium sphaerocephalon L.	G	I			III	IV		III
Onosma microcarpum DC.	H	II						
***Secalion cereale segetale* Zohary 1973**								
Glycyrrhiza glabra L.	G	I	IV	V	IV	V	III	IV
Lactuca serriola L.	T	I		IV	I	III	II	V
Zoegea leptaurea L.	T	III	III	II		II		II
Hyoscyamus reticulatus L.	G			II	III	III		
Medicago monantha (C.A.Mey.) Trautv.	T	I	I	I	I	III	I	
Consolida orientalis (J.Gay) Schrödinger	T			I		III		I
Medicago persica (Boiss.) E.Small	T	I	I				II	I
Sisymbrium irio L.	T				I	II		II
Alcea kurdica Alef.	H		III	I				
Centaurea virgata subsp. *squarrosa* (Boiss.) Gugler	H						I	I
Convolvulus argyracanthus Rech. f., Aellen & Esfand.	Ch	II						
Galium tricornutum Dandy	T							I
Sisymbrium irio L.	T	I				I		
Turgenia latifolia (L.) Hoffm.	T							II
Anchusa azurea Mill.	H						I	
Matthiola chenopodiifolia Fisch. & C.A. Mey.	T					I		
Reseda lutea L.	H		I					
***Hulthemion persicae segetale* Zohary 1973 and *Tricetalia iranica* Zohary 1973**								
Bromus tectorum L.	T	II	I	II			III	III
Gundelia tournefortii L.	H	IV	I					
Papaver argemone L.	T					I		III
Carthamus oxyacantha M.Bieb.	T	I						II
Koelpinia linearis Pall.	T				II			I
Valerianella szovitsiana Fisch. & C.A. Mey.	T					III		
Tragopogon graminifolius DC.	H					II		
Camelina hispida Boiss.	T	I						II
Lepidium draba L.	G	I					I	
***Prosopidion farctae segetale* Zohary 1973, *Triticetalia orientalia* Zohary 1949–50 and *Secalinetea orientalia* Zohary 1973**								
Alhagi maurorum Medik.	Ch		IV	V		II	I	
Alhagi pseudalhagi (M. Bieb.) Desv. ex B. Keller & Shap.	H		I	II	I		I	I
Centaurea solstitialis L.	H			I				III
Falcaria vulgaris Bernh.	T			I			I	IV
Gypsophila pilosa Huds.	G							II
Bongardia chrysogonum (L.) Spach	H				I			
Chenopodietea Oberd. (1957)	T							
Hordeum murinum subsp. *glaucum* (Steud.)	T			V	III	III	III	V
Cyanus depressus (M.Bieb.) Soják	H	II		II	III	II	II	V
Onopordum leptolepis DC.	G	III	I	IV	III	II	I	IV
Cynodon dactylon (L.) Pers.	H	I	I	II	V		I	I
Convolvulus arvensis L.	T			II		II	III	
Chardinia orientalis (L.) Kuntze	T	III	I	I				II
Erodium cicutarium (L.) ''Hér.	G			I				II
Scorzonera tunicata Rech.f. & Köie	H				II			
Tragopogon collinus DC	T			II	I		II	
Companions	T							
Nigella oxypetala Boiss.	T	I	III	II	II	V		II
Eremopyrum bonaepartis (Spreng.) Nevski	H		I	III	IV	IV	II	II
Festuca arundinacea Schreb.	H	III	I			III		I
Marrubium crassidens Boiss.	T	I						II

Table 2. *Cont.*

Clusters	Life Form	Synoptic Table						
		1	2	3	4	5	6	7
Crepis sancta subsp. *nemausensis* (P.Fourn.) Babc.	T			I	I			II
Tragopogon porrifolius subsp. *longirostris* (Sch.Bip.) Greuter	H						I	II
Chorispora tenella (Pall.) DC.	T				I			II
Descurainia sophia (L.) Webb ex Prantl	T			I				II
Hordeum spontaneum K.Koch	T			I	I			I
Marrubium vulgare L.	G		I			I		
Nonea caspica (Willd.) G.Don	T	I						II
Siebera nana (DC.) Bornm.	T	I	I					
Agrostis gigantea Roth	T	II						
Filago pyramidata L.	T	II						
Garhadiolus hedypnois Jaub. & Spach	T							II
Plantago lanceolata L.	H				I			
Rochelia disperma (L.f.) K.Koch	T					II		
Scabiosa persica Boiss.	T				II			
Barbarea plantaginea DC.	H				I			
Hordeum vulgare L.	T	I						
Leopoldia tenuiflora (Tausch) Heldr.	G		I					
Minuartia meyeri (Boiss.) Bornm.	T		I					
Moltkia gypsacea Rech.f. & Aellen	H	I						
Muscari neglectum Guss. ex Ten.	G				I			
Prunus arabica (Olivier) Meikle	P	I						
Rhaponticum repens (L.) Hidalgo	H		I					
Scirpoides holoschoenus (L.) Soják	G				I			
Solanum villosum Mill. (heterotypic synonym)	T			I				
Zeravschania membranacea (Boiss.) Pimenov	H						I	

The dominant species for each cluster, as evidenced by Tables 1 and S1–S5, were: for cluster 1, *Launaea acanthodes, Helichrysum leucocephalum,* and *Dianthus crinitus subsp. kermanensis*; for cluster 2, *Stipa barbata*; for cluster 3, *Alhagi maurorum*; for cluster 4, *Bellevalia saviczii*; for cluster 5, *Glycyrrhiza glabra*; for cluster 6, *Convolvulus arvensis* and *Tragopogon collinus;* and for cluster 7, *Hordeum murinum.* Such dominant species enhance the following bioindication values for the different clusters:

1. dry and windy rocky slopes of lands abandoned after extensive grazing [72–74];
2. dry and stony soils of semi-natural habitats [75];
3. disturbed areas and extreme dry conditions [72,76–78];
4. clayey soils [78];
5. silty-sandy alluvial deposits, subject to grazing and post-cultivation in steppe areas [79,80];
6. ruderal areas [17,81];
7. trampled areas [82].

Vegetation found in steppe soils was predominantly co-dominated by species from the genus *Astragalus,* and is commonly associated with extensive pastures undergoing post-abandonment succession.

3.2. Naturalistic Interest of the Species and the Distribution in the Area

Fifteen endemic species of conservation interest were found at the site, and their distribution and conservation status are detailed in Table 3. These endemic species fall into three distribution groups (Table 3). Most are distributed both in Fars and other Iranian regions (12 species); *Cousinia nekarmanica* and *Astragalus cemerinus* were not reported before [6] for the Fars region. Therefore, this site represents their only regional station; *Acantholimon serotinum* is endemic to the Fars region; *Cousinia gracilis* Boiss. Represents a new discovery within the site, which was not reported by Hosseini et al. [6].

Table 3. Endemic species to Iran recorded in the site and their conservation status, defined following Rechinger [56] and the Red Data Book of Iran [63].

Family	Species	Rechinger	Distribution	Conservation Status
Apiaceae	*Zeravschania membranacea* (Boiss.) Pimenov		Fars and other Iran regions	DD
Asteraceae	*Centaurea balsamita* subsp. *kermanensis* (Bornm.) Wagenitz		Fars and other Iran regions	LR
Asteraceae	*Cousinia gracilis* Boiss.	Iran Endm	Fars and other Iran regions	DD
Asteraceae	*Cousinia nekarmanica* Rech.f.		Other Iran regions	LR
Asteraceae	*Helichrysum leucocephalum* Boiss.		Fars and other Iran regions	LR
Asteraceae	*Hertia angustifolia* Kuntze		Fars and other Iran regions	LR
Boraginaceae	*Moltkia gypsacea* Rech.f. & Aellen	Iran Endm	Fars and other Iran regions	LR
Fabaceae	*Astragalus cemerinus* Beck		Other Iran regions	LR
Fabaceae	*Astragalus fasciculifolius* Boiss. subsp. fasciculifolius	Iran Endm	Fars and other Iran regions	LR
Fabaceae	*Medicago persica* (Boiss.) E.Small		Fars and other Iran regions	LR
Lamiaceae	*Phlomis aucheri* Boiss.	Iran Endm	Fars and other Iran regions	LR
Lamiaceae	*Phlomis persica* Boiss.	Iran Endm	Fars and other Iran regions	LR
Plumbaginaceae	*Acantholimon serotinum* Rech.f. & Schiman-Czeika	Iran Endm	Fars	DD
Polygonaceae	*Polygonum hyrcanicum* Rech.f.		Fars and other Iran regions	LR
Solanaceae	*Hyoscyamus kotschyanus* Pojark.	Iran Endm	Fars and other Iran regions	LR

The IUCN Conservation Status of each species obtained from [63] is reported in Table 3, showing that most of the species are considered at Low Risk (LR), but also three species are data deficient (DD), including *Acantholimon serotinum*, which is Fars endemic.

Table 4 shows the distribution of the endemic species with conservation interest among the different clusters. Cluster 1, which corresponds to the hilly areas furthest from visitor paths, is the one that holds the highest number of species with conservation interest, with a total of nine species, including five found exclusively here. Cluster 2, i.e., grasslands dominated by *Stipa barbata*, occurring sporadically within the site, comes next, with five species, two of which were exclusive. Combining the previous data, we can also note that among the semi-natural grasslands, *Artemisietalia iranica tragacantha* Zohary 1973, within the class *Artemisietea herbae-albae iranica* Zohary 1973, resulted in the richest in terms of species composition, particularly within clusters 1 and 2.

Clusters 3, 4, 6, and 7 appear similar, with two or three species, one of which was exclusive, while Cluster 5, which comprises grasslands dominated by *Glycyrrhiza glabra*, exhibited the lowest richness, with only one species and no exclusive ones.

By analyzing the distribution of plant communities, as derived from QGIS 3.36 software, we can note the significant heterogeneity in the distribution and size of the clusters (Figure 5). Clusters 5, 1, and 7 have the largest size, with the first predominantly in the central part of the site, the second covered almost the entire northernmost part, and the third was mainly found in the southernmost part. The remaining clusters exhibit a more scattered presence: clusters 2 and 3 are sparsely distributed within the site, cluster 4 stretches across a central strip, corresponding to a dry riverbed, and cluster 6 was mainly found in a small area of the central part, around the remnants of the Royal Garden watercourses (Figure 5A).

Table 4. Distribution of the endemic species with conservation interest among the different clusters. For each cluster, the species were assigned frequency classes using Roman numerals I to V, except for cluster 6, which contains three reliefs, in which Arabic numerals 1 to 3 were used. The average coverage values are given in superscript. 1. hilly grasslands; 2. grasslands dominated by *Stipa barbata* Desf.; 3. shrublands dominated by *Alhagi maurorum* Medik; 4. Grasslands dominated by *Bellevalia saviczii* Woronow; 5. grasslands dominated by *Glycyrrhiza glabra* L.; 6. ruderal vegetation of the Royal Garden watercourses; and 7. grasslands dominated by *Hordeum murinum* L.

	Clusters						
Species	**1**	**2**	**3**	**4**	**5**	**6**	**7**
Acantholimon serotinum Rech.f. & Schiman-Czeika	I^{+}						
Astragalus cemerinus Beck	II^{1}	I^{+}					
Astragalus fasciculifolius Boiss. subsp. fasciculifolius	I^{5}						
Centaurea balsamita subsp. *kermanensis* (Bornm.) Wagenitz	I^{+}	II^{+}	II^{+}				I^{+}
Cousinia gracilis Boiss.				II^{+}			
Cousinia nekarmanica Rech.f.	I^{+}						
Helichrysum leucocephalum Boiss.	IV^{2}						
Hertia angustifolia Kuntze		II^{1}					
Hyoscyamus kotschyanus Pojark.			II^{+}	III^{+}	III^{+}		
Medicago persica (Boiss.) E.Small	I^{+}	I^{+}				2^{+}	I^{+}
Moltkia gypsacea Rech.f. & Aellen	I^{+}						
Phlomis aucheri Boiss.		II^{+}					
Phlomis persica Boiss.	I^{+}					2^{+}	
Polygonum hyrcanicum Rech.f.							I^{+}
Zeravschania membranacea (Boiss.) Pimenov						1^{+}	
Total	9	5	2	2	1	3	3
Species found exclusively in the specific cluster	5	2	0	1	0	1	1

3.3. Warming and Desiccation of the Area as Threats to Plant Biodiversity

The climatic data from 2006 to 2022 revealed an increasing trend in temperature and a fluctuation in precipitation values, marked by a significant alteration in average precipitation (222.8 mm), with variations from 120 mm to 314 mm. More recently, despite a peak in average rainfall in 2019 (Figure 6a), the overall trend indicated a relevant decrease in precipitation during traditionally rainy months, especially in Spring and in a more limited way throughout the Fall season (Figure 6b–e). This decline is evident when comparing the monthly rain-temperature chart for the last four years, and the lowest precipitation amounts were recorded in 2008, 2017, and 2022.

Adding to the climatic challenges, the Polvar River, like all watercourses in the province of Fars, has been experiencing a gradual desiccation influenced by a combination of increasing warming and other human interventions, such as the construction of dams and the considerable exploitation for irrigation purposes, that further exacerbated the dryness [67,68,70]. Since the mid-1990s, episodes of drought have become increasingly frequent, significantly affecting the river's flow [67,68,83]. Currently, this river is completely dried up, marking a significant shift from its past status as a perennial river.

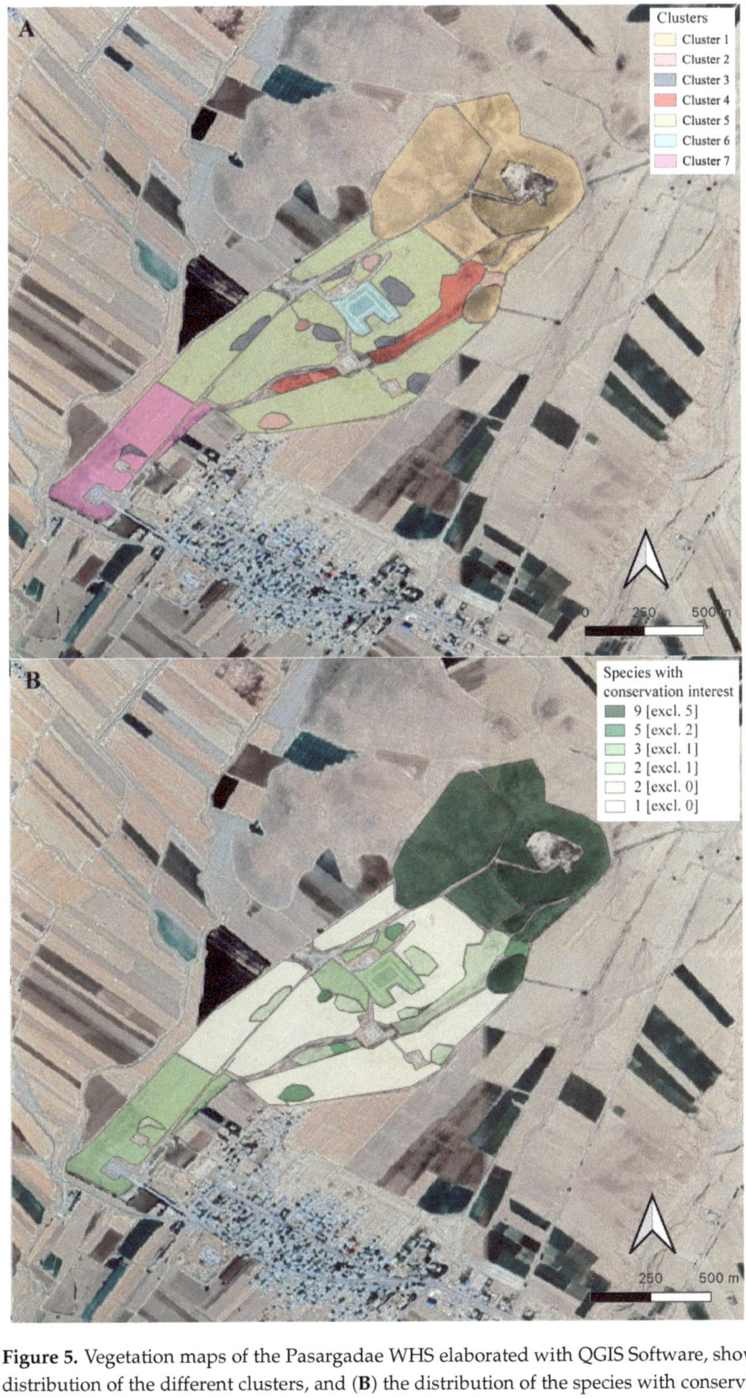

Figure 5. Vegetation maps of the Pasargadae WHS elaborated with QGIS Software, showing: (**A**) the distribution of the different clusters, and (**B**) the distribution of the species with conservation interest that occurred within the clusters considering both the total amount and the presences exclusively found in the specific cluster (in parenthesis).

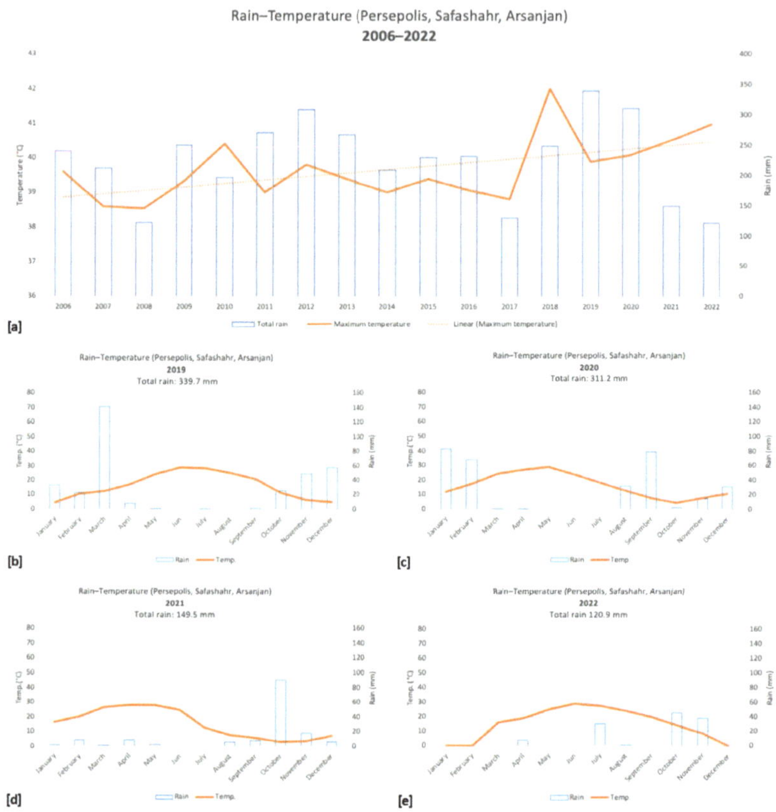

Figure 6. The chart of total rain and maximum temperature of three synoptic Fars stations (Persepolis, Safashar, and Arsenjan): (**a**) the annual chart during 2006–2022; the monthly bioclimatic variation in (**b**) 2019, (**c**) 2020, (**d**) 2021, and (**e**) 2022.

4. Discussion

This pioneering research on vegetation in archaeological sites of Iran contributes to the knowledge of plant communities in archaeological areas, particularly in arid or semi-arid environments, such as in Iran.

Previous studies on the flora of the Pasargadae site and its interaction with stone monuments revealed the importance of understanding the microhabitat of plant colonization on the remaining structures as a tool for controlling biodeterioration phenomena [17,49]. Additionally, the plant diversity of the site [6] and the distribution of the endemic species with conservation interest emphasized the importance of conservation strategies that consider both the natural and cultural values of the site [5,18,24,33,84]. This is particularly important in the context of plant diversity, as the conservation of these sites can help to ensure the protection of plant diversity [39,85].

In this contribution, we have expanded the syntaxonomical attribution of species missing in Zohary [61] to their respective syntaxa, achieved through comparisons with species groupings made by Khodagholi [62].

The presence of the class *Astragaletea iranica* Zohary 1973 is especially noteworthy, as it is typically associated with mountainous environments. Its occurrence in this context likely reflects dynamics related to ecological refuges [61].

The vegetation primarily influenced by human activity is predominantly represented by the *Secalinetea Orientalia* Zohary 1973 class, one of the most prevalent vegetation classes in the Middle East. This class represents the weed communities commonly observed in non-irrigated winter and summer crops. It often occupies abandoned fields that have become partially overgrown by tragacanth astragals. Consequently, this community appears to represent an early post-agricultural phase in the succession towards *Artemisietea herbae-albae iranica* Zohary 1973 communities [61]. This class was more frequent within clusters 3, 4, and 5. The *Chenopodietea* Oberd. (1957) class, notably prevalent in clusters 6 and 7, is characterized by synanthropic vegetation dominated by annual and biennial nitrophilous and semi-nitrophilous species thriving in ruderal and disturbed environments [86]. This component is predominantly found within the site in areas closest to the most trafficked visitor paths and subjected to trampling.

This also explains the similarity in species composition (Figures 3 and 4) and the distribution of the endemic species with conservation interest among the different clusters. Indeed, the highest number of species with conservation interest held by clusters 1 and 2 (Table 4) is in accord with findings made by several authors, namely that when grazing is conducted in a non-intensive manner, it tends to promote greater biodiversity, and the effects of such practices can persist for an extended period of time [87–92].

Additionally, the higher number of species with conservation interest observed in clusters 1 and 2 can be attributed to the microhabitats formed by variations in bedrock and erosion processes. These conditions reduce the competition of dominant species and foster the growth of therophyte species that require environments with lower nutrient and water availability [93,94].

Plant bioindication values have provided significant insights into historical human activities and land-use practices, such as cultivation, corresponding to cluster 5, located in the central and flat parts of the site (see also Figure 5). While the activities proposed by bioindication lack certainty, the likelihood of their occurrence in the region is notably high, particularly in light of the land use management pattern observed in several rangelands of Iran [95]. The land use with extensive grazing has primarily impacted the hilly areas (cluster 1), which are rockier and more sloped, making them less suitable for cultivation. The clayey soils in cluster 4, which likely correspond to a dry riverbed, and the silty-sandy alluvial deposits in areas of cluster 5, indicated fertile soils suitable for cultivation. The ruderal communities in clusters 6 and 7 are instead due, respectively, to the limited availability of soil for the rock outcrop in the remnant of the Royal Garden watercourses and to the influences given from visitor trampling.

Furthermore, the study has highlighted the importance of vegetation maps as a tool for managing different habitats within the site. These maps are instrumental in identifying areas that require protection from anthropogenic pressures such as trampling and mowing, as well as areas hosting invasive species that need to be managed [18,96]. In fact, human activities, including livestock grazing, mowing, and the use of herbicides, also contribute significantly to vegetation distribution [40,97], as seen in the differential coverage around Cyrus the Great's Tomb and the more natural northern hills. This distinction is crucial for developing targeted conservation strategies that respect the site's dual natural and historical significance.

Integrating naturalistic and cultural values in conservation planning ensures not only the preservation of biodiversity, but also the continuity of the landscape's historical narrative [98,99]. Additionally, for the valorization of the Royal Garden of Cyrus the Great, great information could be obtained from vegetation surveys, and analyzing the plant species within and around the presumed location of the lost garden can contribute to the valorization scenario by respecting both ancient and current landscapes [96]. The phytosociological syntaxa could also support the understanding of ancient (cereal and pulse) crop husbandry regimes [100,101].

Finally, the observed recent shifts in precipitation and temperature patterns may have profound implications for the conservation of biodiversity and cultural heritage. The decrease in rainfall of recent years, particularly during critical growth phases, not only may impact phenological patterns and distribution of plants [102–104], but also may threaten species that already have small ranges [105–108], increasing the risk of extinction for many species [109–113]. This is especially concerning for the endemic and restricted-range species found in the study area, which exhibit a heightened vulnerability to environmental change [114]. Such species are vital not only for their intrinsic ecological value, but also as indicators of historical land-use practices, which are part of the cultural narrative of the site.

When looking at our previous floristic data collected in 2019 [6], we can now underline some changes in the dominant species, and an anticipation of the flowering period, such data underscores the increasing stresses induced by the warming and desiccation of the area, posing challenges to the naturalistic value of the site. In general, the impact of climate change in northern Fars will determine a reduction of precipitation in the future [71] and an increase in the duration and severity of drought [115]. Based on the climate scenarios by 2025, the Fars region (in Tashk, Bakhtegan, and Maharlu lakes) will experience a 5.67–15.15% reduction in runoff [116,117].

5. Conclusions

This research has advanced our understanding of vegetation in Iran's archaeological sites, shedding light on plant communities and their interactions with human activity and the natural environment, which have been relatively underexplored in the country. Through vegetation analyses and classification within a syntaxonomic framework, the research has yielded significant findings, particularly in identifying rich species compositions within different vegetation classes. Notably, it highlighted the importance of understanding both natural and human-induced influences on vegetation distribution, emphasizing the critical role of human activities in shaping plant communities.

Furthermore, the bioclimatic analysis in recent years dramatically confirmed the impact of climate change on vegetation patterns and biodiversity, underscoring the urgent need for adaptive conservation strategies. The observed shifts in temperature patterns and the increasing dryness conditions pose significant challenges to both plant diversity and cultural heritage preservation, necessitating proactive measures to mitigate these impacts within the broader context of cultural heritage preservation and sustainable management practices.

Moreover, the work emphasized the importance of integrating ecological and cultural values in conservation planning. By considering both naturalistic and cultural aspects, conservation efforts can ensure the preservation of biodiversity while safeguarding historical monuments.

Supplementary Materials: The following supporting information can be downloaded at: https://www.mdpi.com/article/10.3390/su16093784/s1. Table S1: Analytic table of cluster 1, hilly grasslands; Table S2: Analytic table of cluster 2, grasslands dominated by *Stipa barbata* Desf.; Table S3 Analytic table of cluster 3, shrublands dominated by *Alhagi maurorum* Medik.; Table S4: Analytic table of clusters 4–5, respectively, grasslands dominated by *Bellevalia saviczii* Woronow and grasslands dominated by *Glycyrrhiza glabra* L.; Table S5: Analytic table of clusters 6–7, respectively, ruderal vegetation of remnant of the Royal garden watercourses and grasslands dominated by *Hordeum murinum* L.

Author Contributions: Conceptualization, G.C.; methodology, G.Z., Z.H. and G.C.; investigation G.Z., Z.H. and G.C.; data curation, G.Z. and Z.H.; writing—review and editing, G.Z., Z.H. and G.C. All authors have read and agreed to the published version of the manuscript.

Funding: This research was conducted as part of a PhD study and funded as a PhD scholarship by the University Roma Tre, Science Department. The Grant of Excellence Departments, MIUR, Italy (Art. 1, commi 314–337 legge 232/2016), is gratefully acknowledged.

Institutional Review Board Statement: Not applicable.

Informed Consent Statement: Not applicable.

Data Availability Statement: The data present in this study are available within the text and in the Supplementary Materials.

Acknowledgments: We would like to thank Hamid Fadaei, site manager, and other colleagues at Pasargadae World Heritage Site for their support during field surveys. We would like to mention that any errors are our own and should not tarnish the reputations of these esteemed persons.

Conflicts of Interest: The authors declare no conflicts of interest.

References

1. Mercuri, A.M.; Florenzano, A.; Massamba N'siala, I.; Olmi, L.; Roubis, D.; Sogliani, F. Pollen from Archaeological Layers and Cultural Landscape Reconstruction: Case Studies from the Bradano Valley (Basilicata, Southern Italy). *Plant Biosyst.* **2010**, *144*, 888–901. [CrossRef]
2. Pauknerová, K.; Salisbury, R.B.; Baumanová, M. Human-Landscape Interaction in Prehistoric Central Europe: Analysis of Natural and Built Environments. *Anthropologie* **2013**, *51*, 131–142.
3. Sadori, L.; Allevato, E.; Bellini, C.; Bertacchi, A.; Boetto, G.; Di Pasquale, G.; Giachi, G.; Giardini, M.; Masi, A.; Pepe, C. Archaeobotany in Italian Ancient Roman Harbours. *Rev. Palaeobot. Palynol.* **2015**, *218*, 217–230. [CrossRef]
4. Ackermann, O.; Greenbaum, N.; Bruins, H.; Ayalon, A.; Bar-Matthews, M.; Cabanes, D.; Horwitz, L.K.; Neumann, F.H.; Osband, M.; Porat, N. Ancient Environment and Human Interaction at Tell Eṣ-Ṣâfi/Gath. *Near East. Archaeol.* **2017**, *80*, 244–246. [CrossRef]
5. Caneva, G.; Benelli, F.; Bartoli, F.; Cicinelli, E. Safeguarding Natural and Cultural Heritage on Etruscan Tombs (La Banditaccia, Cerveteri, Italy). *Rend. Lincei* **2018**, *29*, 891–907. [CrossRef]
6. Hosseini, Z.; Bartoli, F.; Cicinelli, E.; Lucchese, F. First Floristic Investigation in Archaeological Sites of Iran: Features and Plant Richness of the Pasargadae World Heritage Site. *Plant Biosyst. Int. J. Deal. All Asp. Plant Biol.* **2023**, *157*, 605–621. [CrossRef]
7. Caneva, G.; Nugari, M.P.; Salvadori, O. *Plant Biology for Cultural Heritage: Biodeterioration and Conservation*; Getty Publications: Los Angeles, CA, USA, 2008; ISBN 0892369396.
8. Cozzolino, A.; Adamo, P.; Bonanomi, G.; Motti, R. The role of lichens, mosses, and vascular plants in the biodeterioration of historic buildings: A review. *Plants* **2022**, *24*, 3429. [CrossRef]
9. Almeida, M.T.; Mouga, T.; Barracosa, P. The Weathering Ability of Higher Plants. The Case of *Ailanthus altissima* (Miller) Swingle. *Int. Biodeterior. Biodegrad.* **1994**, *33*, 333–343. [CrossRef]
10. Caneva, G.; Ceschin, S.; De Marco, G. Mapping the Risk of Damage from Tree Roots for the Conservation of Archaeological Sites: The Case of the Domus Aurea, Rome. *Conserv. Manag. Archaeol. Sites* **2006**, *7*, 163–170. [CrossRef]
11. Motti, R.; Bonanomi, G. Vascular Plant Colonisation of Four Castles in Southern Italy: Effects of Substrate Bioreceptivity, Local Environment Factors and Current Management. *Int. Biodeterior. Biodegrad.* **2018**, *133*, 26–33. [CrossRef]
12. Isola, D.; Bartoli, F.; Langone, S.; Ceschin, S.; Zucconi, L.; Caneva, G. Plant DNA Barcode as a Tool for Root Identification in Hypogea: The Case of the Etruscan Tombs of Tarquinia (Central Italy). *Plants* **2021**, *10*, 1138. [CrossRef] [PubMed]
13. Elgohary, Y.M.; Mansour, M.M.; Salem, M.Z. Assessment of the potential effects of plants with their secreted biochemicals on the biodeterioration of archaeological stones. *Biomass Convers. Biorefinery* **2022**, *26*, 1–5. [CrossRef]
14. Caneva, G.; Galotta, G.; Cancellieri, L.; Savo, V. Tree Roots and Damages in the Jewish Catacombs of Villa Torlonia (Roma). *J. Cult. Herit.* **2009**, *10*, 53–62. [CrossRef]
15. Tiano, P. Biodeterioration of stone monuments: A critical review. In *Recent Advances in Biodeterioration and Biodegradation: Volume 1: Biodeterioration of Cultural Heritage*; Naya Prokash: Calcutta, India, 1993; pp. 301–321.
16. Jain, K.K.; Mishra, A.K.; Singh, T. Biodeterioration of stone: A review of mechanisms involved. In *Recent Advances in Biodeterioration and Biodegredation: Volume 1: Biodeterioration of Cultural Heritage*; Naya Prokash: Calcutta, India, 1993; pp. 323–354.
17. Hosseini, Z.; Zangari, G.; Carboni, M.; Caneva, G. Substrate Preferences of Ruderal Plants in Colonizing Stone Monuments of the Pasargadae World Heritage Site, Iran. *Sustainability* **2021**, *13*, 9381. [CrossRef]
18. Caneva, G. A Botanical Approach to the Planning of Archaeological, Parks in Italy. *Conserv. Manag. Archaeol. Sites* **1999**, *3*, 127–134. [CrossRef]
19. Carcangiu, G.; Casti, M.; Desogus, G.; Meloni, P.; Ricciu, R. Microclimatic Monitoring of a Semi-Confined Archaeological Site Affected by Salt Crystallisation. *J. Cult. Herit.* **2015**, *16*, 113–118. [CrossRef]
20. Fabbri, K.; Canuti, G.; Ugolini, A. A Methodology to Evaluate Outdoor Microclimate of the Archaeological Site and Vegetation Role: A Case Study of the Roman Villa in Russi (Italy). *Sustain. Cities Soc.* **2017**, *35*, 107–133. [CrossRef]
21. Chiusoli, A. *La Scienza Del Paesaggio*; Clueb: Bologna, Italy, 1999; ISBN 8849111762.
22. Bussotti, F.; Grossoni, P.; Batistoni, P.; Ferretti, M.; Cenni, E. Preliminary Studies on the Ability of Plant Barriers to Capture Lead and Cadmium of Vehicular Origin. *Aerobiologia* **1995**, *11*, 11–18. [CrossRef]
23. Caneva, G.; Langone, S.; Bartoli, F.; Cecchini, A.; Meneghini, C. Vegetation Cover and Tumuli's Shape as Affecting Factors of Microclimate and Biodeterioration Risk for the Conservation of Etruscan Tombs (Tarquinia, Italy). *Sustainability* **2021**, *13*, 3393. [CrossRef]
24. Cicinelli, E.; Salerno, G.; Caneva, G. An Assessment Methodology to Combine the Preservation of Biodiversity and Cultural Heritage: The San Vincenzo al Volturno Historical Site (Molise, Italy). *Biodivers. Conserv.* **2018**, *27*, 1073–1093. [CrossRef]
25. Batista, T.; de Mascarenhas, J.M.; Mendes, P.; Pinto-Gomes, C. Assessing Vegetation Heritage Value: The Alentejo Central (Portugal) as a Case Study. *Land* **2021**, *10*, 307. [CrossRef]

26. Batista, T.; Mascarenhas, J.M.D.; Mendes, P.; Pinto-Gomes, C. Methodological proposal for the assessment of vegetation heritage value: Application in Central Alentejo (Portugal). In Proceedings of the ECLAS Conference of Landscape a Place of Cultivation, Porto, Portugal, 21 September 2014; pp. 266–270.
27. Merola, P.; Allegrini, A.; Guglietta, D.; Sampieri, S. Study of Buried Archaeological Sites Using Vegetation Indices. In Proceedings of the Remote Sensing for Environmental Monitoring, GIS Applications, and Geology VI; SPIE, Stockholm, Sweden, 20 October 2006; Volume 6366, pp. 64–75. [CrossRef]
28. Rosen, B.; Galili, E.; Weinstein-Evron, M. Thorny Burnet (*Sarcopoterium spinosum* L.) in a Roman Shipwreck off the Israeli Coast and the Role of Non-Timber Shrubs in Ancient Mediterranean Ships. *Environ. Archaeol.* **2009**, *14*, 163–175. [CrossRef]
29. Bennett, R.; Welham, K.; Hill, R.A.; Ford, A.L.J. The Application of Vegetation Indices for the Prospection of Archaeological Features in Grass-Dominated Environments. *Archaeol. Prospect.* **2012**, *19*, 209–218. [CrossRef]
30. Ceschin, S.; Caneva, G. Plants as Bioindicators for Archaeological Prospection: A Case of Study from Domitian's Stadium in the Palatine (Rome, Italy). *Environ. Monit. Assess.* **2013**, *185*, 5317–5326. [CrossRef]
31. Minissale, P.; Sciandrello, S. The Wild Vascular Flora of the Archaeological Park of Neapolis in Syracuse and Surrounding Areas (Sicily, Italy). *Biodivers. J.* **2017**, *8*, 87–104.
32. Abdelkreem, M.I.M.; Ibrahim, D.A. Check List of Flora and Vegetation of an Archaeological Habitat in North Sudan. *Pyrex J. Biodivers. Conserv.* **2016**, *1*, 1–9.
33. Zangari, G.; Bartoli, F.; Lucchese, F.; Caneva, G. Plant Diversity in Archaeological Sites and Its Bioindication Values for Nature Conservation: Assessments in the UNESCO Site Etruscan Necropolis of Tarquinia (Italy). *Sustainability* **2023**, *15*, 16469. [CrossRef]
34. Pignatti Wikus, E.; Giomi Visentin, M. Ostia Antica and Its Vegetation. *Braun-Blanquetia* **1989**, *3*, 271–278.
35. Lucchese, F.; Pignatti, E. La Vegetazione Nelle Aree Archeologiche Di Roma e Della Campagna Romana. *Quad. Bot. Ambient. Appl.* **2009**, *20*, 3–89.
36. Heneidy, S.Z.; Al-Sodany, Y.M.; Bidak, L.M.; Fakhry, A.M.; Hamouda, S.K.; Halmy, M.W.A.; Alrumman, S.A.; Al-Bakre, D.A.; Eid, E.M.; Toto, S.M. Archeological Sites and Relict Landscapes as Refuge for Biodiversity: Case Study of Alexandria City, Egypt. *Sustainability* **2022**, *14*, 2416. [CrossRef]
37. Albani Rocchetti, G.; Bartoli, F.; Cicinelli, E.; Lucchese, F.; Caneva, G. Linking Man and Nature: Relictual Forest Coenosis with *Laurus nobilis* L. and *Celtis australis* L. in Antica Lavinium, Italy. *Sustainability* **2022**, *14*, 56. [CrossRef]
38. Kanellou, E.; Papafotiou, M.; Saitanis, C.; Economou, G. Ecological Analysis and Opportunities for Enhancement of the Archaeological Landscape: The Vascular Flora of Seven Archaeological Sites in Greece. *Environments* **2024**, *11*, 16. [CrossRef]
39. Minissale, P.; Trigilia, A.; Brogna, F.; Sciandrello, S. Plants and Vegetation in the Archaeological Park of Neapolis of Syracuse (Sicily, Italy): A Management Effort and Also an Opportunity for Better Enjoyment of the Site. *Conserv. Manag. Archaeol. Sites* **2015**, *17*, 340–369. [CrossRef]
40. Matthiesen, H.; Fenger-Nielsen, R.; Harmsen, H.; Madsen, C.K.; Hollesen, J. The Impact of Vegetation on Archaeological Sites in the Low Arctic in Light of Climate Change. *Arctic* **2020**, *73*, 141–152. [CrossRef]
41. Caneva, G.; Pacini, A.; Cutini, M.; Merante, A. The Colosseum Floras as Bio-Indicators of the Climatic Changes in Rome. *Clim. Chang.* **2005**, *70*, 431–443. [CrossRef]
42. Chesson, P.; Gebauer, R.L.E.; Schwinning, S.; Huntly, N.; Wiegand, K.; Ernest, M.S.K.; Sher, A.; Novoplansky, A.; Weltzin, J.F. Resource Pulses, Species Interactions, and Diversity Maintenance in Arid and Semi-Arid Environments. *Oecologia* **2004**, *141*, 236–253. [CrossRef] [PubMed]
43. Cowling, R.M.; Esler, K.J.; Midgley, G.F.; Honig, M.A. Plant Functional Diversity, Species Diversity and Climate in Arid and Semi-Arid Southern Africa. *J. Arid Environ.* **1994**, *27*, 141–158. [CrossRef]
44. Karakuş, Ş.; Tuna, A. Flora of Archaeological Landscape: Case Study of Arslantepe Mound and Its Territory. In *Innovating Strategies and Solutions for Urban Performance and Regeneration*; Springer: Berlin/Heidelberg, Germany, 2022; pp. 303–328. [CrossRef]
45. Boucharlat, R. Pasargadae. *Iran* **2002**, *40*, 279–282. [CrossRef]
46. Stronach, D. Excavations at Pasargadae: First Preliminary Report. *Iran* **1963**, *1*, 19–42. [CrossRef]
47. Chambrade, M.-L.; Gondet, S.; Laisney, D.; Mehrabani, M.; Mohammadkhani, K.; Zareh-Kordshouli, F. The Canal System of Ju-i Dokhtar: New Insight into Water Management in the Eastern Part of the Pasargadae Plain (Fars, Iran). *Water Hist.* **2020**, *12*, 449–476. [CrossRef]
48. Rigot, J.-B.; Gondet, S.; Chambrade, M.-L.; Djamali, M.; Mohammadkhani, K.; Thamó-Bozsó, E. Pulvar River Changes in the Pasargadae Plain (Fars, Iran) during the Holocene and the Consequences for Water Management in the First Millennium BCE. *Quat. Int.* **2022**, *635*, 83–104. [CrossRef]
49. Hosseini, Z.; Caneva, G. Evaluating Hazard Conditions of Plant Colonization in Pasargadae World Heritage Site (Iran) as a Tool of Biodeterioration Assessment. *Int. Biodeterior. Biodegrad.* **2021**, *160*, 105216. [CrossRef]
50. Papafotiou, M.; Bertsouklis, K.F.; Martini, A.N.; Vlachou, G.; Akoumianaki-Ioannidou, A.; Kanellou, E.; Kartsonas, E.D. Evaluation of the Establishment of Native Mediterranean Plant Species Suggested for Landscape Enhancement in Archaeological Sites of Greece. *Acta Hortic.* **2017**, *1189*, 177–180. [CrossRef]
51. Bahrami, B.; Aminzadeh, B. Discovering the Ancient Lakelet of the Archeological Site of Pasargadae in Iran Using a Multi-Layer Technique. *Environ. Geol.* **2007**, *53*, 123–133. [CrossRef]
52. Emami, M.; Eslami, M.; Fadaei, H.; Karami, H.R.; Ahmadi, K. Mineralogical–Geochemical Characterization and Provenance of the Stones Used at the Pasargadae Complex in Iran: A New Perspective. *Archaeometry* **2018**, *60*, 1184–1201. [CrossRef]

53. Sami, A. *Pasargadae: The Oldest Imperial Capital of Iran*; Musavi Printing Office: Shiraz, Iran, 1956.
54. Djamali, M.; Akhani, H.; Khoshravesh, R.; Andrieu-Ponel, V.; Ponel, P.; Brewer, S. Application of the Global Bioclimatic Classification to Iran: Implications for Understanding the Modern Vegetation and Biogeography. *Ecol. Mediterr.* **2011**, *37*, 91–114. [CrossRef]
55. Braun-Blanquet, J. *Plant Sociology. The Study of Plant Communities*, 1st ed.; McGraw-Hill Book Co., Inc.: New York, NY, USA, 1932.
56. Rechinger, K.H.; Rechinger, W. *Flora Iranica: Flora des Iranischen Hochlandes und der Umrahmenden Gebirge: Persien, Afghanistan, Teile von West-Pakistan, Nord-Iraq, Azerbaidjan, Turkmenistan*; Akademische Druck: Graz, Austria, 1978.
57. WFO. World Flora Online. 2020. Available online: http://www.worldfloraonline.org (accessed on 5 November 2023).
58. Rousseeuw, P.J. Silhouettes: A Graphical Aid to the Interpretation and Validation of Cluster Analysis. *J. Comput. Appl. Math.* **1987**, *20*, 53–65. [CrossRef]
59. Cáceres, M.D.; Legendre, P. Associations between Species and Groups of Sites: Indices and Statistical Inference. *Ecology* **2009**, *90*, 3566–3574. [CrossRef]
60. Chahouki, M.A.Z. *Multivariate Analysis Techniques in Environmental Science*; INTECH Open Access Publisher: London, UK, 2011; ISBN 953307468X.
61. Zohary, M. *Geobotanical Foundations of the Middle East*; Gustav Fischer Verlag Press: Stuttgart, Germany, 1973; Volume 1–2.
62. Khodagholi, M. *Ecological Regions of Iranvegetation Types of Fars Province*; Research Institute of Forests and Rangelands: Tehran, Iran, 2017; 337p.
63. Jalili, A.; Jamzad, Z. *Red Data Book of Iran: A Preliminary Survey of Endemic, Rare and Endangered Plant Species in Iran*; Research Institute of Forests and Rangelands: Tehran, Iran, 1999.
64. Sharififar, F.; Koohpayeh, A.; Motaghi, M.M.; Amirkhosravi, A.; Puormohseni Nasab, E.; Khodashenas, M. Study the Ethnobotany of Medicinal Plants in Sirjan, Kerman Province, Iran. *J. Med. Herbs* **2010**, *1*, 19–28.
65. Jalali, M.; Sharifi-Tehrani, M.; Shirmardi, H.-A. Flora of Jahanbin Mountain Area: A Contribution to Flora of the Central Zagros Region of Iran. *J. Genet. Resour.* **2016**, *2*, 26–40. [CrossRef]
66. Parvizi, M.; Sharifi-Tehrani, M.; Jafari, A. Plant Species Diversity in Jokhaneh Plain and Southern Slope of the Nil Mt. in Kohgilouyeh va Boyerahmad Province (Central Zagros Region of Iran). *J. Genet. Resour.* **2016**, *2*, 67–80. [CrossRef]
67. Heydari, M.; Othman, F.; Noori, M. A review of the Environmental Impact of Large Dams in Iran. *Int. J. Adv. Civ. Struct. Environ. Eng. IJACSE* **2013**, *1*, 4.
68. Hajibagheri, R.; Bazaee, A.; Aghamajidi, R. Statistical Estimation of Natural Flood and Discharge Estimation with Different Return Periods (Case Study, Sivand River). *Civ. Proj.* **2022**, *4*, 11–25.
69. Ahmadi, M.H.; Yousefi, H.; Farzin, S.; Rajabpour, R. Management of water resources and demands in Mulla Sadra, Doroodzan and Sivand Dams located in Bakhtegan-Maharlou watershed. *Iran. J. Watershed Manag. Sci. Eng.* **2018**, *12*, 31–41.
70. Samani, N.; Jamshidi, Z. Climate change trend in Fars Province, Iran and its effect on groundwater crisis. In Proceedings of the International Conference of Recent Trends in Environmental Science and Engineering RTESE'17, Toronto, ON, Canada, 23–25 August 2017; p. 133. [CrossRef]
71. Naderi, M. Extreme climate events under global warming in northern Fars Province, southern Iran. *Theor. Appl. Climatol.* **2020**, *142*, 1221–1243. [CrossRef]
72. Mohseni, M.R.; Rad, S.P. The Effect of Edaphic Factors on the Distribution and Abundance of Ants (*Hymenoptera*: *Formicidae*) in Iran. *Biodivers. Data J.* **2021**, *9*, e54843. [CrossRef]
73. Noroozi, J.; Talebi, A.; Doostmohammadi, M.; Bagheri, A. The Zagros Mountain Range. In *Plant Biogeography and Vegetation of High Mountains of Central and South-West Asia*; Springer: Berlin/Heidelberg, Germany, 2020; pp. 185–214.
74. Nadaf, M.; Ejtehadi, H.; Mesdaghi, M.; Farzam, M. Recognizing Ecological Species Groups and Their Relationships with Environmental Factors at Chamanbid-Jozak Protected Area, North Khorasan Province, Iran. *J. Rangel. Sci.* **2017**, *7*, 253–264.
75. Khajeddin, S.J.; Yeganeh, H. Plant Communities of the Karkas Hunting-Prohibited Region, Isfahan-Iran. *Plant Soil Environ.* **2008**, *54*, 347–358. [CrossRef]
76. Iqbal, U.; Ali, A.; Daad, A.; Aslam, M.U.; Rehman, F.U.; Farooq, U.; Gul, M.F. Unraveling the Defensive Strategies of Camel Thorn Alhagi Maurorum Medik. For Thriving in Arid and Semi-Arid Environments. *J. Arid Environ.* **2023**, *219*, 105076. [CrossRef]
77. Salama, F.M.; Abd El-Ghani, M.M.; Gaafar, A.E.; Hasanin, D.M.; Abd El-Wahab, D.A. Adaptive Eco-Physiological Mechanisms of Alhagi Graecorum in Response to Severe Aridity in the Western Desert of Egypt. *Plant Biosyst. Int. J. Deal. All Asp. Plant Biol.* **2022**, *156*, 528–537. [CrossRef]
78. Ghahraman, A.; Heydari, J.; Atar, F.; Hamzehei, B. A Floristic Study of the Southwestern Slopes of Binaloud Elevations (Iran: Khorassan Province). *J. Sci.* **2006**, *32*, 1–12.
79. Rahchamani, R.; Faramarzi, M.; Moslemipor, F.; Bayat Kohsar, J. Effect of Supplementing Sheep Diet with *Glycyrrhiza glabra* and *Urtica dioica* Powder on Growth Performance, Rumen Bacterial Community and Some Blood Biochemical Constituents. *Iran. J. Appl. Anim. Sci.* **2019**, *9*, 95–103.
80. Roozitalab, M.H.; Siadat, H.; Farshad, A. *The Soils of Iran*; Springer: Berlin/Heidelberg, Germany, 2018; ISBN 3319690485.
81. Pourrezaei, J.; Khajeddin, S.J.; Karimzadeh, H.R.; Vahabi, M.R.; Mozaffarian, V.A.; Esfahani, M.T. Phytogeographical Distribution of Roadside Flora along the Plain to Mountainous Natural Areas (Northern Khorasan Province, Iran). *Flora* **2017**, *234*, 92–105. [CrossRef]
82. Davison, A.W. The Ecology of *Hordeum murinum* L.: II. The Ruderal Habit. *J. Ecol.* **1971**, *59*, 493–506. [CrossRef]

83. Brisset, E.; Djamali, M.; Bard, E.; Borschneck, D.; Gandouin, E.; Garcia, M.; Stevens, L.; Tachikawa, K. Late Holocene hydrology of Lake Maharlou, southwest Iran, inferred from high-resolution sedimentological and geochemical analyses. *J. Paleolimnol.* **2019** *61*, 111–128. [CrossRef]
84. Taylor, K.; Lennon, J. Cultural Landscapes: A Bridge between Culture and Nature? *Int. J. Herit. Stud.* **2011**, *17*, 537–554. [CrossRef]
85. Davis, S.D.; Heywood, V. *Centres of Plant Diversity: A Guide and Strategy for Their Conservation: Volume 1: Europe, Africa, South West Asia and the Middle East*; World Conservation Union: London, UK, 1994.
86. Mucina, L.; Bültmann, H.; Dierßen, K.; Theurillat, J.; Raus, T.; Čarni, A.; Šumberová, K.; Willner, W.; Dengler, J.; García, R.G Vegetation of Europe: Hierarchical Floristic Classification System of Vascular Plant, Bryophyte, Lichen, and Algal Communities *Appl. Veg. Sci.* **2016**, *19*, 3–264. [CrossRef]
87. Fóti, S.; Bartha, S.; Balogh, J.; Pintér, K.; Koncz, P.; Biró, M.; Süle, G.; Petrás, D.; De Luca, G.; Mészáros, Á. Fluctuations and Trends in Spatio-temporal Patterns of Plant Species and Diversity in a Sandy Pasture. *J. Veg. Sci.* **2023**, *34*, e13190. [CrossRef]
88. Kun, R.; Babai, D.; Csathó, A.I.; Erdélyi, A.; Hartdégen, J.; Lengyel, A.; Kálmán, N.; Mártonffy, A.; Hábenczyus, A.A.; Szegleti, Z Effects of Management Complexity on the Composition, Plant Functional Dominance Relationships and Physiognomy of High Nature Value Grasslands. *Nat. Conserv.* **2024**, *55*, 1–19. [CrossRef]
89. Marini, L.; Fontana, P.; Scotton, M.; Klimek, S. Vascular Plant and Orthoptera Diversity in Relation to Grassland Management and Landscape Composition in the European Alps. *J. Appl. Ecol.* **2008**, *45*, 361–370. [CrossRef]
90. Marini, L.; Scotton, M.; Klimek, S.; Isselstein, J.; Pecile, A. Effects of Local Factors on Plant Species Richness and Composition of Alpine Meadows. *Agric. Ecosyst. Environ.* **2007**, *119*, 281–288. [CrossRef]
91. Orlandi, S.; Probo, M.; Sitzia, T.; Trentanovi, G.; Garbarino, M.; Lombardi, G.; Lonati, M. Environmental and Land Use Determinants of Grassland Patch Diversity in the Western and Eastern Alps under Agro-Pastoral Abandonment. *Biodivers. Conserv.* **2016**, *25*, 275–293. [CrossRef]
92. Rahmanian, S.; Hejda, M.; Ejtehadi, H.; Farzam, M.; Pyšek, P.; Memariani, F. Effects of Livestock Grazing on Plant Species Diversity Vary along a Climatic Gradient in Northeastern Iran. *Appl. Veg. Sci.* **2020**, *23*, 551–561. [CrossRef]
93. Grace, J.B. The Factors Controlling Species Density in Herbaceous Plant Communities: An Assessment. *Perspect. Plant Ecol. Evol Syst.* **1999**, *2*, 1–28. [CrossRef]
94. Wellstein, C.; Campetella, G.; Spada, F.; Chelli, S.; Mucina, L.; Canullo, R.; Bartha, S. Context-Dependent Assembly Rules and the Role of Dominating Grasses in Semi-Natural Abandoned Sub-Mediterranean Grasslands. *Agric. Ecosyst. Environ.* **2014**, *182*, 113–122. [CrossRef]
95. Moradi, E.; Heshmati, G.A.; Ghilishli, F.; Mirdeilami, S.Z.; Pessarakli, M. Grazing Intensity and Environmental Factors Effects on Species Composition and Diversity in Rangelands of Iran. *J. Plant Nutr.* **2016**, *39*, 2002–2014. [CrossRef]
96. Hosseini, Z.; Caneva, G. Lost Gardens: From Knowledge to Revitalization and Cultural Valorization of Natural Elements *Sustainability* **2022**, *14*, 2956. [CrossRef]
97. Dai, A. Drought under global warming: A review. *Wiley Interdiscip. Rev. Clim. Chang.* **2011**, *2*, 45–65. [CrossRef]
98. Crawford, G.W. Palaeoethnobotanical Contributions to Human-Environment Interaction. In *Environmental archaeology: Current Theoretical and Methodological Approaches*; Springer: Berlin/Heidelberg, Germany, 2018; pp. 155–180. [CrossRef]
99. Pavlik, B.M.; Louderback, L.A.; Vernon, K.B.; Yaworsky, P.M.; Wilson, C.; Clifford, A.; Codding, B.F. Plant Species Richness at Archaeological Sites Suggests Ecological Legacy of Indigenous Subsistence on the Colorado Plateau. *Proc. Natl. Acad. Sci. USA* **2021**, *118*, e2025047118. [CrossRef]
100. Halbrooks, M.C. The English Garden at Stan Hywet Hall and Gardens: Interpretation, Analysis, and Documentation of a Historic Garden Restoration. *Horttechnology* **2005**, *15*, 196–213. [CrossRef]
101. Jones, G. Weed Phytosociology and Crop Husbandry: Identifying a Contrast between Ancient and Modern Practice. *Rev. Palaeobot Palynol.* **1992**, *73*, 133–143. [CrossRef]
102. Guo, X.Y.; Zhang, H.Y.; Wang, Y.Q. Comparison of the spatio-temporal dynamics of vegetation between the Changbai Mountains of eastern Eurasia and the Appalachian Mountains of eastern North America. *J. Mt. Sci.* **2018**, *15*, 1–12. [CrossRef]
103. Deng, S.F.; Yang, T.B.; Zeng, B.; Zhu, X.F.; Xu, H.J. Vegetation cover variation in the Qilian Mountains and its response to climate change in 2000–2011. *J. Mt. Sci.* **2013**, *10*, 1050–1062. [CrossRef]
104. Kayiranga, A.; Ndayisaba, F.; Nahayo, L.; Karamage, F.; Nsengiyumva, J.B.; Mupenzi, C.; Nyesheja, E.M. Analysis of climate and topography impacts on the spatial distribution of vegetation in the Virunga Volcanoes Massif of East-Central Africa. *Geosciences* **2017**, *7*, 17. [CrossRef]
105. Manne, L.L.; Pimm, S.L. Beyond Eight Forms of Rarity: Which Species Are Threatened and Which Will Be Next? In *Animal Conservation Forum*; Cambridge University Press: Cambridge, UK, 2001; Volume 4, pp. 221–229.
106. Mawdsley, J.R.; O'Malley, R.; Ojima, D.S. A Review of Climate-change Adaptation Strategies for Wildlife Management and Biodiversity Conservation. *Conserv. Biol.* **2009**, *23*, 1080–1089. [CrossRef] [PubMed]
107. Behroozian, M.; Ejtehadi, H.; Townsend Peterson, A.; Memariani, F.; Mesdaghi, M. Climate Change Influences on the Potential Distribution of Dianthus Polylepis Bien. Ex Boiss. (Caryophyllaceae), an Endemic Species in the Irano-Turanian Region. *PLoS ONE* **2020**, *15*, e0237527. [CrossRef]
108. Velásquez-Tibatá, J.; Salaman, P.; Graham, C.H. Effects of Climate Change on Species Distribution, Community Structure, and Conservation of Birds in Protected Areas in Colombia. *Reg. Environ. Chang.* **2013**, *13*, 235–248. [CrossRef]

109. Nabout, J.C.; Magalhães, M.R.; de Amorim Gomes, M.A.; Da Cunha, H.F. The Impact of Global Climate Change on the Geographic Distribution and Sustainable Harvest of Hancornia Speciosa Gomes (Apocynaceae) in Brazil. *Environ. Manag.* **2016**, *57*, 814–821. [CrossRef]
110. Xu, X.; Zhang, H.; Xie, T.; Xu, Y.; Zhao, L.; Tian, W. Effects of Climate Change on the Potentially Suitable Climatic Geographical Range of Liriodendron Chinense. *Forests* **2017**, *8*, 399. [CrossRef]
111. Cuena-Lombraña, A.; Fois, M.; Fenu, G.; Cogoni, D.; Bacchetta, G. The Impact of Climatic Variations on the Reproductive Success of Gentiana Lutea L. in a Mediterranean Mountain Area. *Int. J. Biometeorol.* **2018**, *62*, 1283–1295. [CrossRef]
112. Guisan, A.; Holten, J.I.; Spichiger, R.; Tessier, L. *Potential Ecological Impacts of Climate Change in the Alps and Fennoscandian Mountains. An Annex to the Intergovernmental Panel on Climate Change Second Assessment Report, Working Group II-C (Impacts of Climate Change on Mountain Regions)*; Conservatoire et Jardin Botaniques: Geneva, Switzerland, 1995.
113. Aguirre-Gutiérrez, J.; van Treuren, R.; Hoekstra, R.; van Hintum, T.J.L. Crop Wild Relatives Range Shifts and Conservation in Europe under Climate Change. *Divers. Distrib.* **2017**, *23*, 739–750. [CrossRef]
114. Casazza, G.; Giordani, P.; Benesperi, R.; Foggi, B.; Viciani, D.; Filigheddu, R.; Farris, E.; Bagella, S.; Pisanu, S.; Mariotti, M.G. Climate Change Hastens the Urgency of Conservation for Range-Restricted Plant Species in the Central-Northern Mediterranean Region. *Biol. Conserv.* **2014**, *179*, 129–138. [CrossRef]
115. Modarres, R.; Sarhadi, A.; Burn, D.H. Changes of extreme drought and flood events in Iran. *Glob. Planet. Chang.* **2016**, *144*, 67–81. [CrossRef]
116. Soltani, M.; Laux, P.; Kunstmann, H.; Stan, K.; Sohrabi, M.M.; Molanejad, M.; Sabziparvar, A.A.; Ranjbar SaadatAbadi, A.; Ranjbar, F.; Rousta, I.; et al. Assessment of climate variations in temperature and precipitation extreme events over Iran. *Theor. Appl. Climatol.* **2016**, *126*, 775–795. [CrossRef]
117. Karimi, V.; Karami, E.; Keshavarz, M. Climate change and agriculture: Impacts and adaptive responses in Iran. *J. Integr. Agric.* **2018**, *17*, 1–15. [CrossRef]

sustainability

MDPI

Article

Spatio-Temporal Variation in Landforms and Surface Urban Heat Island in Riverine Megacity

Namita Gorai [1], Jatisankar Bandyopadhyay [1], Bijay Halder [2,3], Minhaz Farid Ahmed [4], Altaf Hossain Molla [5,*] and Thomas M. T. Lei [6,*]

1 Department of Remote Sensing and GIS, Vidyasagar University, Midnapore 721102, India; namitagorai2000@gmail.com (N.G.); jatib@mail.vidyasagar.ac.in (J.B.)

2 Department of Earth Sciences and Environment, Faculty of Sciences and Technology, Universiti Kebangsaan Malaysia (UKM), Bangi 43600, Selangor, Malaysia; halder06bijay@gmail.com

3 New Era and Development in Civil Engineering Research Group, Scientific Research Center, Al-Ayen University, Nasiriyah 64001, Iraq

4 Institute for Environment and Development (LESTARI), Universiti Kebangsaan Malaysia (UKM), Bangi 43600, Selangor, Malaysia; minhaz@ukm.edu.my

5 Department of Mechanical and Manufacturing Engineering, Faculty of Engineering and Built Environment, Universiti Kebangsaan Malaysia (UKM), Bangi 43600, Selangor, Malaysia

6 Institute of Science and Environment, University of Saint Joseph, Macau, China

* Correspondence: altafhossain1994@gmail.com (A.H.M.); thomas.lei@usj.edu.mo (T.M.T.L.)

Abstract: Rapid urbanization and changing climatic procedures can activate the present surface urban heat island (SUHI) effect. An SUHI was considered by temperature alterations among urban and rural surroundings. The urban zones were frequently warmer than the rural regions because of population pressure, urbanization, vegetation insufficiency, industrialization, and transportation systems. This investigation analyses the Surface-UHI (SUHI) influence in Kolkata Municipal Corporation (KMC), India. Growing land surface temperature (LST) may cause an SUHI and impact ecological conditions in urban regions. The urban thermal field variation index (UTFVI) served as a qualitative and quantitative barrier to the SUHI susceptibility. The maximum likelihood approach was used in conjunction with supervised classification techniques to identify variations in land use and land cover (LULC) over a chosen year. The outcomes designated a reduction of around 1354.86 Ha, 653.31 Ha, 2286.9 Ha, and 434.16 Ha for vegetation, bare land, grassland, and water bodies, correspondingly. Temporarily, from the years 1991–2021, the built-up area increased by 4729.23 Ha. The highest LST increased by around 7.72 °C, while the lowest LST increased by around 5.81 °C from 1991 to 2021. The vegetation index and LST showed a negative link, according to the correlation analyses; however, the built-up index showed an experimentally measured positive correlation. This inquiry will compel the administration, urban planners, and stakeholders to observe humanistic activities and thus confirm sustainable urban expansion.

Keywords: climate change; ecological disturbance; heat island; urban environment; remote sensing and GIS

check for updates

Citation: Gorai, N.; Bandyopadhyay, J.; Halder, B.; Ahmed, M.F.; Molla, A.H.; Lei, T.M.T. Spatio-Temporal Variation in Landforms and Surface Urban Heat Island in Riverine Megacity. *Sustainability* **2024**, *16*, 3383. https://doi.org/10.3390/su16083383

Academic Editors: Jun Qin and Hou Jiang

Received: 7 March 2024
Revised: 13 April 2024
Accepted: 15 April 2024
Published: 18 April 2024

1. Introduction

With increasing industrialization, urbanization, and population growth, the biological landscapes have been deformed into impermeable surfaces related to building structures, apartments, roads, parking lots, and urban infrastructure [1,2]. It is established that land surface imperviousness was a dangerous constituent influence on the quality of the urban environment [3,4]. Consequently, a significant portion of Kolkata's natural catchment parts have been developed into urban areas. This urbanization process has altered the physical features of the land's surface, including soil moisture, heat transfer efficiency, thermodynamic irradiance characteristics, and albedo. Variations in the land surface temperature (LST) have the greatest ecological impact since they directly affect human comfort levels, air quality, building energy demand, convective, and latent heat transfer procedures [5–8]. The LST has measured the surface temperature values of the Earth's crust, where reflectance values are observed through space-based observation. Many investigators worldwide have used LST widely in LULC modification assessments to ascertain environmental issues of a particular area [9–12].

Due to the dynamics of green spaces, water scarcity, soil moisture loss, low infiltration rate, and erosion, the urban heat differential has a significant influence on how susceptible an urban environment is [13–15]. Along with environmental degradation, human activities and global climate conditions have significantly affected resident and global land adjustment [16–18]. In the urban region, health subjects like lung cancer, asthma, skin disorders, respiratory ailments, and other health-related challenges were also common [19]. Urban areas are also often the site of additional pollution-related issues as well as the SUHI result [20–22]. The SUHI is an index to identify the heat island effects on the Earth's surface with the help of LST and some methods. This technique is applied to build a heat island effects study and implement awareness and strategies. The SUHI causes pollutants and lowers the air quality in the urban region, which has an impact on the surrounding natural environment and local urban ecology. Due to SUHI stress, SUHIs not only impair human health but also have the potential to increase death rates [23]. In urban areas like Kolkata, the transportation and public industries contributed to air pollution [24]. In this scenario, proper urban management and planning provided further advantages for sustainable growth.

According to some investigation results, the single channel (SC) technique has the lowest accuracy, and the split window (SW) algorithm has reasonable accuracy. Still, LST, overturned from the radiative transfer equation-based technique applying band 10, has the highest inaccuracy with an RMSE lower than 1K [25]. By cross-referencing and linking the Moderate Resolution Imaging Spectroradiometer (MODIS) with Geostationary Ocean Color Imager (GOCI)-derived NDVIs in addition to in situ NDVI dimensions, the researcher evaluated the Landsat 8 OLI/TIRS-derived NDVI characteristics in contrast to Landsat 7 ETM+ [26]. The reproductions of surface reflectance and Top of Atmosphere (TOA) reflectance of broadleaf water and trees are shown for the Landsat 8 OLI/TIRS, MODIS, and Landsat 7 ETM^{+}, to appraise the influence of bandpass alteration on the NDVI calculation [27]. The NDVI is a space-based vegetation monitoring method to identify the green space available on the Earth's surface through different bands in the satellite.

Another examination outcome designates how the stretched (or 'universal') triangle can be applied to understand pixel outlines inside the triangle, and presents how the temporal trajectories of opinions exclusively designate decorations of the LULC change [28]. Finally, we conclude the research with a succinct assessment of the limitations of the method. According to some scientists, over the last 27 years, the built-up zone of Ismailia has gradually increased. The areas with the highest surface radiant temperatures are the barren land (37.34 °C in 1984 and 42.801 °C in 2011) and the built-up region (37.65 °C in 1984 and 43.876 °C in 2011). Vegetated surfaces (28.73 °C in 1984 and 32.96 °C in 2011) have the lowest surface radiant temperatures [29]. The investigation revealed a strong agreement between the real-time ground and satellite RS datasets, with correlation coefficient (R^2) values of 0.90 [30]. The 28-year period (1984–2011) saw significant variations in LST in the Salt Lake Basin area, with an estimated 2 °C shift. This was revealed using themed catalogue diagrams created from remotely sensed and modified satellite imagery.

Urban regions progressively established the amenity's expansion with high reduction buildings. Those belongings were mostly additionally impactful for activating the SUHI effects over the examining area. The UTFVI was also applied to calculate the ecological diversification identified through LST datasets and the notified formulation. Kolkata Municipal Corporation (KMC) progressively documented high green space losses because of urban expansion. Those belongings are likewise triggering land subsidence-related difficulties with high groundwater shortages. Consequently, the SUHI examination and green space examination are essential for improved urban planning. Current results also showed that, during the summer, Delhi's heavily populated metropolitan area has a greater thermal inertia than the nearby rural areas [31]. The main research gap in this area is in the details of SUHI analysis with ecological variation. Most of the studies applied a ten-year gap of UHI analysis or only LULC-related analysis. In addition, this examination applied a five-year gap-wise landforms analysis with surface temperature measurement and heat island studies. The accuracy assessment of the classification maps and year-wise change detection was also applied to identify the landform alteration in the KMC area. Gradually, different geo-spatial indices with five-year information provide more landform alteration-related evidence. The built-up land, water, soil moisture, bareness analysis, and vegetation scenarios are useful for planning details, management, and novel adaptation strategies toward sustainable urban development. The spatial fluctuation in thermal inertia over the region is starting to make sense of the presence of the Cool Island during the day. The primary goal of this study is to use statistical modelling and RS-based datasets to determine the decadal LULC modification of Kolkata's megacity. The outcomes of this investigation include (a) imageries (LULC) classification from 1991 to 2021 with built-up expansion and vegetation losses; (b) the LST approximation for the thermal condition measurement with SUHI and UTFVI alteration examination; (c) certain geo-spatial indicators for ecological and the environmental impact assessment. These examinations may help the local planners with forthcoming management planning and adaptation strategies to shelter the megacity's environment.

2. Materials and Methods

2.1. Study Area

The British Empire and the East India Company founded the imperial metropolis of Kolkata. The British Indian Empire had its capital at the megacity of Kolkata until 1911 when it was moved to Delhi. To grow the second metropolis of the British Indian Empire, this megacity was created quickly in the 19th century. This was addressed through the fusion of Arabian tradition with Indian philosophy in cultural development. Kolkata is also well known for its creative past, which spans from India to trade the union, and Marxist Naxalite movements. Kolkata, also known as the "City of Palaces", "Cultural Capital of India", "City of Joy", and "The City of Processions", has also been the residence of prominent politicians and yogis. The pandemic that struck Kolkata in 1653 brought with it challenges associated with hasty urbanization, and the city's remnants serve as a model for the urbanization initiatives of less developed countries. The monthly mean temperatures are 19–30 °C (66–86 °F); the annual mean temperature is roughly 26.8 °C (80.2 °F). The summer months of March through June are hot and muggy, with highs in the low 30s; during periods of drought, the highest temperatures ever recorded were higher than 40 °C (104 °F) in May and June (Figure 1). The highest known temperature is around 43.9 °C (111.0 °F), and the lowest is around 5 °C (41 °F). The southwest summer monsoon's Bay of Bengal branch brought rainfall to Kolkata between June and September, giving the city its highest annual precipitation total of roughly 1850 mm (73 in). The provisional Census of India says that 4,496,694 people were living in this megacity in 2011; 2,356,766 of those people were female and 2,139,928 were male. Despite having a megacity population of 4,496,694, Kolkata has an urban/metropolitan population of around 14,035,959, with 6,784,051 women and 7,251,908 men. The oldest disruptive mass transit system in India is the Kolkata Metro, which was developed in 1984. Kolkata, a megacity has five long-distance lines.

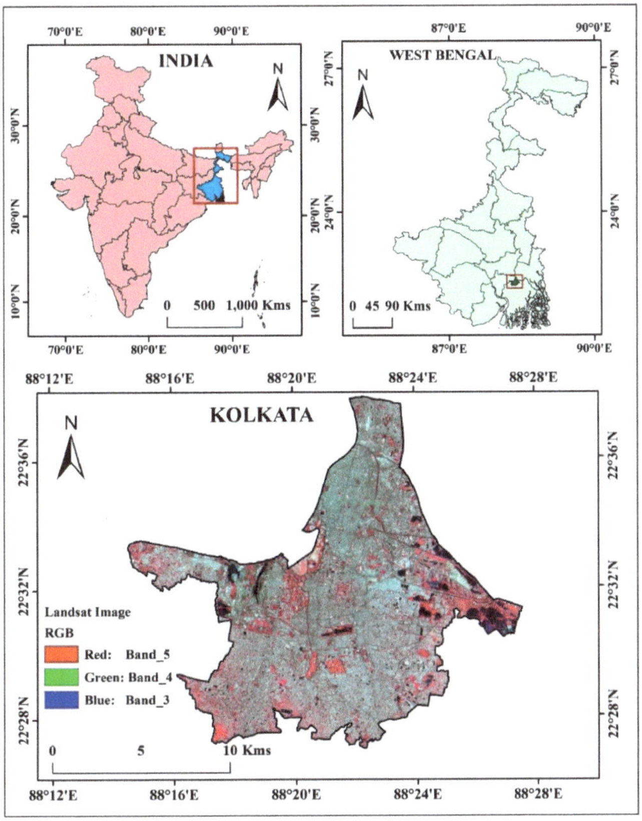

Figure 1. The location map of the case study.

2.2. Applied Datasets

The RS datasets are derived from the website USGS Earth Explorer (https://earthexplorer.usgs.gov/, accessed on 12 March 2023). For this analysis, six satellite datasets are used (Path 138 and row 044). The Landsat 5 TM datasets from 1991, 1996, 2001, and 2006, as well as the Landsat 8 OLI/TIRS datasets from 2016 and 2021, are used to classify the LULC of the study region and SUHI study using LST and certain spectral indicators. Table 1 is utilized for the gathering of the details of datasets (Table 1). Six satellite datasets are used in this analysis (Path 138 and row 044). Classifying the LULC of the inquiry region and SUHI study using LST and specific spectral indicators is done using Landsat 5 TM datasets from 1991, 1996, 2001, and 2006, as well as the 2016 and 2021 Landsat 8 OLI/TIRS datasets. Table 1 is used for dataset gathering and details.

Table 1. Details of satellite datasets.

Satellite	Sensor	Date of Acquisition	Path/Row	Website
Landsat 5	TM	6 March 1991 20 March 1996 17 March 2001 19 June 2006	138/044	https://earthexplorer.usgs.gov/, accessed on 12 March 2023
Landsat 8	OLI/TIRS	11 April 2016 25 April 2021		

2.3. Image Pre-Processing and Classification

Satellite photos are initially pre-processed using RS software to perform topological, geometric, and atmospheric adjustments. ArcGIS software version 10.8 is applied for the layer stacking, masking, and ultimately the clipping of the region of interest under inspection. The human involvement in the portion of land utilized for commercial activity is known as the Land Use Change (LU). The term "Land Cover" (LC) refers to the physical features of the Earth's surface, including vegetation, soil, water bodies, and other actual land shares [32]. The most effective approach for classifying images is through digital image processing, or DIP (Figure 2). The supervised image classification approach and a maximum likelihood algorithm are utilized for the pixel-based LULC classification There are five classes in this examining area. Using a maximum likelihood algorithm and supervised classification technique, vegetation, built-up land, bare land, and grassland are classified(Table 2).

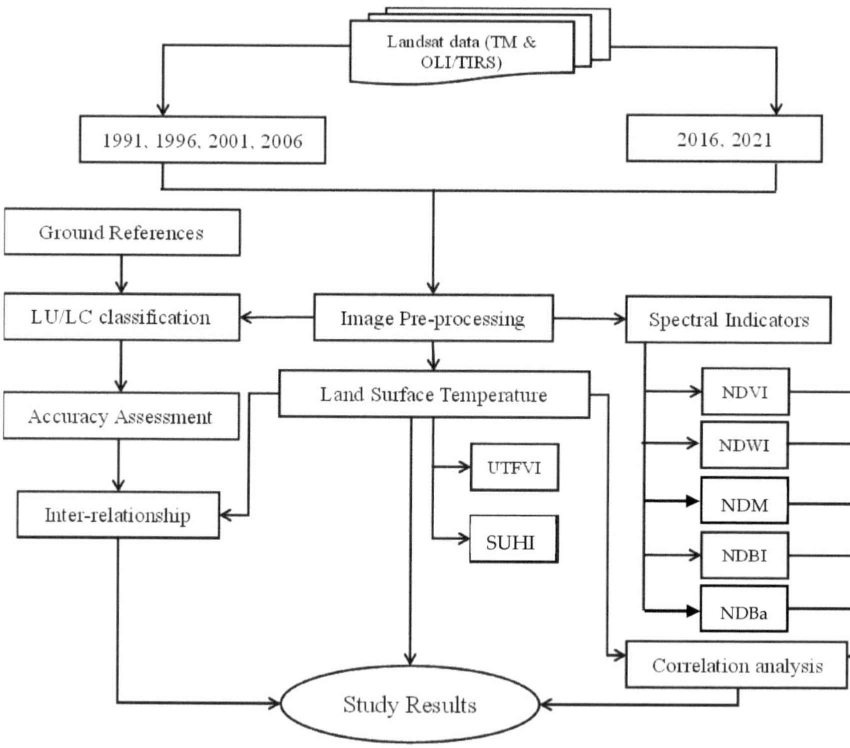

Figure 2. The modelling framework of the adopted methodology.

Table 2. LULC classes applied in this study classification.

Built-up land	Residential area, commercial area, industrial area, transportation, roads, and construction area.
Vegetation	Evergreen forest, deciduous Forest Land, Mixed Forest Land, Shrub/degraded vegetation.
Water Bodies	River, Ponds, lakes, and open water bodies.
Bare land	These types of classes are mainly playgrounds, open area, and many others.
Grass Land	Many types of trees, Grass area, open vegetated area

2.4. Accuracy Assessment and Kappa Statistic

Accuracy assessment comes in the post-classification stage. The urban areas consist of a diverse range of features, such as residential areas, water bodies, roadways, green spaces, and railroads [33]. Since this approach indicates the correctness of the classification outcomes, the accuracy assessment has the biggest important influence after the classification [34]. This method is used to associate both categorized and ground truth datasets. Google Earth Pro and field surveys make up the ground truth datasets. The accuracy of classification is assessed using the Non-parametric Kappa test. In addition to being a diagonal element, the Kappa coefficient also forms the basis of the misperception matrix [34]. Following this equation is the calculation of the Kappa coefficient (Table 3).

$$k = \frac{N\sum_{i=1}^{r} X_{ii} - \sum_{i=1}^{r}(X_{i+}X_{+i})}{N^2 - \sum_{i=1}^{r}(X_{i+}X_{+i})} \tag{1}$$

Table 3. The scale of Kappa coefficient.

SL. No	Value of K	Strength of Agreement
1	<0.20	Poor
2	0.21–0.40	Fair
3	0.41–0.60	Moderate
4	0.61–0.80	Good
5	0.81–1.00	Very Good

2.5. Geo-Spatial Indices

2.5.1. NDVI

Green space was the most significant feature of Earth's surface because it regulates temperature swings, surface runoff, infiltration rate, soil erosion losses, drought control, and water level over the land's surface [35,36]. Many areas were losing their green space land due to urbanization, which also caused droughts, temperature fluctuations, and increased evapotranspiration [6,37]. The land transformation also influenced the green space situation. The monitoring of the vegetation state made considerable use of the Landsat 5 TM and 8 OLI/TIRS datasets [38,39]. The NDVI, which can be expressed by applying Equation (2), was used in the current investigation to assess the health status of green spaces in Kolkata, West Bengal.

$$NDVI = \frac{(\rho_{NIR} - \rho_R)}{(\rho_{NIR} + \rho_R)} \tag{2}$$

where *R* represents the red band of satellite datasets and *NIR* indicates the near-infrared band of the Landsat imageries. The remaining LULC classifications are denoted by 0 to −1 in the *NDVI* standards, while a region's healthy green space is indicated by 0 to +1.

2.5.2. NDBI

Urban growth affects environmental deprivation and localized climatic change [40,41]. In the megacity of Kolkata, population density has a significant impact on both urban development and the expansion of built-up territory. Urban planning is more important for sustainable urban growth, although conditions are eliminated by excessive population density [42,43]. The megacity of Kolkata has seen significant infrastructure growth in the past. Such urban expansion was observed using the *NDBI*.

$$NDBI = \frac{(\rho_{SWIR1} - \rho_{NIR})}{(\rho_{SWIR1} + \rho_{NIR})} \tag{3}$$

In satellite imagery collections, the acronym *SWIR* refers to shortwave infrared bands, while *NIR* denotes near-infrared bands. The *NDBI* guidelines range from −1 to +1. Positive standards are built-up areas, while negative standards represent other LULC traits.

2.5.3. NDMI

Urbanization generally lowers relative humidity (RH) because of the higher temperatures and less surface water that can evaporate, in addition to aggregating the temperature over the megacity. The Landsat TM, ETM+, and OLI/TIRS near-infrared (NIR) and shortwave infrared bands were used to calculate the NDMI (https://www.usgs.gov/core-science-systems/nli/landsat/normalized-difference-moisture-index, accessed on 12 March 2023). The NDMI can be determined by using Equation (4).

$$NDMI = \frac{(\rho_{NIR} - \rho_{SWIR1})}{(\rho_{NIR} + \rho_{SWIR1})} \tag{4}$$

The areas with problems with water stress will be categorized using the normalized differential moisture index standards. Typically, the *NDMI* norms fall between −1 and 1.

2.5.4. NDBal

The normalized difference bareness index is typically used to determine the exterior hardness of the infertile ground. Applying the *SWIR* and thermal infrared (*TIR*) bands as forecasted by Zhao and Chen yields the *NDBaI* [44].

$$NDBaI = \frac{(\rho_{SWIR1} - \rho_{TIR})}{(\rho_{SWIR1} + \rho_{TIR})} \tag{5}$$

2.5.5. NDWI

With the NDWI applied to satellite imagery, open water landscapes were brought to light, making a body of water "pop out" in contrast to the surrounding soil and vegetation [45].

$$NDWI = \frac{(\rho_{G} - \rho_{NIR})}{(\rho_{G} + \rho_{NIR})} \tag{6}$$

2.6. LST Estimation

The thermal difference and the heat change of a region are prejudiced by the LST [46] The megacity of Kolkata's land surface temperature (LST) is monitored using the Landsat 5 TM (Band 6) and the Landsat 8 OLI/TIRS (Band 10). To monitor the LST, data from Landsat OLI/TIRS for 2016 and 2021, as well as Landsat TM datasets for 1991, 1996, 2001, and 2006, are used. Landsat 8 has two thermal bands, namely 10 and 11. But band 11 was not treated in this evaluation because of the possibility that it will rise in the LST approximation due to the satellite orbit's tilt. As a result, only the Landsat band 10 is used to approximate the LST imagery in the megacity of Kolkata. Las Vegas and Baghdad are two different cities in the world whose temperature changes are calculated using the LST of four decadal Earth observation datasets. The land surface temperature (Band 10) is designed using Landsat TM (Band 6) and Landsat OLI/TIRS [47]. The process that follows is utilized to advance the LST maps of certain research areas [48]. Thermal fluctuation is greater in urban areas than in rural ones. The metropolitan areas that are green or blue have a lower temperature than the surrounding areas. In the area under research, the location and effect of temperature are displayed by the LST charge. The complete computing process for Landsat 5 and 8 LST is defined in the literature. The SUHI remained resolute through the LST. A shift in surface temperature was connected to a modification in LULC. Regular updates are made to the SUHI requirements.

2.7. UTFVI

The SUHI's effects were often described using the *UTFVI*. The *SUHI* and *UTFVI* singularities are the result of several factors influencing land surface temperature, psychometrics, and light intensity, such as heat waves and Earth surface modification [49]. The formula below was used to determine the *UTFVI*.

$$UTFVI = \left(\frac{T_s + T_{mean}}{T_{mean}}\right) \tag{7}$$

where T_s signifies the LST in kelvin, and T_{mean} is used to display the mean LST in kelvin. The *UTFVI* is categorized into six groups according to the reflected changes in the urban thermal field: none, moderate, intermediate, strong, stronger, and strongest.

2.8. SUHI

The investigation of *SUHI* is significant for the study of urban heat balance. The *SUHI* diagram is projected to regulate the heat variation in the Kolkata megacity area:

$$SUHI = \left(\frac{T_s + T_{mean}}{SD}\right) \tag{8}$$

where T_s stands for the LST (K), T mean is the mean LST (K), and *SD* is the standard deviation of the estimated LST map.

3. Results and Discussion

The three-colour bands' composition (blue, green, and red) is applied for classification based on six altered years of the Landsat 5 TM and Landsat 8 OLI/TIRS datasets. Between KMC and environs regions, there has been a foremost LULC change in the aforementioned 30 years as an outcome of urban growth. At large, population growth is controlled, but the built-up region is quickly improved, while water bodies, vegetation, and grassland are deceptively reduced in maximum portions of the investigation region. The diminuendos of LULCC are predictable from the year 1991 to 2021. The outcomes will be deliberated in the subsequent subdivisions.

3.1. Areal Change of LULC

The supervised classification technique with a maximum likelihood algorithm is utilized to recognize the LULCC outlines from the year 1991 to 2021. Five categories of LULC are identified in LULC diagrams: built-up land, grassland, bare ground, vegetation, and aquatic bodies. Due to the disastrous process of urbanization and population increase, the entire region has displayed a water body, a decline in vegetation, bare ground, and grassland. The percentages of the vegetation region are acknowledged as 21.82% (1991), 35.52% (1996), 7.40% (2001), 15.62% (2006), 26.07% (2016), and 14.52% (2021), correspondingly (Table 4), over the investigation zone. The water body variations regions over the years were 875.43 Ha (1991), 614.43 Ha (1996), 523.89 Ha (2001), 682.56 Ha (2006), 785.97 Ha (2016), and 441.27 Ha (2021), correspondingly. Due to the urbanization of such areas, a 434.16 Ha water body region has shrunk throughout the last 30 years (Figure 3). According to the land modification examination, many vegetated lands have changed into various LULC features that can be experienced in terms of infiltration rate, soil moisture content, and slope stability. The residential zone close to the industrial region has significantly intensified in the KMC area. High temperatures and air pollution consistently plagued the populated zones, and the SUHI result was also observed in this location (Table 4). The grassland has shown a reduction from 27.63% (1991) to 22.63% (1996), 31.95% (2001), 29.97% (2006), 9.13% (2016), and 15.31% (2021), and a total grassland of 2286.9 Ha has been reduced in the past 30 years (Figure 4). The classified diagrams show enormous variations in the built-up lands, which enlarged from 41.61% (1991) to 37.60% (1996), 55.19% (2001), 50.69% (2006), 54.27% (2016), and 67.10% (2021), respectively (Table 5). Along with lowering the temperature, these built-up locations have also significantly lowered the rate of infiltration and surface runoff escalation.

Table 4. Area calculation of the different years' classification.

Class Name	Area in Ha					
	1991	**1996**	**2001**	**2006**	**2016**	**2021**
Water body	875.43	614.43	523.89	682.56	785.97	441.27
Vegetation	4050.27	6591.51	1374.03	2899.44	4837.77	2695.41
Grass Land	5127.93	4198.95	5928.21	5561.55	1694.25	2841.03
Built-up Land	7721.55	6978.15	10,241.73	9406.98	10071	12,450.78
Bare Land	779.04	171.18	486.36	3.69	1165.23	125.73
Class Name	**Area in Percentage (%)**					
	1991	**1996**	**2001**	**2006**	**2016**	**2021**
Water body	4.71	3.31	2.82	3.67	4.23	2.37
Vegetation	21.82	35.52	7.4	15.62	26.07	14.52
Grass Land	27.63	22.63	31.95	29.97	9.13	15.31
Built-up Land	41.61	37.6	55.19	50.69	54.27	67.1
Bare Land	4.19	0.92	2.62	0.01	6.28	0.67

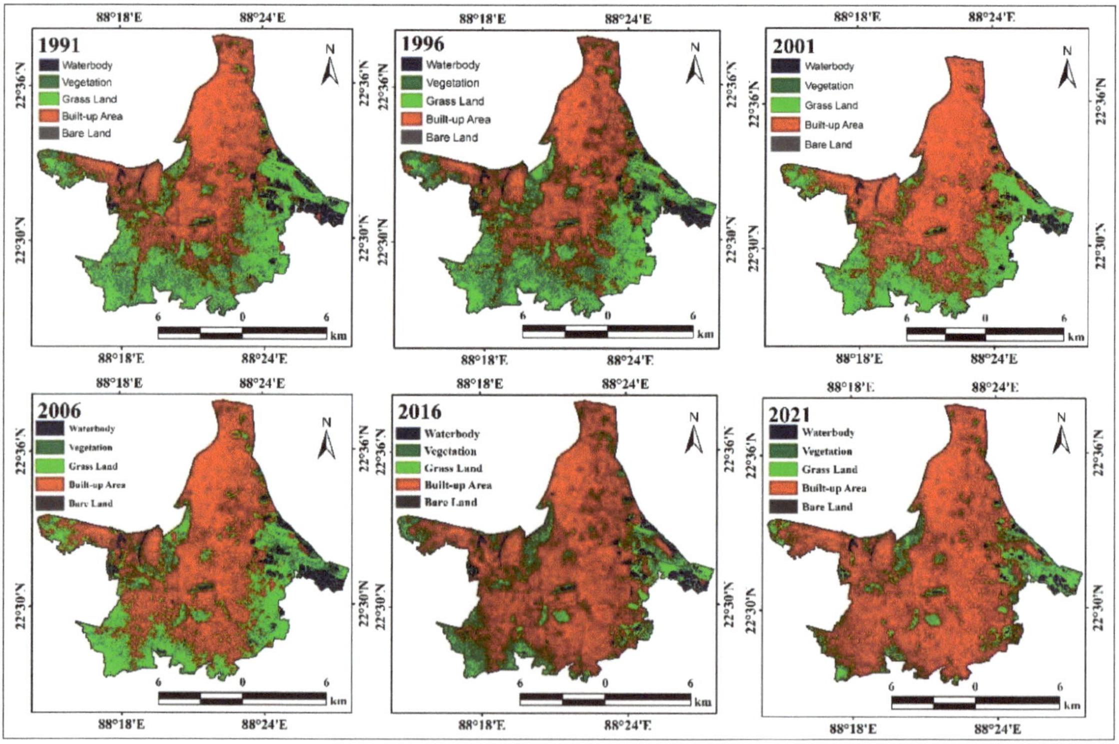

Figure 3. Maps of the land cover/use of the studied Kolkata district.

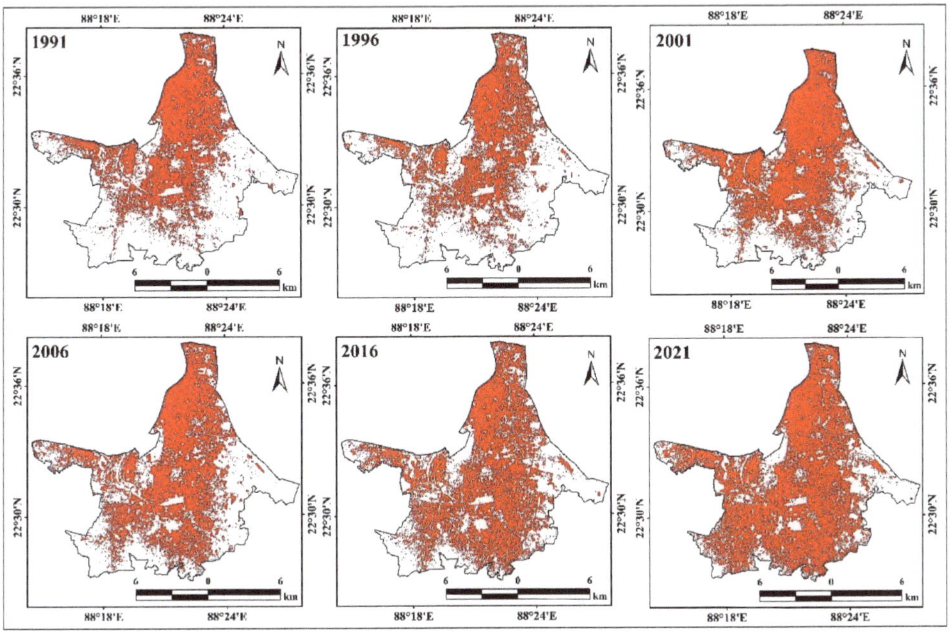

Figure 4. Maps of built-up land in different years (1991–2021).

Table 5. Loss/Gain analysis of different periods.

Class Name	Area Increased/Decreased (Ha)					
	(1991–1996)	(1996–2001)	(2001–2006)	(2006–2016)	(2016–2021)	(1991–2021)
Water body	−261	−90.54	158.67	103.41	−344.7	−434.16
Vegetation	2541.24	−5217.48	1525.41	1938.33	−2142.36	−1354.86
Grass Land	−928.98	1729.26	−366.66	−3867.3	1146.78	−2286.9
Built-up Land	−743.4	3263.58	−834.75	664.02	2379.78	4729.23
Bare Land	−607.86	315.18	−482.67	1161.54	−1039.5	−653.31

Notwithstanding the LULC elements listed above, the remaining LULC elements also share a striking amount of dynamic similarity. Bare land (0.55), built-up (0.52), and fallow land (0.56) dynamicity are close to each other in agricultural land dynamics. It is discovered that homestead dynamics with a plantation are almost identical to those of bare land (0.53). The similarity index values of 0.53 and 0.57, respectively, indicate that the dynamics of built-up areas are nearly identical to those of homesteads with plantations and water bodies [50]. Over the previous 20 years, there has been a 108.94 km^2 spatial growth due to the growing urban population. Moreover, 88.71 km^2 in 1989, 144.64 km^2 in 2006, and 197.65 km^2 in 2010 make up the urban built-up area within and surrounding the city. These modifications have raised the study region's surface temperature. Biophysical parameter analysis reveals a negative association between NDBI and NDWI, a negative correlation between LST and NDVI, and a positive correlation between LST and NDBI [51]. Urban built-up, open terrain, vegetation, agricultural land, and aquatic bodies are the five classes into which the multi-temporal satellite data are classified using the supervised Maximum Likelihood Classification technique. The findings showed that new road construction, flyovers, settlement building, etc., caused the urban built-up area to gradually rise by roughly 21.17% (239.097 km^2) throughout the study period. Other geographical features have gradually decreased, including open space, flora, agricultural land, and bodies of water [52].

A built-up zone was developed on 4729.23 hectares of natural land after the categorized LULCC diagrams from 1991 to 2021 were examined. The differences in the bare land zone are recorded as 4.19% (1991), 0.92% (1996), 2.62% (2001), 0.01% (2006), 6.28% (2016), and 0.67% (2021), and a total decrement of 653.31 Ha (1991–2021) of bare land is detected above the investigation regions (Figure 5). The overall accuracy assessment outcomes for LULC classification are 93% (1991), 95.59% (1996), 92.97% (2016), and 92.94% (2021), respectively. Apart from that, the kappa statistics for the years 1991, 1996, 2016, and 2021 are obtained, respectively, and they are 0.91, 0.94, 0.91, and 0.91. Large-scale land changes in Kolkata's megacity have intentionally impacted the local ecology and natural environmental conditions (Figure 6). The vegetation decrease and high thermal variation have increased the SUHI effects in the investigation region [53].

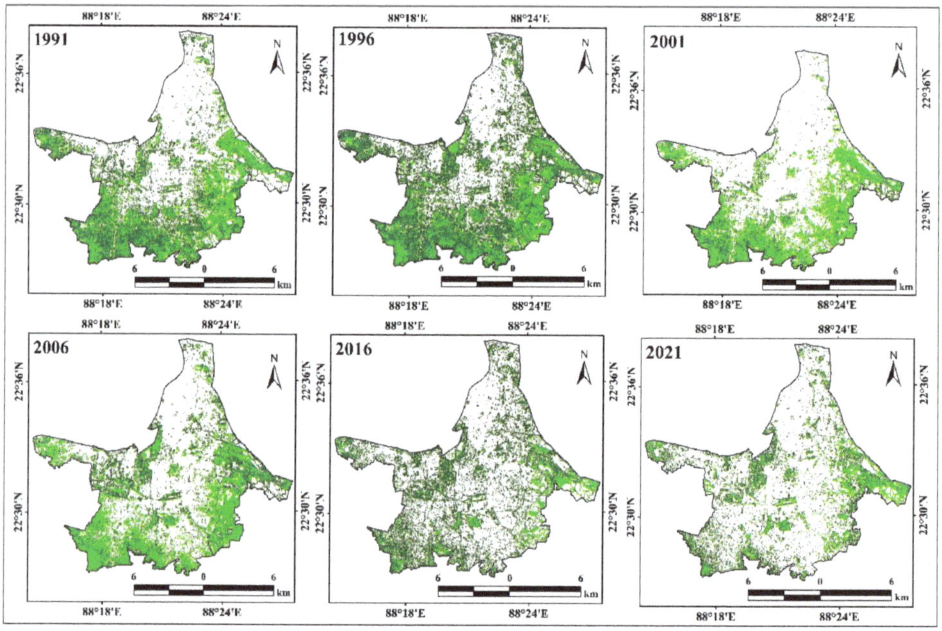

Figure 5. Maps of vegetation land in different years (1991–2021).

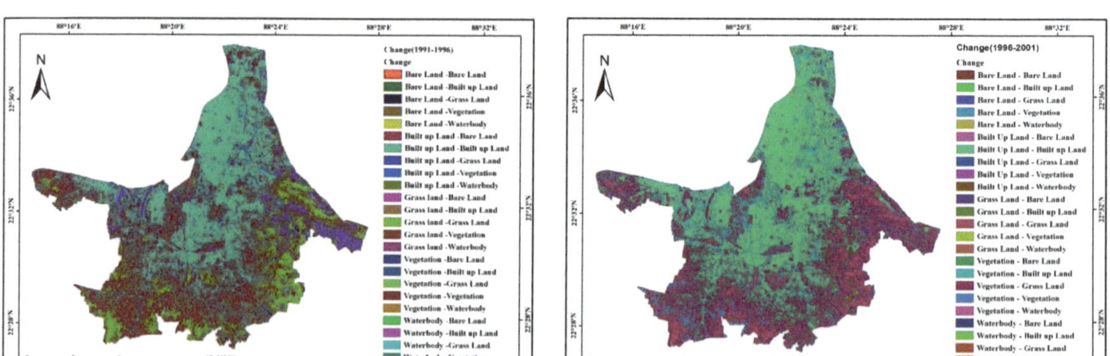

(a) (b)

Figure 6. *Cont.*

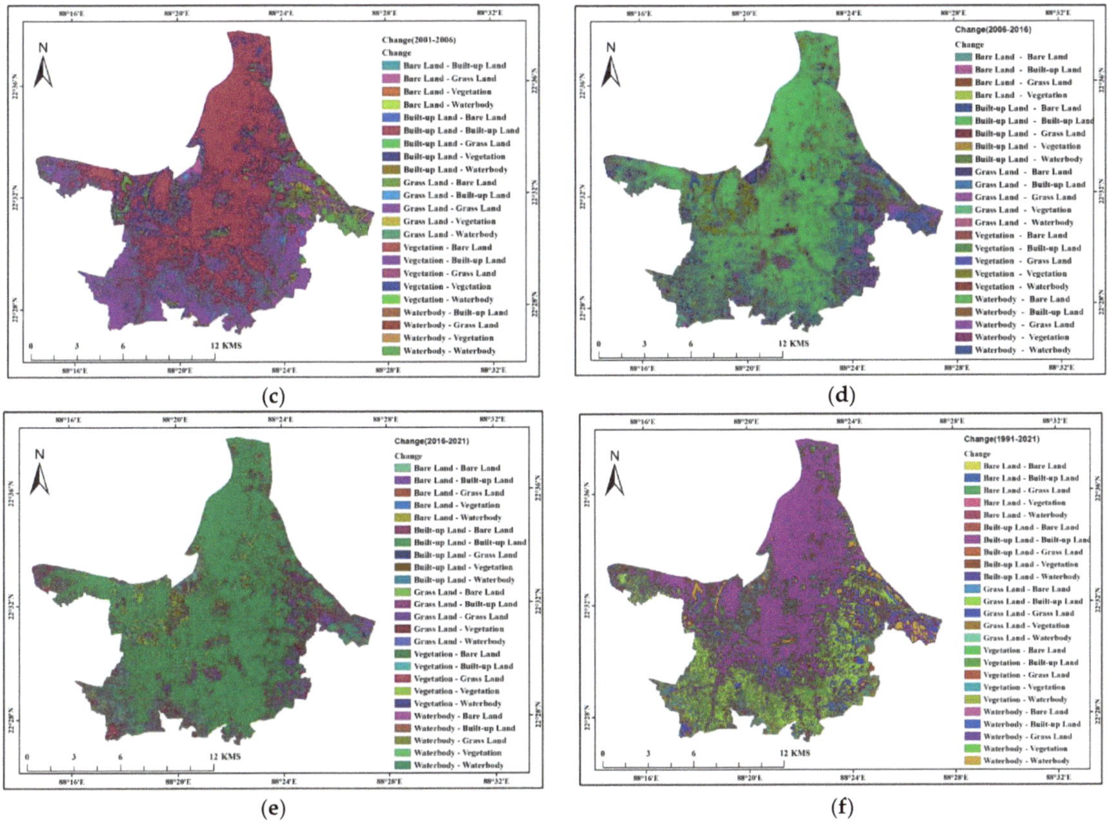

Figure 6. Change detection of LULC (**a**) 1991–1996; (**b**) 1996–2001; (**c**) 2001–2006; (**d**) 2006–2016; (**e**) 2016–2021; (**f**) 1991–2021.

3.2. LST Variation

This area experienced wide variations in temperature in different areas due to the growth of the cities. Due to human activity, the center sections are situated in areas with significant temperature variations. Verifications of the Earth's surface temperature variations and high points throughout 1991–2021 correspond with the spatial circulation of the regions under examination. In various areas of the research site, open space and grasslands are converted into developed areas. The temperature in any given area determines the type of vegetation. The application of the known equation to remote sensing (RS) data from Landsat 5 TM and 8 OLI thermal bands produced the spatio-temporal disseminations of land surface temperature (LST) for the years 1991–2021. This diagram's red hue indicated the highest temperature, while the blue colour indicated the lowest temperature in real-time (Figure 7). The LST varies in the following years: 1991, 20.62 °C to 30.83 °C with a mean temperature of 25.72 °C; 1996, 22.97 °C to 31.26 °C with a mean temperature of 27.11 °C; 2001, 23.94 °C to 32.05 °C with a mean temperature of 27.99 °C; 2006, 23.97 °C to 33.05 °C with a mean temperature of 27.995 °C; 2016, 24.67 °C to 35.59 °C with a mean temperature of 30.13 °C; and 2021, 25.32 °C to 38.82 °C with a corresponding mean temperature of 32.07 °C. The annual differences in the hottest and lowest temperatures between 1991 and 2021 are around 5.81 °C and 7.72 °C, respectively. The annual temperature rose every year between 1995 and 2020, and there was a notable rise in the mean land surface temperature (LST) due to the conversion of agricultural land and trees outside forests (TOF) to built-up areas. Compared to the suburbs, the mean LST over Kolkata City was rather high.

The average land surface temperature (LST) increased by 4.32 °C in the winter and by approximately 8.43 °C in the summer between 1995 and 2020. Over built-up areas (7.06 °C), agricultural land without crops (5.55 °C), and open land (5.54 °C), the rate of increase in LST was found to be relatively high. Over TOF (4.66 °C) and water bodies, however, it was quite low (3.68 °C) [54].

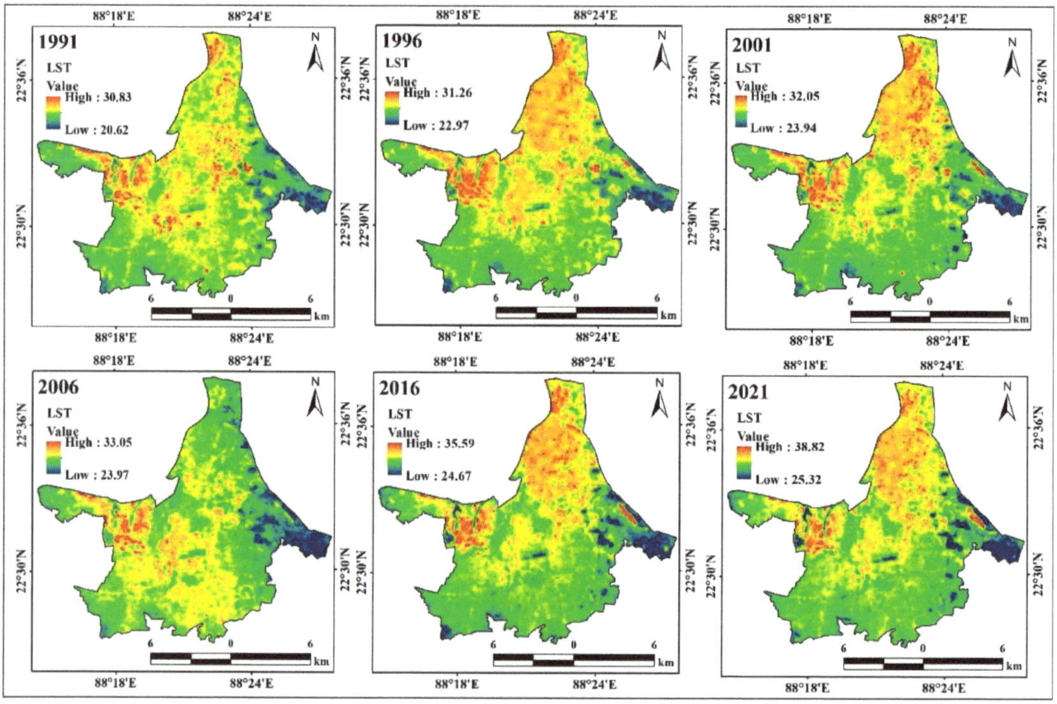

Figure 7. Maps of the LST in different years (1991–2021).

3.3. Geo-Spatial Indices

The SUHI study is classified using RS-based dissimilar spectral indices. The NDWI, NDBI, NDMI, NDVI, and NDBaI are computed using the Landsat 5 TM and Landsat 8 OLI/TIRS bands (NDBaI). The urban zone grew over the years, according to the NDBI graphic. The primary cause of the devastating fall in heat differential is likewise the decline in the vegetated zone in such zones. This inquiry region's altered flora is depicted in the NDVI graphic. Many areas of this region had a healthy green space in 1991; however, those areas were aware of the degradation of the green space due to population pressure and urbanization. Because there are differences in land values in different areas of the region under examination, the urban area has grown. The land value in KMC and its surrounding areas is generally high, while there are occasional times of low land value in other places. As a result, the societies were able to relocate to the nearby KMC and benefit from the development and convenience of the Kolkata metropolitan area. The south, west, and northeast regions of this study area have had substantial changes to the green space during the past 20 years, according to the NDVI diagram of various years The locations of Rajarhat-Newtown, Khidirpur, Amta, Sonarpur-Rajpur municipality, and Pujali have reduced the amount of green space. The SUHI influence on the area under investigation is also produced via the Urban Thermal Field Variance Index (UTFVI) figure Urban expansion is causing a significant rate of value alteration in the RS technique-based UTFVI. There were also significant thermal field variations between 1991 and 2021 in the areas surrounding Khidirpur, Kolkata port, Dharmotala, Sealdha station, and Netaji Subhash Chandra Bose International Airport. Urbanization caused a 0.015 Urban Thermal Field Variance to increase everywhere throughout those years. This study highlights how urbanization has significantly impacted the area under investigation and shows a persistent pattern of green space conversion into built-up areas during the investigation. As a result of this change, there is less greenery and a rise in surface temperatures. Via NDVI, NDBI, and LST studies, this study shows strong relationships and patterns, highlighting the necessity for urban planners, environmentalists, and ecologists to give this issue their full attention [55]. The regulatory amplitudes for the NDBI and albedo were highest because the marginal effect values had the biggest range. When the BCR changed from 0.3 to roughly 0.5, there was an increase in the positive link between NDBI and BCR and LST. When it was more than 60%, there was a substantial negative correlation between GS and LST. There was some complexity in the correlations between LST, SVF, and NDVI, respectively. When the NDVI values were greater than 0.6, the relationship between the NDVI and LST turned negative. Otherwise, there was a generally positive correlation between the two variables. After surpassing 0.8, the connection between SVF and LST turned positive. Between 0.2 and 0.8, it was negative [56].

To control the area of vegetation deterioration between 1991 and 2021, the NDVI was computed (Figure 8). Due to the region under study's ongoing urbanization and deforestation, the NDVI interpretations based on the various years demonstrated a significant reduction in the green space region. The NDVI's experiential interpretations from 1991 are 0.63 and −0.34, respectively, representing the highest and lowest interpretations. However, the NDVI interpretations have reduced unexpectedly with the maximum value individually validated as 0.44, while the lowest reading is −0.10. Because there is less greenery or grassland over KMC regions, the NDVI values in urban and industrial zones are often low. the years 1991 to 2021. Everywhere 0.015 Urban Thermal Field Variance amplified throughout those years due to urbanization.

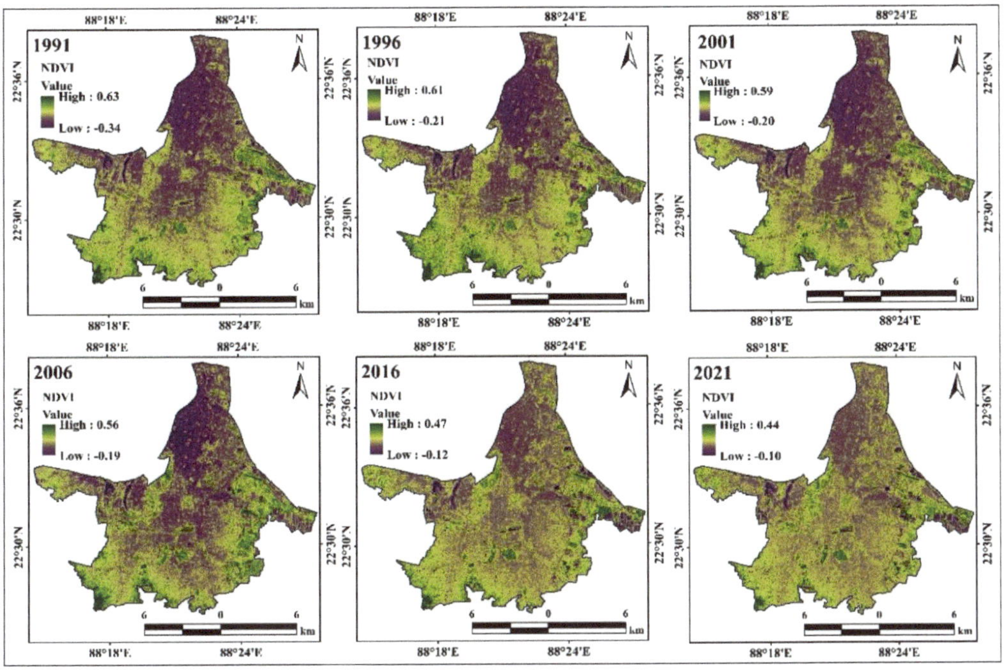

Figure 8. Maps of the NDVI in different years (1991–2021).

Satellite imagery was used to track this metropolitan megacity's built-up expansion over several years. To compute the built-up indicators for six distinct years, two bands were used. According to the NDBI results, over the last 30 years, the greatest significance rose from 0.38 (1991) to 0.49 (2021). This discovery indicates that the examined region's built-up land and urban expansion became appealing locations quickly (Figure 9). The NDMI scale runs from −1 to +1, with the lowest requirements representing low water content in green spaces and the highest criteria representing high water content. Stated differently, a decrease in the NDMI will be indicative of water stress, whereas abnormally high NDMI values may suggest waterlogging. The NDMI's high requirements are 0.62 (1991), 0.52 (1996), 0.48 (2001), 0.40 (2006), 0.35 (2016), and 0.32 (2021). In some areas of the region under examination, the NDMI's standards have decreased. Over the last 30 years, the overall NDMI norm of 0.3 has been lowered (Figure 10). The entire inspecting area's bareness level is categorized using the bareness index. The maximum standards varied between 0.17 (1991) and 0.08 (2021), continually. The substantial built-up growth over the examination zone caused the NDBal to be gradually reduced in the same manner (Figure 11). The high standards of NDWI are 0.38 (1991), 0.23 (1996), 0.21 (2001), 0.20 (2006), 0.13 (2016), and 0.12 (2021). The NDWI standards reduced in certain portions of this examining region. A total NDWI standard of 0.18 has been condensed in the past 30 years (Figure 12).

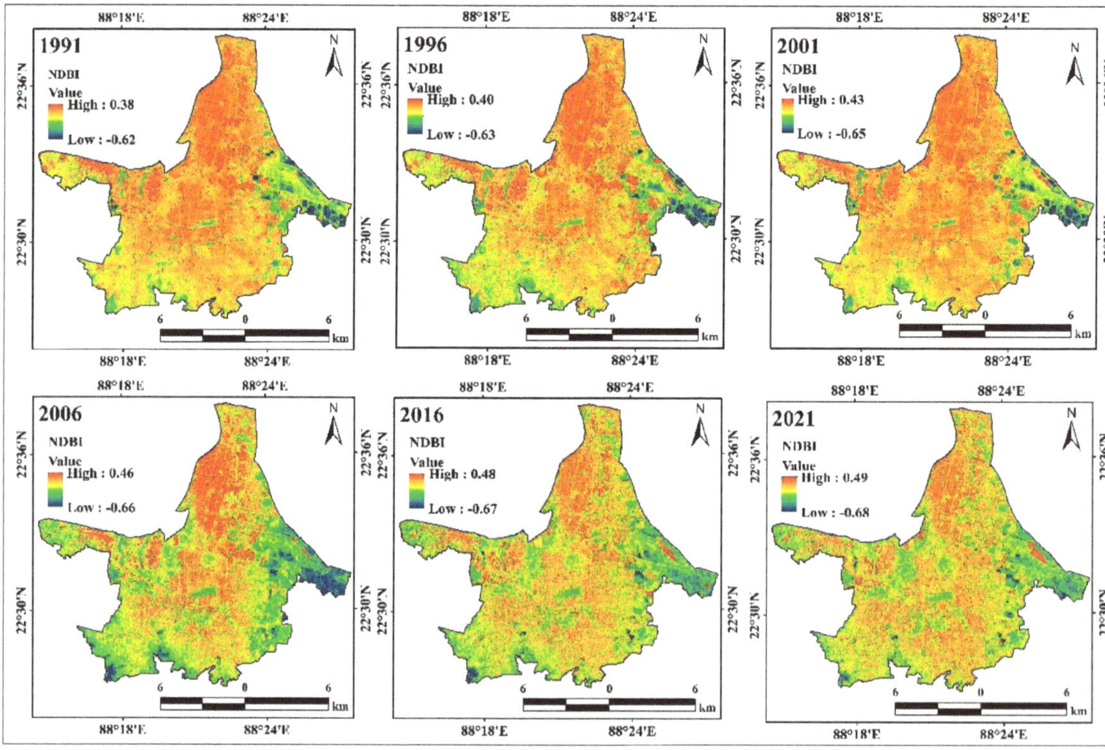

Figure 9. Maps of the NDBI in different years (1991–2021).

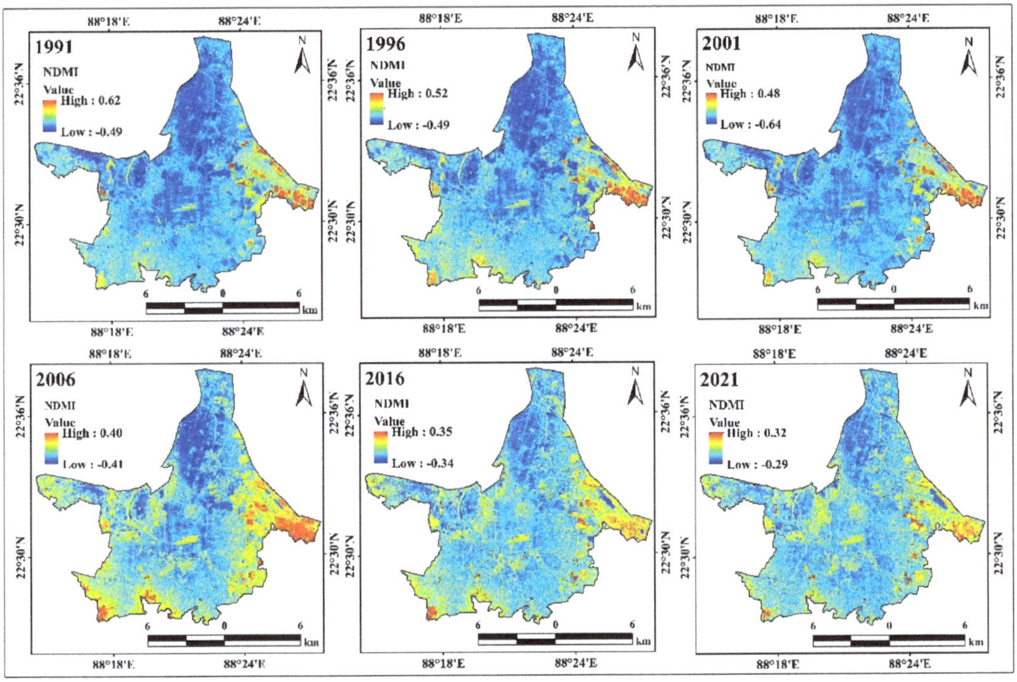

Figure 10. Maps of the NDMI in different years (1991–2021).

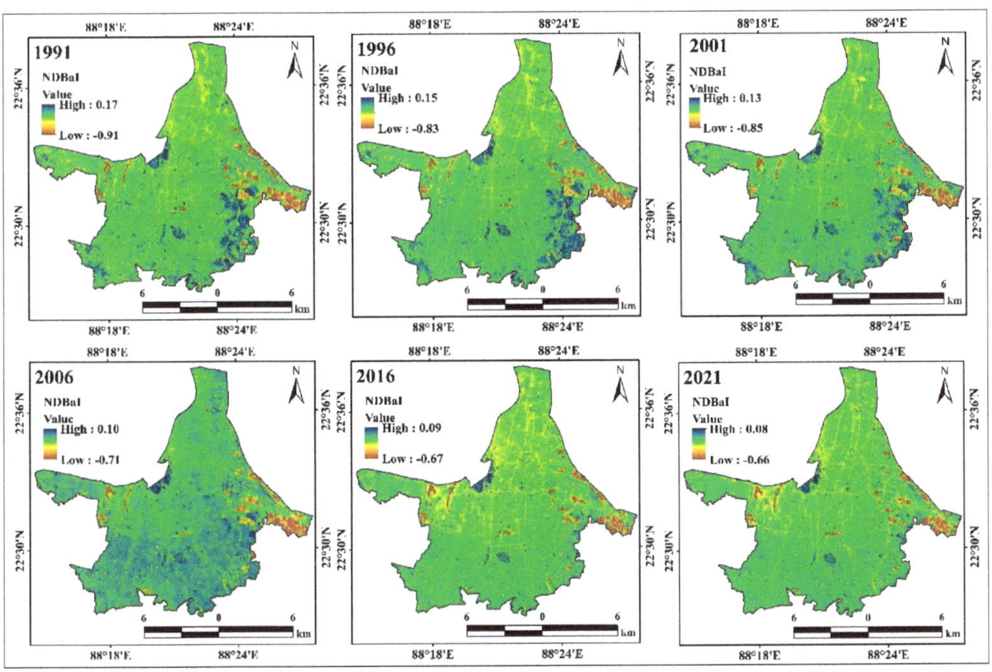

Figure 11. Maps of the NDBaI in different years (1991–2021).

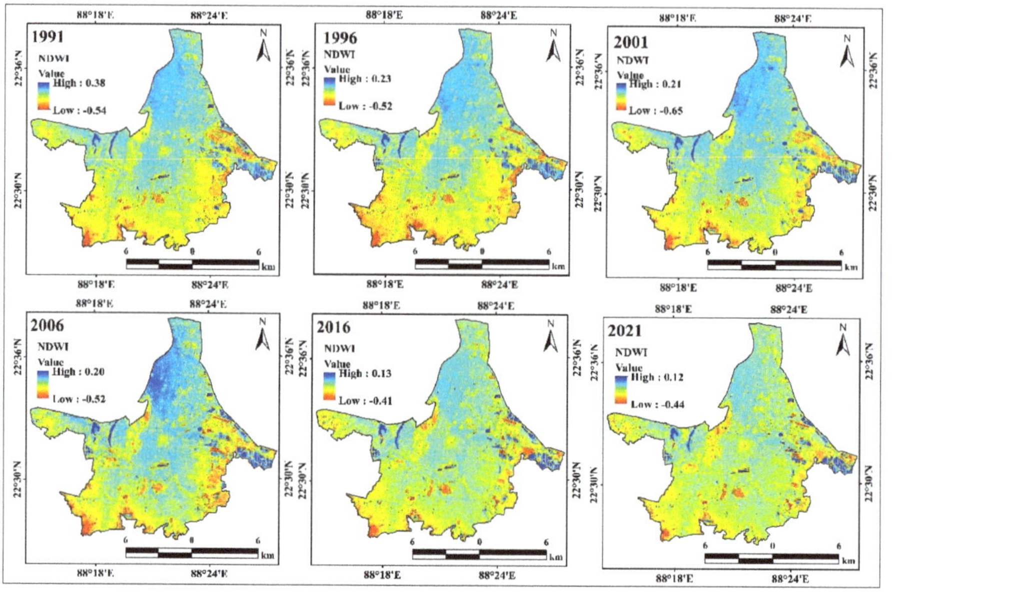

Figure 12. Maps of the NDWI in different years (1991–2021).

3.4. SUHI

The SUHI calculation is additionally significant for environmental situation supervision and preparation for a better future (Figure 13). In every corner of the world, people migrated in the direction of the cities. In a similar vein, the rural–urban fringe area currently transforms into an urban area during designated times. The years 1991–2021 are indicated by a high SUHI by the colour green to blue. Planning for smart cities is now even more important for future environmental progress. The focus of this investigation is to raise the necessary awareness among those who can lessen the challenges (Figure 14). The SUHI measurements are further significant in support of urbanization examination and estimation of the general environmental situation. Consequently, the SUHI dimension and assessment of the thermal difference over the urbanized regions were additionally vital. The global urbanization and changing climate influences of the KMC megacity, settlement, and city sites are a cumulative phenomenon that is activating the environmental variables and health problems. Those circumstances are essential for wide valuation and organization; climatic circumstances affect the general Earth's surface and increase the risk of thermal variation. In addition, unplanned metropolises were frequently affected by the SUHI variation because numerous countries were knowledgeable about unexpected urbanization which activated the heat island influence over the sphere. The highest SUHI values observed were 3.21 (1991), 3.49 (1996), 3.66 (2001), 3.90 (2006), 4.37 (2016), and 4.56 (2021), respectively. The SUHI increased by 1.25 over 30 years, while the affected areas were Khidirpur port, Dharmotala, Kalighat, Sealdha, Sovabazar, and Ultadanga areas. Similarly, for the ecological diversity assessment, UTFVI information was applied. The values of UTFVI were 0.20 (1991), 0.21 (1996), 0.24 (2001), 0.26 (2006), 0.26 (2016), and 0.26 (2021), respectively. Most of the KMC areas were affected by heat effects and thermal variation Therefore, that information is more helpful for decision-making and future management and adaptation strategies. With the diversity and dynamic growth of built-up morphology and urban surface cover, the traditional method of examining the temporal pattern of LST to investigate UHI has lost some of its significance. Rather, the Local Climate Zones (LCZ) system, which divides the city into areas according to building height, density, and forms of land cover that interact differently with the microclimate, emerged as a strong substitute [57]. Meanwhile, urban surfaces act as large heat energy reservoirs due to their high thermal inertia, which could have accelerated the heat flux from the Earth. In Kolkata, the noon UHI effect is not as severe or strong because of the thermal characteristics of the urban surface, which influence the daytime UHI impacts. However, because there is no direct solar heating during the night, the situation is reversed and local-scale convection stops [58].

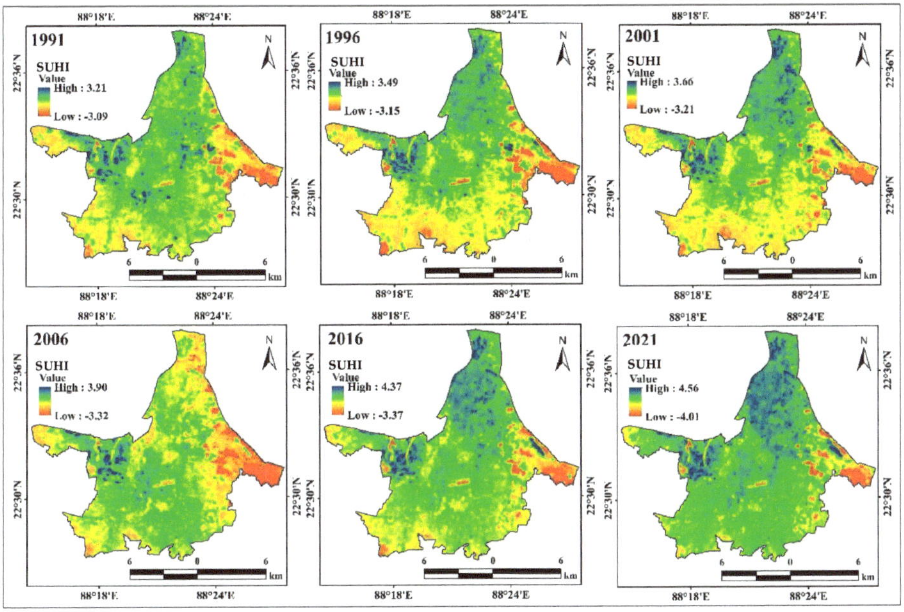

Figure 13. Maps of the SUHI in different years (1991–2021).

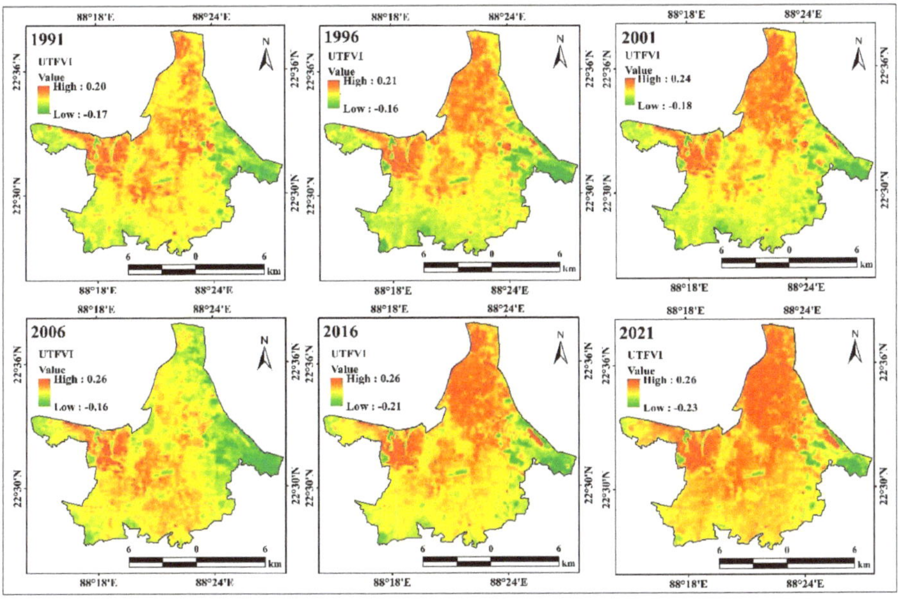

Figure 14. Maps of the UTFVI in different years (1991–2021).

3.5. Correlation Analysis

The association between the LST and LULC diagram is caused by the green space region showing a slightly lower temperature than built-up land. The thermal difference is different for diverse LULC features. The built-up area is hotter than the rest, and the water body is colder. The association is projected to apply the ArcGIS software v10.8 and calculate the situation in different years of this examining part. The built-up area is hotter than the rest of the water body, which is quite cool and the R^2 standards are 0.32, 0.36, 0.32, 0.22, 0.24, and 0.20 in the years 1991, 1996, 2001, 2006, 2016, and 2021 correspondingly. Because of the building degradation, the green space land, and the reasons for the significant temperature difference on this site, there is an additional side correlation with LST and NDVI that is negative. The relationship between LST and NDVI is 0.07, 0.13, 0.13, 0.03, 0.05, and 0.01 in the years 1991, 1996, 2001, 2006, 2016, and 2021, respectively. Because of the high frequency of deforestation and the increase in LST over the examined region, the NDMI standards demonstrated a negative connection with LST. The R2 standards are 0.32, 0.36, 0.32, 0.22, 0.24, and 0.20 in the years 1991, 1996, 2001, 2006, 2016, and 2021, respectively (Figure 15). The correlation with LST and NDWI shows a positive relationship and the R^2 values are 0.03, 0.06, 0.07, 0.004, 0.02, and 0.002 in the years 1991, 1996, 2001, 2006, 2016, and 2021 correspondingly. Additionally, there was a positive association with LST as a result of the correlation with NDBal.

The dynamic relationship between LST and plant cover (NDVI) and built-up (NDBI) area is also investigated, demonstrating how vegetation cools the city's microclimate while the built-up area plays a heating role. A top-down method for verifying the effect of shifting land use on LST is offered: a microscale study with grids. The conversion of natural and agricultural lands into built-up areas is one of the main causes of the significant rise in urban hotspots in the city's southern and central regions in 2019 [59]. In contrast to NIR reflectance, which is only related to leaf structure and dry matter, SWIR reflectance is related to both leaf structure and water content. Therefore, differences in leaf internal structure are cancelled out by spectral indices employing the NIR and SWIR bands, which increases the accuracy of vegetation water content detection. Reduced leaf water content would limit transpiration, resulting in less water evaporating from the leaf surface, decreasing cooling and raising leaf temperature [60,61]. Due to their capacity to detect the water content of vegetation, the indices that use SWIR and LST have strong connections that could indicate drought-like situations during heatwaves. It has been observed that surfaces with sparse vegetation experience water stress during heatwaves; this could prevent or reverse the cooling effect of vegetation since there is less water available for plant transpiration and drying out [62]

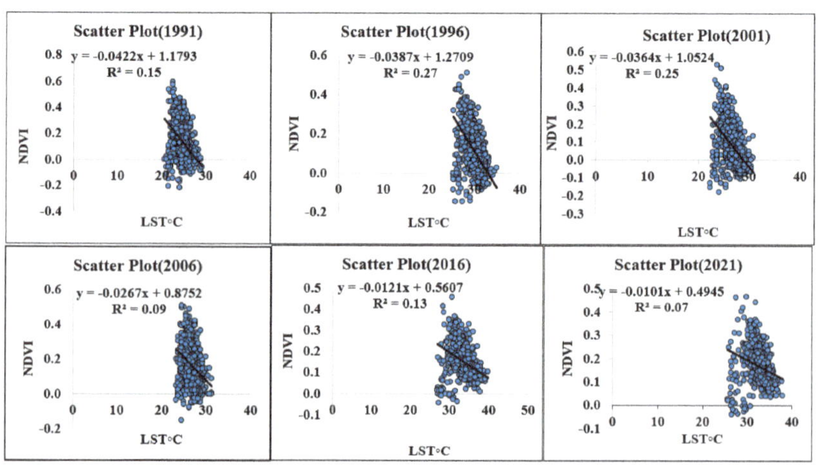

Figure 15. *Cont.*

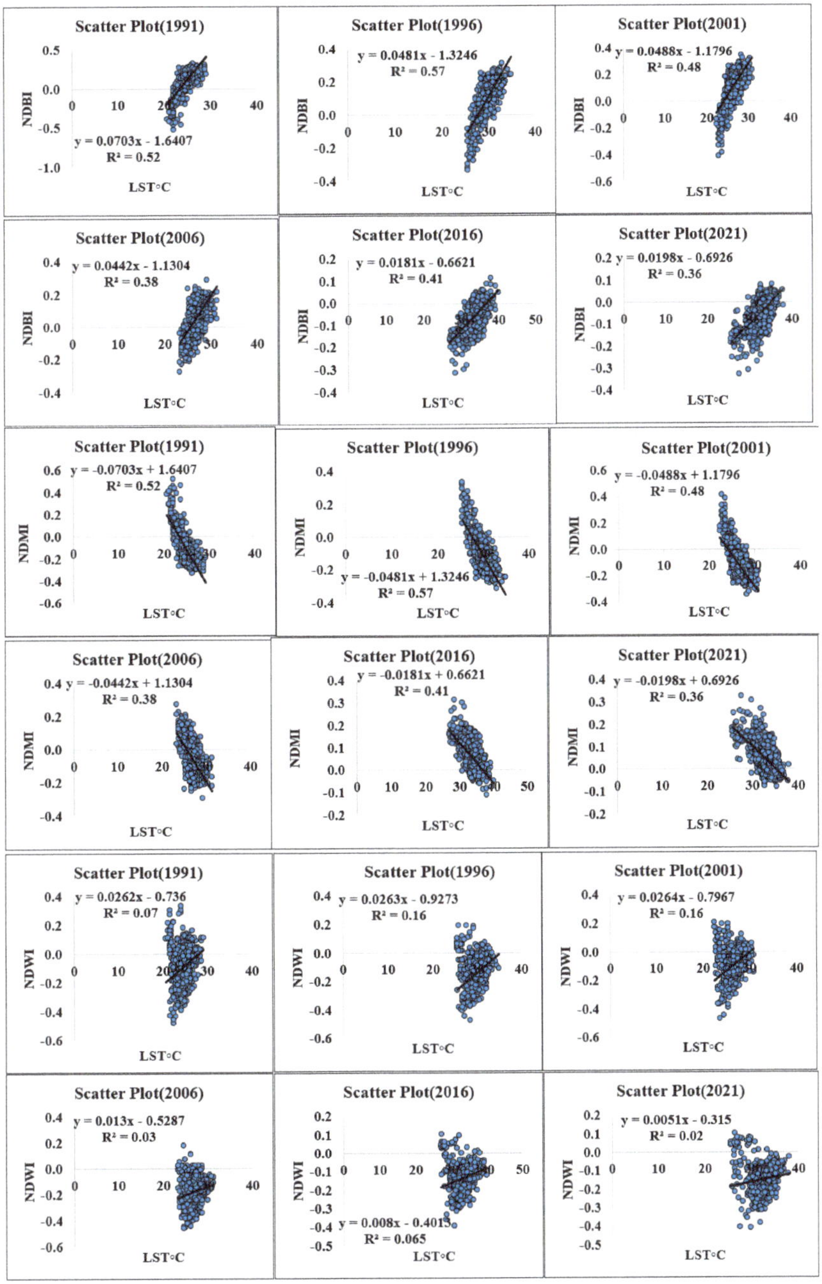

Figure 15. *Cont.*

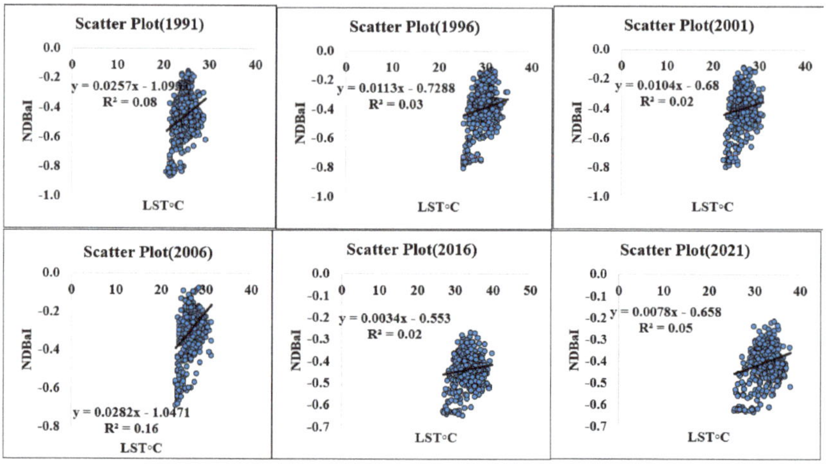

Figure 15. Correlation analysis of LST and some geo-spatial indices in different years (1991–2021).

3.6. Limitations and Recommendation

Because the satellites were capturing the images from satellite orbit, there is a distinction between surface SUHI and atmospheric SUHI. Two other factors that were crucial to examine in the upcoming research are timing and cloud cover. The massive space between the satellite location and the urban area might decrease the sensor's efficacy as there is still a substantial inconsistency between the predicted and real air temperatures [63]. It is possible that the 30 m resolution Landsat files will not be sufficient for accurately classifying the picture. The land alteration investigation's primary challenges are mixed pixels and heterogeneous values, which will make the issues in the urban expansion study worse. The upcoming research establishes the need for real surface temperatures, enhanced radiation adaption and filtering models, and Earth observation satellite systems. Aside from that, the study regions' urban green space (UGS) analysis should be conducted to advance sustainable urban design and management.

4. Conclusions

In this examination, a positive connection is found between LULC features and the LST because of the urbanization growth, industrialization, and population. An increasing LST trend is experiential from the years 1991 to 2021. The outcomes demonstrate that an increase of 4729.23 Ha in built-up areas is observed in the examining region. Temporarily 1354.86 Ha of vegetation land, 653.31 Ha of bare land, 2286.9 Ha of grassland, and 434.16 Ha of water body region reduced from the year 1991 until 2021. Both the RS and GIS methods contribute to the main benefit of the spatio-temporal urban sprawls trend with SUHI, which, when applied as the main important impact on behalf of suitable urbanized and transportation growth, as well as ecological organization difficulties like high LST in the urbanized regions, produces SUHI; consistently, we can differentiate the UTFVI.

- The UHI and LULC results commend the significant strengthening in residential regions, similar to the temperature of the urbanized regions in the last three periods, as supplementary to added LULC features. The local thermal shape of the natural surroundings appears to have been impacted by the urbanization process, according to correlations found between LST and NDBI, NDVI, NDWI, NDMI, and NDBal.
- The significant positive link found between LST and NDBI suggests that rapid urban growth has directly impacted the region under investigation's temperature conditions Moreover, an inverse relationship between the decline in green space and the urban thermal field is suggested by the negative correlation between LST and NDVI.

- The primary regulating factor for SUHI and heat stress in Kolkata and the surrounding area, according to this study, is surface area. Policymakers, administrators, urban planners, and other interested parties can use this analysis for project management and planning that will reduce thermal variance and land modification over the KMC regions.

Author Contributions: Conceptualization, N.G., B.H. and J.B.; methodology, N.G., B.H., A.H.M. and T.M.T.L.; software, N.G. and B.H.; validation, B.H., J.B., M.F.A., A.H.M. and T.M.T.L.; formal analysis, N.G., B.H., J.B., M.F.A., A.H.M. and T.M.T.L.; investigation, N.G., B.H., J.B., M.F.A., A.H.M. and T.M.T.L.; resources, A.H.M. and T.M.T.L.; data curation, N.G., B.H., A.H.M. and M.F.A.; writing—original draft preparation, N.G., B.H., J.B., A.H.M., M.F.A. and T.M.T.L.; writing—review and editing, N.G., B.H., J.B., M.F.A., A.H.M. and T.M.T.L.; visualization, B.H., A.H.M. and T.M.T.L.; supervision, J.B., B.H. and T.M.T.L.; project administration, A.H.M. and T.M.T.L.; funding acquisition, A.H.M. and T.M.T.L. All authors have read and agreed to the published version of the manuscript.

Funding: This research received no external funding.

Institutional Review Board Statement: Not applicable.

Informed Consent Statement: Not applicable.

Data Availability Statement: Data will be supplied upon request from the corresponding author.

Acknowledgments: The authors are thankful to the Vidyasagar University for this research opportunity and truly thankful to the local government body for the field survey and data collection.

Conflicts of Interest: The authors declare no conflict of interest.

References

1. Ishola, K.A.; Okogbue, E.C.; Adeyeri, O.E. Dynamics of Surface Urban Biophysical Compositions and Its Impact on Land Surface Thermal Field. *Model. Earth Syst. Environ.* **2016**, *2*, 1–20. [CrossRef]
2. Atef, I.; Ahmed, W.; Abdel-Maguid, R.H. Modelling of Land Use Land Cover Changes Using Machine Learning and GIS Techniques: A Case Study in El-Fayoum Governorate, Egypt. *Environ. Monit. Assess.* **2023**, *195*, 637. [CrossRef] [PubMed]
3. Armanuos, A.; Ahmed, K.; Sanusi Shiru, M.; Jamei, M. Impact of Increasing Pumping Discharge on Groundwater Level in the Nile Delta Aquifer, Egypt. *Knowl.-Based Eng. Sci.* **2021**, *2*, 13–23. [CrossRef]
4. Zhang, S.; Omar, A.H.; Hashim, A.S.; Alam, T.; Khalifa, H.A.E.-W.; Elkotb, M.A. Enhancing Waste Management and Prediction of Water Quality in the Sustainable Urban Environment Using Optimized Algorithm of Least Square Support Vector Machine and Deep Learning Techniques. *Urban Clim.* **2023**, *49*, 101487. [CrossRef]
5. Voogt, J.A.; Oke, T.R. Thermal Remote Sensing of Urban Climates. *Remote Sens. Environ.* **2003**, *86*, 370–384. [CrossRef]
6. Lu, D.; Weng, Q. Use of Impervious Surface in Urban Land-Use Classification. *Remote Sens. Environ.* **2006**, *102*, 146–160. [CrossRef]
7. Sarrat, C.; Lemonsu, A.; Masson, V.; Guedalia, D. Impact of Urban Heat Island on Regional Atmospheric Pollution. *Atmos. Environ.* **2006**, *40*, 1743–1758. [CrossRef]
8. Abd Alraheem, E.; Jaber, N.A.; Jamei, M.; Tangang, F. Assessment of Future Meteorological Drought under Representative Concentration Pathways (RCP8. 5) Scenario: Case Study of Iraq. *Knowl.-Based Eng. Sci.* **2022**, *3*, 64–82.
9. Mundia, C.N.; James, M.M. Dynamism of Land Use Changes on Surface Temperature in Kenya: A Case Study of Nairobi City. *Int. J. Sci. Res.* **2014**, *3*, 38–41.
10. Avdan, U.; Jovanovska, G. Algorithm for Automated Mapping of Land Surface Temperature Using LANDSAT 8 Satellite Data. *J. Sens.* **2016**, *2016*, 1480307. [CrossRef]
11. Cheruto, M.C.; Kauti, M.K.; Kisangau, D.P.; Kariuki, P.C. Assessment of Land Use and Land Cover Change Using GIS and Remote Sensing Techniques: A Case Study of Makueni County, Kenya. *J. Remote. Sens. GIS* **2016**, *5*, 1000175. [CrossRef]
12. Meer, M.S.; Mishra, A.K. Land Use/Land Cover Changes over a District in Northern India Using Remote Sensing and GIS and Their Impact on Society and Environment. *J. Geol. Soc. India* **2020**, *95*, 179–182. [CrossRef]
13. Estoque, R.C.; Murayama, Y.; Myint, S.W. Effects of Landscape Composition and Pattern on Land Surface Temperature: An Urban Heat Island Study in the Megacities of Southeast Asia. *Sci. Total Environ.* **2017**, *577*, 349–359. [CrossRef] [PubMed]
14. Gohain, K.J.; Mohammad, P.; Goswami, A. Assessing the Impact of Land Use Land Cover Changes on Land Surface Temperature over Pune City, India. *Quat. Int.* **2021**, *575*, 259–269. [CrossRef]
15. Mahmoud, A.A.; Mbengue, M.T.M.; Hussain, S.; Abdullahi, M.A.; Beddal, D.; Abba, S.I. Investigation for Flood Flow Quantification of Porous Asphalt with Different Surface and Subsurface Thickness. *Knowl.-Based Eng. Sci.* **2023**, *4*, 78–89.
16. Zheng, B.; Myint, S.W.; Fan, C. Spatial Configuration of Anthropogenic Land Cover Impacts on Urban Warming. *Landsc. Urban Plan.* **2014**, *130*, 104–111. [CrossRef]
17. Kumar, R.; Raj Gautam, H. Climate Change and Its Impact on Agricultural Productivity in India. *J. Climatol. Weather Forecast.* **2014**, *2*, 1000109. [CrossRef]

18. Saleem, M.; Iqbal, J.; Shah, M.H. Non-Carcinogenic and Carcinogenic Health Risk Assessment of Selected Metals in Soil around a Natural Water Reservoir, Pakistan. *Ecotoxicol. Environ. Saf.* **2014**, *108*, 42–51. [CrossRef]
19. Chakraborty, T.; Hsu, A.; Manya, D.; Sheriff, G. A Spatially Explicit Surface Urban Heat Island Database for the United States: Characterization, Uncertainties, and Possible Applications. *ISPRS J. Photogramm. Remote Sens.* **2020**, *168*, 74–88. [CrossRef]
20. Shahmohamadi, P.; Che-Ani, A.I.; Etessam, I.; Maulud, K.N.A.; Tawil, N.M. Healthy Environment: The Need to Mitigate Urban Heat Island Effects on Human Health. *Procedia Eng.* **2011**, *20*, 61–70. [CrossRef]
21. Gupta, K.; Kumar, P.; Pathan, S.K.; Sharma, K.P. Urban Neighborhood Green Index—A Measure of Green Spaces in Urban Areas *Landsc. Urban Plan.* **2012**, *105*, 325–335. [CrossRef]
22. Veettil, B.K.; Grondona, A.E.B. Vegetation Changes and Formation of Small-Scale Urban Heat Islands in Three Populated Districts of Kerala State, India. *Acta Geophys.* **2018**, *66*, 1063–1072. [CrossRef]
23. Hashim, B.M.; Sultan, M.A.; Attyia, M.N.; Al Maliki, A.A.; Al-Ansari, N. Change Detection and Impact of Climate Changes to Iraqi Southern Marshes Using Landsat 2 Mss, Landsat 8 Oli and Sentinel 2 Msi Data and Gis Applications. *Appl. Sci.* **2019**, *9*, 2016 [CrossRef]
24. Jamei, M.; Ali, M.; Jun, C.; Bateni, S.M.; Karbasi, M.; Farooque, A.A.; Yaseen, Z.M. Multi-Step Ahead Hourly Forecasting of Air Quality Indices in Australia: Application of an Optimal Time-Varying Decomposition-Based Ensemble Deep Learning Algorithm *Atmos. Pollut. Res.* **2023**, *14*, 101752. [CrossRef]
25. Yu, X.; Guo, X.; Wu, Z. Land Surface Temperature Retrieval from Landsat 8 TIRS—Comparison between Radiative Transfer Equation-Based Method, Split Window Algorithm and Single Channel Method. *Remote Sens.* **2014**, *6*, 9829–9852. [CrossRef]
26. Ke, Y.; Im, J.; Lee, J.; Gong, H.; Ryu, Y. Characteristics of Landsat 8 OLI-Derived NDVI by Comparison with Multiple Satellite Sensors and In-Situ Observations. *Remote Sens. Environ.* **2015**, *164*, 298–313. [CrossRef]
27. Sekertekin, A.; Bonafoni, S. Land Surface Temperature Retrieval from Landsat 5, 7, and 8 over Rural Areas: Assessment of Different Retrieval Algorithms and Emissivity Models and Toolbox Implementation. *Remote Sens.* **2020**, *12*, 294. [CrossRef]
28. Carlson, T. An Overview of the "Triangle Method" for Estimating Surface Evapotranspiration and Soil Moisture from Satellite Imagery. *Sensors* **2007**, *7*, 1612–1629. [CrossRef]
29. Omran, E.-S.E. Detection of Land-Use and Surface Temperature Change at Different Resolutions. *J. Geogr. Inf. Syst.* **2012**, *4*, 189–203. [CrossRef]
30. Orhan, O.; Ekercin, S.; Dadaser-Celik, F. Use of Landsat Land Surface Temperature and Vegetation Indices for Monitoring Drought in the Salt Lake Basin Area, Turkey. *Sci. World J.* **2014**, *2014*, 142939. [CrossRef]
31. Berwal, S.; Kumar, D.; Pandey, A.K.; Singh, V.P.; Kumar, R.; Kumar, K. Dynamics of Thermal Inertia over Highly Urban City: A Case Study of Delhi. In *Remote Sensing Technologies and Applications in Urban Environments*; SPIE: Bellingham, WA, USA, 2016, Volume 10008, pp. 108–114.
32. Kayet, N.; Pathak, K.; Chakrabarty, A.; Sahoo, S. Spatial Impact of Land Use/Land Cover Change on Surface Temperature Distribution in Saranda Forest, Jharkhand. *Model. Earth Syst. Environ.* **2016**, *2*, 127. [CrossRef]
33. Jensen Mausel, P.; Dias, N.; Gonser, R.; Yang, C.; Everitt, J.; Fletcher, R. Spectral Analysis of Coastal Vegetation and Land Cover Using AISA+ Hyperspectral Data. *Geocarto Int.* **2007**, *22*, 17–28. [CrossRef]
34. Cohen, J. Weighted Kappa: Nominal Scale Agreement Provision for Scaled Disagreement or Partial Credit. *Psychol. Bull.* **1968** *70*, 213. [CrossRef] [PubMed]
35. Zoungrana, B.J.B.; Conrad, C.; Thiel, M.; Amekudzi, L.K.; Da, E.D. MODIS NDVI Trends and Fractional Land Cover Change for Improved Assessments of Vegetation Degradation in Burkina Faso, West Africa. *J. Arid Environ.* **2018**, *153*, 66–75. [CrossRef]
36. Li, Q.; Zhang, T.; Yu, Y. Using Cloud Computing to Process Intensive Floating Car Data for Urban Traffic Surveillance. *Int. J. Geogr. Inf. Sci.* **2011**, *25*, 1303–1322. [CrossRef]
37. Jin, Z.; Zhang, L.; Lv, J.; Sun, X. The Application of Geostatistical Analysis and Receptor Model for the Spatial Distribution and Sources of Potentially Toxic Elements in Soils. *Environ. Geochem. Health* **2021**, *43*, 407–421. [CrossRef] [PubMed]
38. Sobrino, J.A.; Raissouni, N.; Li, Z.-L. A Comparative Study of Land Surface Emissivity Retrieval from NOAA Data. *Remote Sens Environ.* **2001**, *75*, 256–266. [CrossRef]
39. Guha, S.; Govil, H.; Dey, A.; Gill, N. Analytical Study of Land Surface Temperature with NDVI and NDBI Using Landsat 8 OLI and TIRS Data in Florence and Naples City, Italy. *Eur. J. Remote Sens.* **2018**, *51*, 667–678. [CrossRef]
40. Singh, P.; Kikon, N.; Verma, P. Impact of Land Use Change and Urbanization on Urban Heat Island in Lucknow City, Central India. A Remote Sensing Based Estimate. *Sustain. Cities Soc.* **2017**, *32*, 100–114. [CrossRef]
41. Kedia, S.; Bhakare, S.P.; Dwivedi, A.K.; Islam, S.; Kaginalkar, A. Estimates of Change in Surface Meteorology and Urban Heat Island over Northwest India: Impact of Urbanization. *Urban Clim.* **2021**, *36*, 100782. [CrossRef]
42. Chandler, T.J. Urban Climatology and Urban Planning. *Geogr. J.* **1976**, *142*, 57. [CrossRef]
43. Estoque, R.C.; Murayama, Y. Monitoring Surface Urban Heat Island Formation in a Tropical Mountain City Using Landsat Data (1987–2015). *ISPRS J. Photogramm. Remote Sens.* **2017**, *133*, 18–29. [CrossRef]
44. Zhao, H.; Chen, X. Use of Normalized Difference Bareness Index in Quickly Mapping Bare Areas from TM/ETM+. In Proceedings of the International Geoscience and Remote Sensing Symposium, Seoul, Republic of Korea, 29 July 2005; Volume 3, p. 1666.
45. Guha, S.; Govil, H.; Besoya, M. An Investigation on Seasonal Variability between LST and NDWI in an Urban Environment Using Landsat Satellite Data. *Geomat. Nat. Hazards Risk* **2020**, *11*, 1319–1345. [CrossRef]

46. Sobrino, J.A.; Jiménez-Muñoz, J.C.; Paolini, L. Land Surface Temperature Retrieval from LANDSAT TM 5. *Remote Sens. Environ.* **2004**, *90*, 434–440. [CrossRef]
47. Scarano, M.; Sobrino, J.A. On the Relationship between the Sky View Factor and the Land Surface Temperature Derived by Landsat-8 Images in Bari, Italy. *Int. J. Remote Sens.* **2015**, *36*, 4820–4835. [CrossRef]
48. Mirzaei, P.A.; Haghighat, F. Approaches to Study Urban Heat Island—Abilities and Limitations. *Build. Environ.* **2010**, *45*, 2192–2201. [CrossRef]
49. Abir, F.A.; Ahmmed, S.; Sarker, S.H.; Fahim, A.U. Thermal and Ecological Assessment Based on Land Surface Temperature and Quantifying Multivariate Controlling Factors in Bogura, Bangladesh. *Heliyon* **2021**, *7*, e08012. [CrossRef] [PubMed]
50. Ray, R.; Das, A.; Hasan, M.S.U.; Aldrees, A.; Islam, S.; Khan, M.A.; Lama, G.F.C. Quantitative Analysis of Land Use and Land Cover Dynamics Using Geoinformatics Techniques: A Case Study on Kolkata Metropolitan Development Authority (KMDA) in West Bengal, India. *Remote Sens.* **2023**, *15*, 959. [CrossRef]
51. Sharma, R.; Chakraborty, A.; Joshi, P.K. Geospatial Quantification and Analysis of Environmental Changes in Urbanizing City of Kolkata (India). *Environ. Monit. Assess.* **2015**, *187*, 4206. [CrossRef]
52. Hasnine, M.; Rukhsana. Spatial and Temporal Analysis of Land Use and Land Cover Change in and around Kolkata City, India, Using Geospatial Techniques. *J. Indian Soc. Remote Sens.* **2023**, *51*, 1037–1056. [CrossRef]
53. Das, P.; Vamsi, K.S.; Zhenke, Z. Decadal Variation of the Land Surface Temperatures (LST) and Urban Heat Island (UHI) over Kolkata City Projected Using MODIS and ERA-Interim DataSets. *Aerosol Sci. Eng.* **2020**, *4*, 200–209. [CrossRef]
54. Biswas, S.; Ghosh, S. Estimation of Land Surface Temperature in Response to Land Use/Land Cover Transformation in Kolkata City and Its Suburban Area, India. *Int. J. Urban Sci.* **2022**, *26*, 604–631. [CrossRef]
55. Ali, M.B.; Jamal, S.; Ahmad, M.; Saqib, M. Unriddle the Complex Associations among Urban Green Cover, Built-up Index, and Surface Temperature Using Geospatial Approach: A Micro-Level Study of Kolkata Municipal Corporation for Sustainable City. *Theor. Appl. Climatol.* **2024**, 1–22. [CrossRef]
56. Lu, L.; Fu, P.; Dewan, A.; Li, Q. Contrasting Determinants of Land Surface Temperature in Three Megacities: Implications to Cool Tropical Metropolitan Regions. *Sustain. Cities Soc.* **2023**, *92*, 104505. [CrossRef]
57. Bhowmick, D.; Mukherjee, K.; Dash, P.; Mondal, R. Use of "Local Climate Zones" for Detecting Urban Heat Island: A Case Study of Kolkata Metropolitan Area, India. In *IOP Conference Series: Earth and Environmental Science, Proceedings of the International Conference on Geospatial Science for Digital Earth Observation (GSDEO-2021), Online, 26–27 March 2021*; IOP Publishing: Bristol, UK, 2023; Volume 1164, p. 12007.
58. Khorat, S.; Das, D.; Khatun, R.; Aziz, S.M.; Anand, P.; Khan, A.; Santamouris, M.; Niyogi, D. Cool Roof Strategies for Urban Thermal Resilience to Extreme Heatwaves in Tropical Cities. *Energy Build.* **2024**, *302*, 113751. [CrossRef]
59. Sharma, R.; Pradhan, L.; Kumari, M.; Bhattacharya, P. Assessing Urban Heat Islands and Thermal Comfort in Noida City Using Geospatial Technology. *Urban Clim.* **2021**, *35*, 100751. [CrossRef]
60. Ceccato, P.; Flasse, S.; Tarantola, S.; Jacquemoud, S.; Grégoire, J.-M. Detecting Vegetation Leaf Water Content Using Reflectance in the Optical Domain. *Remote Sens. Environ.* **2001**, *77*, 22–33. [CrossRef]
61. Yao, X.; Yu, K.; Zeng, X.; Lin, Y.; Ye, B.; Shen, X.; Liu, J. How Can Urban Parks Be Planned to Mitigate Urban Heat Island Effect in "Furnace Cities"? An Accumulation Perspective. *J. Clean. Prod.* **2022**, *330*, 129852. [CrossRef]
62. Alexander, C. Normalised Difference Spectral Indices and Urban Land Cover as Indicators of Land Surface Temperature (LST). *Int. J. Appl. Earth Obs. Geoinf.* **2020**, *86*, 102013. [CrossRef]
63. Mohamed, A.H.; Adwan, I.A.I.; Ahmeda, A.G.F.; Hrtemih, H.; Al-MSari, H. Identification of Affecting Factors on the Travel Time Reliability for Bus Transportation. *Knowl.-Based Eng. Sci.* **2021**, *2*, 19–30. [CrossRef]

sustainability

Article

Land-Use-Based Runoff Yield Method to Modify Hydrological Model for Flood Management: A Case in the Basin of Simple Underlying Surface

Chaowei Xu, Hao Fu *, Jiashuai Yang, Lingyue Wang and Yizhen Wang

College of Urban and Environmental Sciences, Peking University, No.5 Yiheyuan Road, Haidian District, Beijing 100871, China
* Correspondence: fuhao@pku.edu.cn; Tel.: +86-18101201990

Abstract: The study of runoff under the influence of human activities is a research hot spot in the field of water science. Land-use change is one of the main forms of human activities and it is also the major driver of changes to the runoff process. As for the relationship between land use and the runoff process, runoff yield theories pointed out that the runoff yield capacity is spatially heterogeneous. The present work hypothesizes that the distribution of the runoff yield can be divided by land use, which is, areas with the same land-use type are similar in runoff yield, while areas of different land uses are significantly different. To prove it, we proposed a land-use-based framework for runoff yield calculations based on a conceptual rainfall–runoff model, the Xin'anjiang (XAJ) model. Based on the framework, the modified land-use-based Xin'anjiang (L-XAJ) model was constructed by replacing the yielding area (f/F) in the water storage capacity curve of the XAJ model with the area ratio of different land-use types (L/F; L is the area of specific land-use types, F is the whole basin area). The L-XAJ model was then applied to the typical cultivated–urban binary land-use-type basin (Taipingchi basin) to evaluate its performance. Results showed great success of the L-XAJ model, which demonstrated the area ratio of different land-use types can represent the corresponding yielding area in the XAJ model. The L-XAJ model enhanced the physical meaning of the runoff generation in the XAJ model and was expected to be used in the sustainable development of basin water resources.

Keywords: runoff generation; saturation-excess runoff generation theory; yielding area; conceptual hydrological model; land-use-based Xin'anjiang model (L-XAJ); cultivated–urban binary land-use-type basin; sustainable development of water resources

check for updates

Citation: Xu, C.; Fu, H.; Yang, J.; Wang, L.; Wang, Y. Land-Use-Based Runoff Yield Method to Modify Hydrological Model for Flood Management: A Case in the Basin of Simple Underlying Surface. *Sustainability* **2022**, *14*, 10895. https://doi.org/10.3390/su141710895

Academic Editors: Jun Qin and Hou Jiang

Received: 5 August 2022
Accepted: 28 August 2022
Published: 31 August 2022

Publisher's Note: MDPI stays neutral with regard to jurisdictional claims in published maps and institutional affiliations.

1. Introduction

The sustainable development of water resources has intimate associations with the quality of human life in modern society [1–3]. A runoff yield calculation is one of the key components in basin water resource management, which plays an important role in hydrologic processes [4–8]. In the past few decades, runoff yield models, which were used for hydrological forecasting and water resources management, have been providing decision-making services for basin management and planning [5,9–13]. Saturation-excess runoff generation theory pointed out that runoff occurs when the soil water content in the unsaturated zone exceeds the field capacity [14]. Based on this theory, numerous rainfall–runoff models have been developed and applied extensively around the world over the past century, which proved the effectiveness of the theory [15–19].

The runoff yield is closely related to the underlying surface condition [20,21]. In a basin, the characteristics of runoff yield are spatially differentiated. Different hydrological models invariably construct a curve to describe this difference in the runoff yield, e.g., the water storage capacity curve of the XAJ model, which has demonstrated widespread utility in most natural basins of humid and semi-humid regions [22,23]. However, intensive human activities are changing the natural basin deeply [24], which has a profound impact on the runoff yield. The natural distribution of the runoff yield and the original runoff yield mechanism are changed along with these drastic land-use changes [25–27]. Zhu et al. [28] and Zheng et al. [29] have found that land-use change is the strongest contributor to a change in the runoff process and may be directly responsible for more than 70%. Studies also found that hydrological models tend to underperform more in artificial basins than in natural basins [25]. Some scholars pointed out that it is mainly due to the land-use changes, which influence the runoff yield [30–32].

The effect of land use on the runoff yield was widely discussed in hydrology [33–35]. The basic consensus is that land-use change is the most important factor affecting the runoff yield [24,36–44]. Among them, lots of studies concentrated on the relationship between different land-use changes and runoff yield (e.g., vegetation [45–48], urbanization [49,50], agricultural activities [51–53]). In general, afforestation will reduce the runoff yield, while deforestation, urbanization and overgrazing will increase the runoff yield; furthermore, the effect of agricultural activities varies with tillage practices [54–59]. Moreover, relevant studies indicated that the land-use structure and spatial layout (e.g., land-use distribution [60,61], land-use pattern and landscape features [62,63]) also have a deep impact on the runoff yield, runoff patterns and the runoff processes. These studies suggest that the land-use type is closely related to the runoff processes, which may be the key factor to determine the runoff yield.

Although many studies concerned the relationship between land-use change and runoff [64–66], few of them took land use as a parameter to integrate into conceptual hydrological models. Inspired by previous scholars, this paper assumes that the distribution of the runoff yield can be divided by the land-use form, i.e., areas with the same land-use form are similar in their runoff yield, while areas of different land uses are significantly different. As for the XAJ model, it can be expressed by replacing the yielding area with the area ratio of land use in the basin. The main objective of this study is to construct a land-use-based Xin'anjiang (L-XAJ) model with the relationship between the yielding area (f/F) and the area of different land uses (L/F) for better runoff generation simulation in a typical cultivated–urban binary land-use-type basin, thus providing a better tool for the sustainable development of water resources at the basin scale. This research is expected to be used for flood management and the sustainable development of water resources in the basin. The remainder of this paper was organized as follows: Section 2 describes the methods and study area, Section 3 summarizes the research results and discussions and, lastly, the conclusions are drawn in Section 4.

2. Materials and Methods

2.1. Runoff Yield in XAJ Model

XAJ model is one of the most famous hydrological models in China, which is widely used in humid and semi-humid areas [14]. XAJ model can be divided into four parts: evapotranspiration, runoff generation, runoff sources partition and runoff concentration (Figure 1). Runoff generation is one of the most important modules and the parabolic curve is used to calculate the runoff yield, which can be represented by the following equation:

$$\frac{f}{F} = 1 - \left(1 - \frac{W'_m}{W'_{mm}}\right)^B \tag{1}$$

where W'_m is the storage capacity of a point in the basin (mm), f is the fraction of the basin area for which the storage capacity is less than W'_m, F is the whole basin area, W'_{mm} is the maximum value of W'_m and B is the shape parameter of the storage capacity distribution.

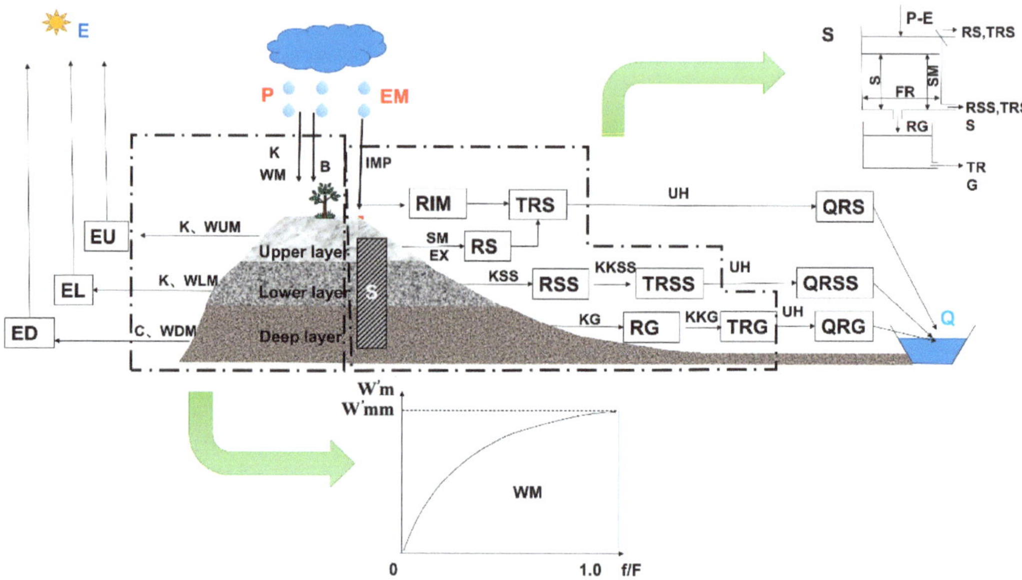

Figure 1. The framework of XAJ model. (**Variables:** P: precipitation; EM: potential evapotranspiration; E: actual evapotranspiration; RIM: runoff from the impervious area; RS: surface runoff; RSS: interflow runoff; RG: groundwater runoff; TRS: outflows from the reservoirs of surface run-off components; TRSS: outflows from the reservoirs of interflow run-off components; TRG: outflows from the reservoirs of groundwater run-off components; QRS: surface runoff inflow to river network; QRSS: interflow to river network; QRG: groundwater inflow to river network; s: water content in free water store reservoir; FR: ratio of runoff-producing area; Upper layer, Lower layer and Deep layer are the three soil layers for evapotranspiration; S: free water storage reservoir; UH: unit hydrograph; WM: the average water storage capacity; W'_m: the water storage capacity; W'_{mm}: the maximum value of W'_m; EU, EL and ED are evaporation from the upper, lower and deepest layer, respectively; Q: the discharge at the outlet of the basin; **Parameters:** the others are parameters and will be introduced in Section 2.4).

Based on Equation (1), the average water storage capacity (WM) of the basin can be obtained:

$$\mathrm{WM} = \int_0^1 W'_m d\left(\frac{f}{F}\right) = \frac{W'_{mm}}{1+B} \tag{2}$$

As shown in Figure 2a, there are two basic initial assumptions for the runoff generation process: (1) the initial soil water content of the basin is W_0 and the maximum field storage capacity is A; (2) the area of a proportion of α_0 over the basin is in saturation state and the rainfall that falls on this area directly produces runoff, on the area of $1-\alpha_0$, it does not Hence, the initial state of the basin is:

$$A = W'_{mm} * \left[1 - \left(1 - \frac{W_0}{\mathrm{WM}}\right)^{\frac{1}{1+B}}\right] \tag{3}$$

If rainfall is P and evapotranspiration is E, when evapotranspiration exceeds rainfall $(P - E < 0)$, runoff is not generated; when rainfall exceeds evapotranspiration $(P - E > 0)$, then runoff is generated.

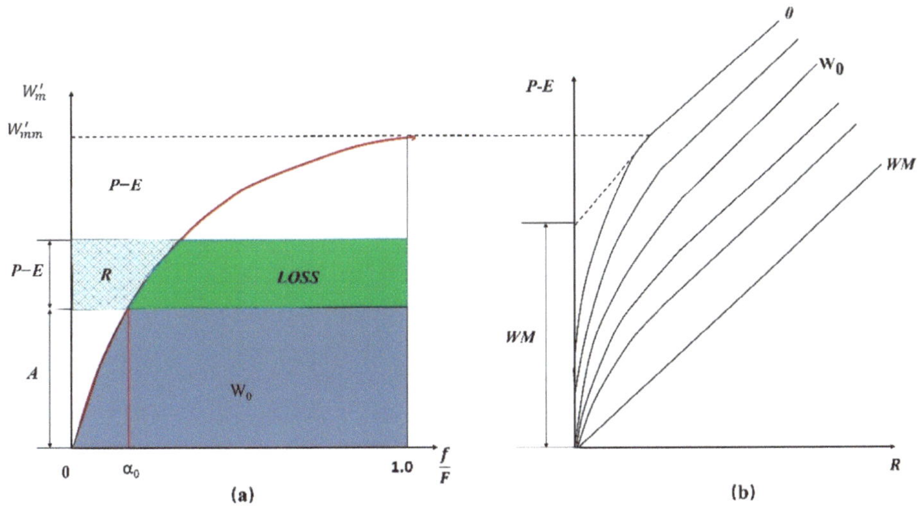

Figure 2. (**a**) The water storage capacity curve; (**b**) Rainfall–runoff relationship.

If $P - E + A < W'_{mm}$ for local runoff generation, the soil water storage is the LOSS part in Figure 2a and the runoff yield R can be obtained by the following equation:

$$\begin{aligned} R &= dPE - \int_A^{A+(P-E)} (1 - \alpha_0) dW'_m \\ &= P - E + WM(1 - \frac{A+P-E}{W'_{mm}})^{B+1} + W_0 - WM \end{aligned} \quad (4)$$

Otherwise, the runoff yield R can be obtained by the equation:

$$R = P - E - (WM - W_0) \quad (5)$$

The runoff generation process can be calculated by Equations (4) and (5). From Equations (4) and (5), the rainfall–runoff relationship can be obtained as Figure 2b, which indicates that runoff yield only is controlled by net rainfall P-E and soil moisture W_0.

2.2. Runoff Yield in L-XAJ Model

The XAJ model provides an effective solution for runoff yield calculation but has no clear physical meaning [67,68]. However, its parameters implicitly represent the influence of underlying surface factors such as land use on runoff yield. To clarify the underlying surface information of the model, we assumed that the distribution of runoff yield in a basin can be divided by land-use form, i.e., areas with the same land-use form are similar in runoff yield, while areas of different land uses are significantly different. L-XAJ model calculates the runoff yield under each land-use type by specific water storage capacity value, accumulates the runoff yield of all land-use types as the basin's runoff yield and then goes into the free water storage reservoir for the partition of runoff sources (Figure 3).

As shown in Figure 4, land-use pattern can be obtained from remote sensing images. This is assuming that the land use of the rectangle can be divided into four regions, A, B, C and D, which is grassland, forest, urban and grassland, respectively. Though the water storage capacity is spatially heterogeneous in this rectangle, it can be roughly distinguished as the four regions. The water storage capacity values of regions A and D, are roughly at the same level; while the values of regions A, B and C are at different levels. So, we can use a mean value a to represent the average water storage capacity of region A and D, a mean value b for region B and a mean value c for region C; a, b and c are not equal to each other.

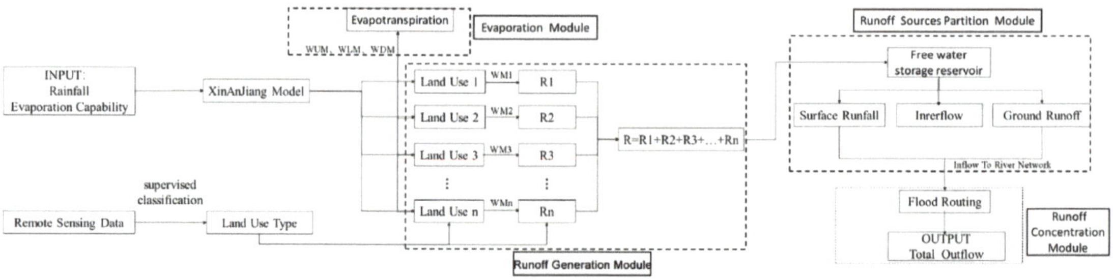

Figure 3. Flow chart of L-XAJ model.

Figure 4. Picture of different land-use types (examples of different land-use types with water storage capacity: (**A**): grassland, (**B**): forest, (**C**): urban and (**D**): grassland).

In this framework, water storage capacity is indeed different at different points within the same land-use type. It is difficult to calculate the water storage capacity of each point, but its statistical law is presented in the water storage capacity curve of XAJ model. The mean value of each land-use segment is used to represent its water storage capacity value.

To describe the land-use-based water storage capacity curve by clear mathematical formulas, we assume that there are *n* different land-use types in a certain basin (Figure 5a) and their area ratios over the basin are s_1, $s_2 \ldots s_n$, respectively, as the abscissa in L-XAJ model by α_i, such as $(\alpha_0, \alpha_1]$, $(\alpha_1, \alpha_2] \ldots (\alpha_{n-1}, \alpha_n]$, where $\alpha_0 = 0$, $\alpha_i - \alpha_{i-1} = s_i$ $(1 \leq i \leq n)$ (Figure 5b). The water storage capacity value of different land-use types are $W_1, W_2 \ldots W_n$, representing the average value of the water storage capacity of different land-use types as the ordinate in L-XAJ model. As shown in Figure 6a, L-XAJ model is a monotone increasing piecewise function and each segment of the function represents one kind of land-use type in the basin, including urban, surface water bodies, grassland, crops, forest, etc. In application, land-use types can be adjusted based on the true condition of different basins. The area of i-th $(1 \leq i \leq n)$ land-use type is $\alpha_i - \alpha_{i-1} = s_i$ and was reflected on the abscissa in L-XAJ model. The water storage capacity value of the ith land-use type is W_i, reflecting the ordinate in L-XAJ model, and the water storage capacity value within $[\alpha_{i-1}, \alpha_i]$ is always W_i. Therefore, the average water storage capacity of the ith land-use type WS_i can be obtained:

$$WS_i = W_i \times (\alpha_i - \alpha_{i-1}) = W_i \times s_i \; (1 \leq i \leq n) \tag{6}$$

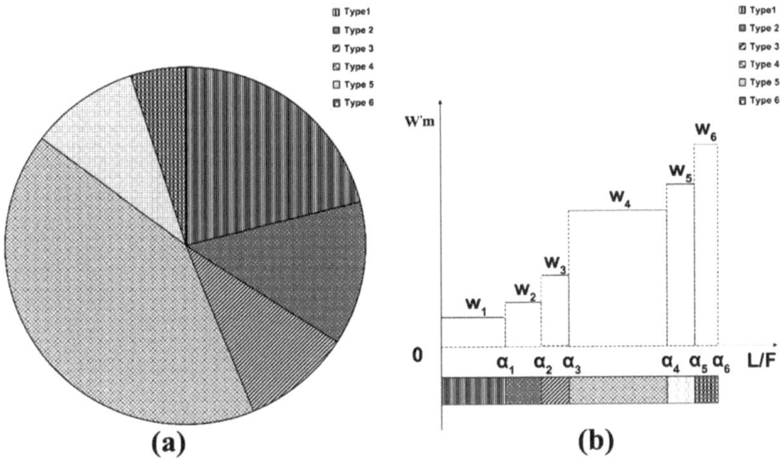

Figure 5. (**a**) Schematic diagram of the area ratio of different land-use types in the basin; (**b**) relationship between storage capacity value and land-use types. (α1, α2, α3, α4, α5, α6 are the area ratio sums of different land-use types; w1, w2, w3, w4, w5, w6 are the average storage capacity values of different land-use types, in which w1 < w2 < w3 < w4 < w5 < w6. L/F is the area ratio of different land-use types; W'_m is water storage capacity; six land-use types for examples).

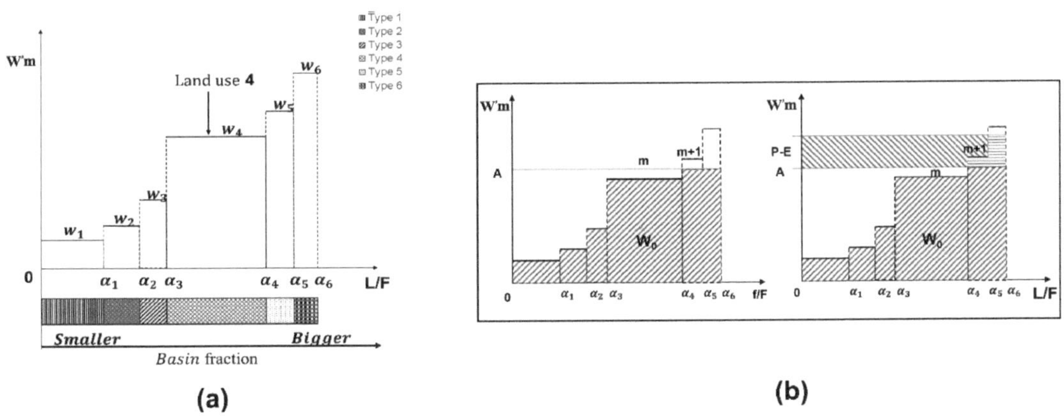

Figure 6. The construction of L-XAJ model (six land-use types for examples; (**a**) the lower water storage capacity values are located in smaller area and higher are in bigger area; (**b**) left: the basin state before rainfall for which soil moisture is W_0; right: the runoff process when the net rainfall is P-E, set (α3, α4] is m-th land-use type and (α4, α5] is (m + 1)th land-use type).

In addition, the average storage capacity, WM_{l-xaj}, of L-XAJ model can be calculated as follows:

$$WM_{l-xaj} = \sum_{i=1}^{n} WS_i = \sum_{i=1}^{n} W_i \times (\alpha_i - \alpha_{i-1}) \; (1 \leq i \leq n) \tag{7}$$

In the quantification, the shape of the L-XAJ model is determined by certain constraints. In this paper, the total water storage capacity of the same basin should be the same. Therefore, there are two constraints of XAJ model and L-XAJ model: (1) the maximum water storage capacity of the basin should be the same and (2) the average water storage capacity also should be the same. So:

$$WM = \frac{W'_{mm}}{1+B} = WM_{l-xaj} = \sum_{i=1}^{n} W_i \times (\alpha_i - \alpha_{i-1}) \tag{8}$$

Additionally,

$$W'_{mm} = W_n \tag{9}$$

Here, W'_{mm} and B are the parameters of XAJ model, which can be obtained by parameter calibration, n and α_i are the land-use parameters, which can be obtained from the land-use analysis of the basin, W_i is the water storage capacity value of ith land-use type and W_n is the maximum water storage capacity value.

After the L-XAJ model is constructed, it is necessary to further analyze the rainfall–runoff relationship and calculate the runoff yield under different rainfall conditions. As shown in Figure 6b, we assume that the basin soil moisture before rainfall is W_0, which is distributed horizontally in L-XAJ model, the first m land-use types in the basin have reached the saturation state, while the m + 1 land-use type has not reached or has just reached it. In this case:

$$A = \frac{W_0 - \sum_{i=1}^{m} W_i \times (\alpha_i - \alpha_{i-1})}{1 - \alpha_m} \tag{10}$$

If rainfall is P, evapotranspiration is E. When evapotranspiration exceeds rainfall ($P - E < 0$), runoff is not generated. When rainfall exceeds evapotranspiration ($P - E > 0$) then, if $P - E + A > W_n$, total runoff generation:

$$R = P - E - \left(WM_{l-xaj} - W_0\right) \tag{11}$$

Otherwise, for local runoff generation, the soil water storage is the horizontal fill part in Figure 6b and runoff yield R is:

$$R = (P - E) - \sum_{i=a}^{b} (W_i - A)\cdot(\alpha_i - \alpha_{i-1}) - \sum_{i=b+1}^{n} (P - E)\cdot(\alpha_i - \alpha_{i-1}) \tag{12}$$

where a and b satisfy the constraints:

$$W_a > A\,P - E + A \geq W_b \tag{13}$$

where a takes the smallest integer value that satisfies Equation (13) and b takes the largest integer value.

So, the runoff yield of L-XAJ model can be calculated by Equations (11) and (12) and the rainfall–runoff interactions can be studied. When set, the soil moisture content is S_i ($S_0 = 0$) and at the same time, the soil moisture is just enough to make the land-use type i reach the storage-full state. In this condition, the rainfall–runoff relationship (Figure 7) is different to XAJ model; it is segmented form, but there is continuity between adjacent segments. When $W_0 = S_0$, the line is n segments with different slopes, which in turn are $1/\alpha i$ ($1 \leq i \leq n$); when $W_0 = S_1$, the line is $n - 1$ segments with different slopes, which in turn are $1/\alpha i$ ($2 \leq i \leq n$), and so on. When $S_{i-1} < W_0 < S_i$, the segment is similar to the $W_0 = S_{i-1}$, but the position will be changed. When $W_0 = WM_{l-xaj}$, then $R = P - E$ and the line which is straight line from the origin with slope is 1, which means all rainfall generated runoff yield. So, the L-XAJ model satisfies the principle of the saturation-excess runoff mechanism.

2.3. Study Area and Data Set

In the piecewise-function-described L-XAJ model, the more diverse the land-use types of the basin are, the more segments the curve is divided into and the more accurate it is to describe the water storage capacity with a continuous curve. As a result, a basin with simple land-use form and homogeneous soil is more suitable to verify the L-XAJ model. Therefore, Taipingchi basin was chosen as the study area.

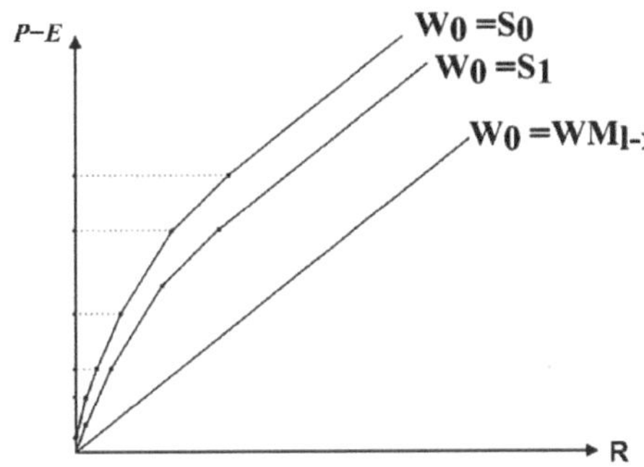

Figure 7. Rainfall–runoff relationship in L-XAJ model.

Taipingchi basin, located in northeast China (Figure 8a), has an area of 1706 km^2. Two tributaries, the Wengke River and the Xinkai River, flow into the mainstream of Taipingchi (Figure 8b). The elevation of the whole basin from the northwest to the southeast gradually increases, from 178 m to 552 m, with gentle fluctuations (Figure 8a). The main landform type in the basin is valley plains. Taipingchi basin is a typical human activity-dominated basin. Almost all of the basin has been built into towns or reclaimed as farmland. Urban and agricultural land account for above 95% of the total area of the basin (urban about 10% and agricultural land about 85%). According to the L-XAJ model, we guess the continuous water storage capacity curve would not be accurate enough to describe the real situation accurately as possible and the XAJ model would not perform well in this basin.

The basin is dominated by a typical semi-arid and semi-humid climate. The average annual temperature in the basin is about 4.9 °C. The annual average precipitation is 515.7 mm. Seventy percent of rainfall is in the summer from July to September. In the main flood season, July and August are prone to short-term heavy rainfall, which can easily lead to heavy flood disasters. The inter-annual variability of precipitation is large and the distribution is extremely uneven during the year. The annual average evaporation in the basin is about 947 mm.

The hydrological data mainly include the 21 flood events of 12 hydrological observation stations in the Taipingchi basin from 2009 to 2012 and 1 runoff observation station (Figure 8b), including average rainfall and runoff data ($\Delta t = 6$ h).

Based on 30 m Landsat TM image data (the data set is provided by Geospatial Data Cloud site, Computer Network Information Center, Chinese Academy of Sciences (http://www.gscloud.cn, (accessed on 24 February 2022))) and in consideration of the impact of land characteristics on runoff, the supervised classification method (by the maximum likelihood classification in ArcGIS) was adopted to classify the land-use types into five categories: forest, crops, grassland, surface water bodies, and urban (Figure 9). See Table 1 for statistical information.

2.4. Modeling Set

In this study, 21 flood events from 2009 to 2012 that took place in the Taipingchi basin were used for model calibration and verification at a 6-h time step. Fifteen flood events were chosen to calibrate the model parameters and six events to verify the model. Calibration and optimization of XAJ model parameters were based on the parameter estimation algorithm (PEST) with MATLAB environment [69]. Thirteen parameters related to evapotranspiration, runoff generation, runoff source partition and runoff routing (Table 2) were calibrated. There are three main factors to consider in the calibration process: the lower and upper boundaries [23], the objective function and termination condition. The objective function can be updated as below:

$$OF = \sum_{i=1}^{n} (Q_s - Q_o)^2 \tag{14}$$

where OF is objective function, i is the time order, n is the time step, Q_s is the simulated discharge and Q_o is the observed discharge.

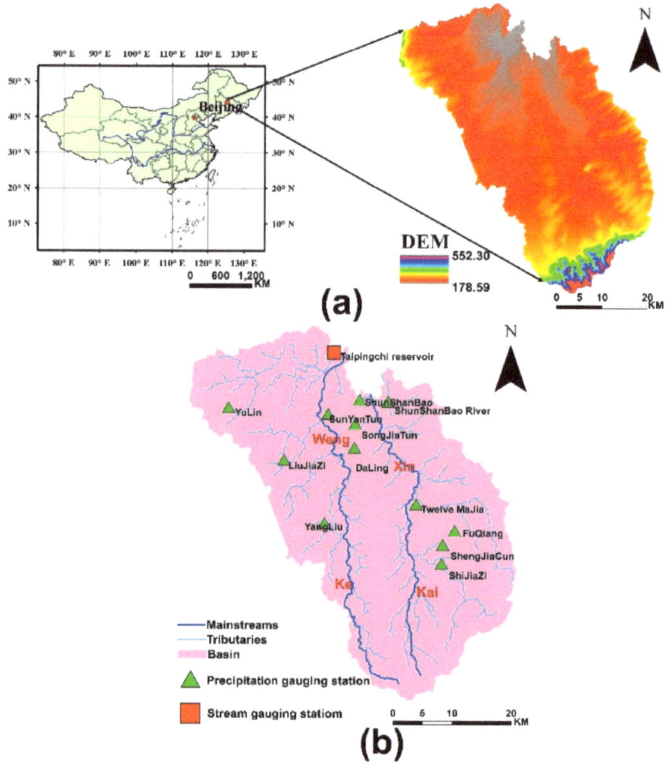

Figure 8. Study area information ((**a**): geographical location and DEM; (**b**) mainstream, branch and hydrological stations network).

Table 1. Area of different land-uses during the studied years.

Land-Use Type	2009		2010		2011		2012	
	Area (km^2)	Ratio (%)	Area (km^2)	Ratio (%)	Area (km^2)	Ratio (%)	Area (km^2)	Ratio (%)
Urban	341.33	19	352.28	20	386.65	22	353.42	21
Crops	1308.77	77	1301.01	76	1267.46	74	1301.07	75
Grassland	4.64	1	4.01	1	3.52	1	3.11	1
Forest	23.4	1	21.7	1	21.6	1	21.71	1
Surface water bodies	28.53	2	27.67	2	27.44	2	27.36	2

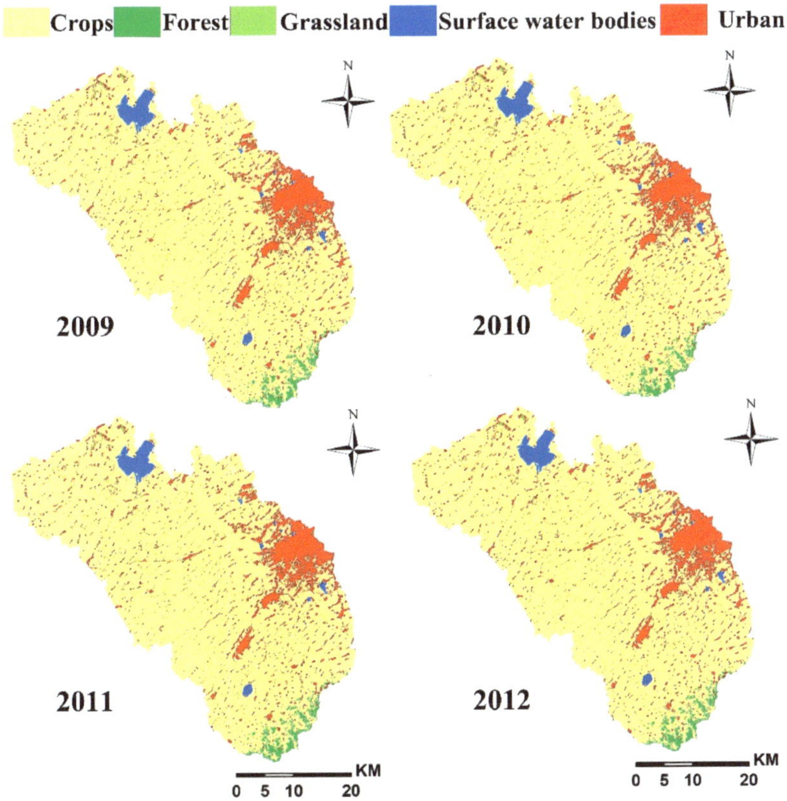

Figure 9. Land-use map in the Taipingchi basin.

Table 2. The Parameters of XAJ and the L-XAJ.

XAJ			L-XAJ		
Parameters	**Value**	**Physical Meaning**	**Parameters**	**Value**	**Physical Meaning**
WUM	15.2	Averaged soil moisture storage capacity of the upper layer	W_1	0	Urban land soil moisture storage capacity
WLM	78.6	Averaged soil moisture storage capacity of the lower layer	W_2	153.3	Cultivated land soil moisture storage capacity
WDM	29.5	Averaged soil moisture storage capacity of the deep layer	Ra1	0.2	Area ratio of urban land
B	0.35	Exponential of the distribution to tension water capacity	Ra2	0.8	Area ratio of cultivated land
K	0.71	Conversion coefficient of evaporation	K	-	-
C	0.2	Coefficient of the deep layer	C	-	-
IMP	0.02	Percentage of impervious and saturated areas in the basin	IMP	-	-
SM	32.5	Areal mean free water capacity of the surface soil layer	SM	-	-

Table 2. *Cont.*

XAJ			L-XAJ		
Parameters	**Value**	**Physical Meaning**	**Parameters**	**Value**	**Physical Meaning**
EX	1.02	Exponent of the free water capacity curve influencing the development of the saturated area	EX	-	-
KG	0.06	Outflow coefficients of the free water storage to groundwater relationships	KG	-	-
KSS	0.11	Outflow coefficients of the free water storage to interflow relationships	KSS	-	-
KKG	0.98	Recession constants of the groundwater storage	KKG	-	-
KKSS	0.71	Recession constants of the lower interflow storage	KKSS	-	-

Note: represent the parameters of L-XAJ are the same as XAJ.

L-XAJ model and XAJ model are slightly different in the parameter calibration process of runoff yield module: the WUM, WLM, WDM and B are the parameters by the XAJ, however, these parameters are replaced by the area ratios of specific land-use type (Ra1 and Ra2) and their corresponding water storage capacity values (W_1 and W_2) in L-XAJ. The remaining parameters are the same in both XAJ and L-XAJ. See Table 2 for parameter results.

2.5. Statistical Criteria

According to the accuracy standard for hydrological forecasting in China, the results were evaluated by three statistical criteria including the Nash–Sutcliffe efficiency coefficient (NSE), the flood volume error (FVE) and the flood peak error (FPE). The detailed equations are expressed as follows:

$$\text{NSE} = 1 - \frac{\sum (Q_o - Q_s)^2}{\sum (Q_o - \overline{Q_o})^2} \tag{15}$$

$$\text{FVE} = \frac{\sum Q_s - \sum Q_o}{\sum Q_o} \times 100\% \tag{16}$$

$$\text{FPE} = \frac{\text{MAX}(Q_s) - \text{MAX}(Q_o)}{\text{MAX}(Q_o)} \times 100\% \tag{17}$$

where Q_o is the observed discharge (m^3/s), Q_s is the simulated discharge (m^3/s), $\overline{Q_o}$ is the mean value of the observed discharge (m^3/s), MAX(Q_s) is the simulated peak discharge and MAX(Q_o) is the observed peak discharge.

According to the accuracy standard, when NSE exceeds 0.9, it is considered to meet standard A and when $0.7 \leq \text{NSE} \leq 0.9$, it meets standard B.

3. Results and Discussions

3.1. Simulated Results and Global Analysis

Both the L-XAJ model and XAJ model were applied in the Taipingchi basin. From 2009 to 2012, a total of 21 flood events occurred in the Taipingchi basin and they were all simulated by these two models. The performances of the two models were tested by the statistical indicators (NSE, FVE, FPE) mentioned in Section 2.5, and showed in Table 3.

Table 3. Simulation results by the XAJ and L-XAJ model.

Period	Flood Event ID	Date	L-XAJ NSE	L-XAJ FVE (%)	L-XAJ FPE (%)	XAJ NSE	XAJ FVE (%)	XAJ FPE (%)	P
Calibration	1	28 May 2009	0.81	5.36	16.33	0.74	−4.55	−29.16	+
	2	28 June 2009	0.92	−15.13	−19.38	0.80	−23.46	−25.92	+
	3	16 July 2009	0.93	4.66	10.84	0.83	26.27	23.04	+
	4	27 August 2009	0.84	−17.14	−10.73	0.74	−19.49	−23.56	+
	5	3 May 2010	0.82	−18.23	−9.75	0.66	−27.99	−19.87	+
	6	1 July 2010	0.84	−5.41	−11.15	0.71	−7.37	−24.68	+
	7	19 July 2010	0.87	−3.73	−13.15	0.85	−6.59	−16.68	○
	8	4 August 2010	0.76	14.25	−5.30	0.75	18.92	5.25	○
	9	10 October 2010	0.69	5.49	10.49	0.69	5.10	11.01	○
	10	11 November 2010	0.84	4.17	6.53	0.68	6.54	10.69	+
	11	18 May 2011	0.79	3.11	5.52	0.69	2.52	−12.39	+
	12	29 May 2011	0.91	1.62	−12.19	0.82	−12.93	−17.57	+
	13	30 June 2011	0.88	10.07	7.90	0.81	17.32	18.48	+
	14	20 July 2011	0.89	−7.57	9.73	0.70	−9.45	26.95	+
	15	30 July 2011	0.91	3.79	9.38	0.85	2.87	16.85	+
Validation	16	29 June 2012	0.88	−19.48	−17.83	0.80	25.28	20.26	+
	17	22 July 2012	0.93	−3.32	−13.45	0.88	3.18	−18.04	+
	18	18 August 2012	0.86	15.42	−8.02	0.64	24.97	16.41	+
	19	27 August 2012	0.92	−9.47	19.09	0.91	−12.83	17.01	○
	20	27 September 2012	0.88	9.74	7.42	0.62	18.20	28.31	+
	21	10 November 2012	0.89	9.20	8.82	0.61	18.75	20.22	+

Note: P is a sign of whether L-XAJ is better than XAJ, + represents that L-XAJ is better than XAJ model and ○ is not.

As shown in Table 3, the NSE of the L-XAJ model ranged from 0.69 to 0.93, with the average being 0.86. Meanwhile, the NSE of the XAJ model ranged from 0.61 to 0.91, with the average being 0.75. For the FVE and FPE, all 21 flood events of the L-XAJ model were within 20% and the qualified rate was 100%. However, there were only 12 flood events within 20% of the FPE in the XAJ model and the qualified rate only was 57.14%. The L-XAJ model reduced the average FPE from 19.16% to 11.10% and the FVE from 14.03% to 8.87%. Overall, only 10 flood events were simulated accurately in the XAJ model, while all flood events were accurately simulated by the L-XAJ model; the simulation results of the L-XAJ model were better than the XAJ model under all the three statistical criteria. This showed that the L-XAJ model was successfully used in the Taipingchi basin.

The distributions of the FVE, FPE and NSE statistics for all simulations (both calibration and validation events) were showed in Figure 10. The NSE of the L-XAJ model was higher than the XAJ model (except 20101010) and the FPE and FVE distribution of the L-XAJ model was lower than the XAJ model. This showed that the performance of the L-XAJ model in the Taipingchi basin was comprehensively better than that of XAJ model.

Several rainfall–runoff processes were shown in Figure 11. The discharge process of the XAJ model and the L-XAJ model were basically similar and there was only a certain difference in the flood volume, indicating that the two models had the same runoff sources partition and runoff concentration and differ only in the runoff generation. The discharge processes of the two models were similar with the observed discharge (OBQ), indicating that the two models can reflect the runoff process.

Focusing on the calibration period, as the parameters of the XAJ model were calibrated by these 15 floods, it should be expected to perform well in this period. However, none of the NSE achieved standard A and three flood events of FVE and four flood events of FPE exceeded 20%, which indicates that the XAJ model is not accurate enough to reproduce the rainfall–runoff process. This is in line with our prediction in Section 2.3. In contrast, the L-XAJ model performed well during the calibration period, though its parameters were calibrated by the XAJ model. The NSE of four flood events achieved standard A and all the FVE and FPE are within 20%. The results indicated that the L-XAJ model could reflect the rainfall–runoff process more accurately; or more precisely, the land-use-based water storage capacity curve is more accurate in illustrating the runoff yield in the Taipingchi basin. Furthermore, it indicated that the land-use area ratio (L/F) is substantially associated with the yielding area (f/F) of the XAJ model, which validates the hypothesis of a corresponding relationship between the different land-use types and the yield area.

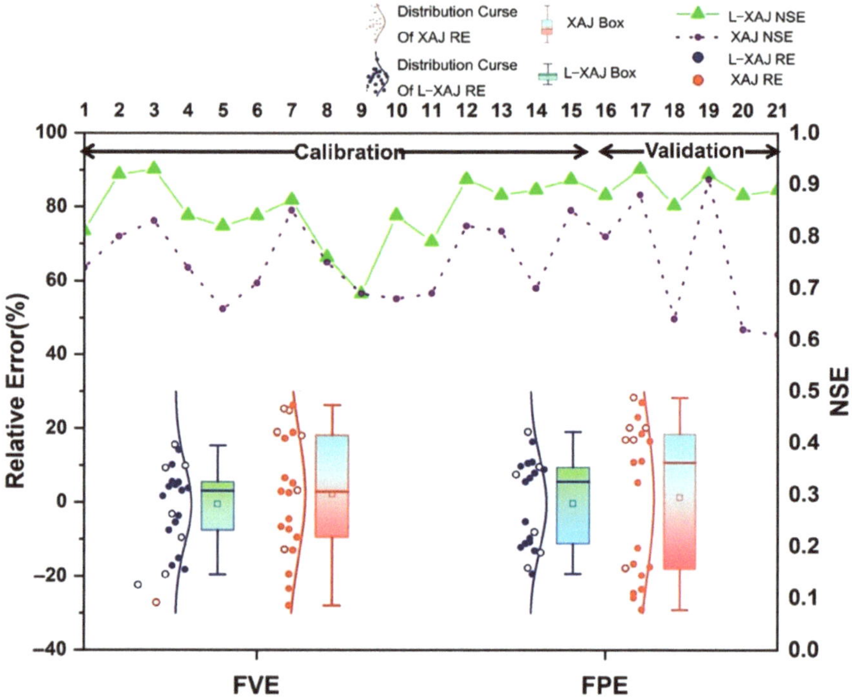

Figure 10. Comparison of the simulation results with XAJ model and L-XAJ model (the number is the order of the flood event, open circles are calibration events and filled circles are validation events).

As for the validation period, the NSE of the L-XAJ model had two flood events for standard A and four for standard B in all of the six floods, while all the FVE and FPE were within 20%. The validation results performed well.

Generally, the performance of the XAJ model in the Taipingchi basin is mediocre, which indicated that the XAJ model would not perform well in a simple land-use basin. On the other hand, the L-XAJ model outperformed in 17 of 21 floods, not only in the validation period, but also in the calibration period. This indicated that the L-XAJ model is more suitable for simulating the hydrological process of the Taipingchi basin than the XAJ model or that the land-use-based water storage capacity curve can describe the runoff yield more accurately than the original water storage capacity curve of the XAJ model.

3.2. Simulation Results in Different Yielding Area

During severe rainfall events or high-intensity rainfalls, the runoff yield is not synchronized everywhere in the basin [70]. Generally, an impervious surface usually yields earlier than the other areas; farmland with low vegetation usually has less interception than forests with tall vegetation, so it yields earlier than the forest area. Under the same underlying surface type, the higher the soil moisture is, the earlier the flow is produced. That is to say, in different flood events, the actual yielding area and yielding process are different.

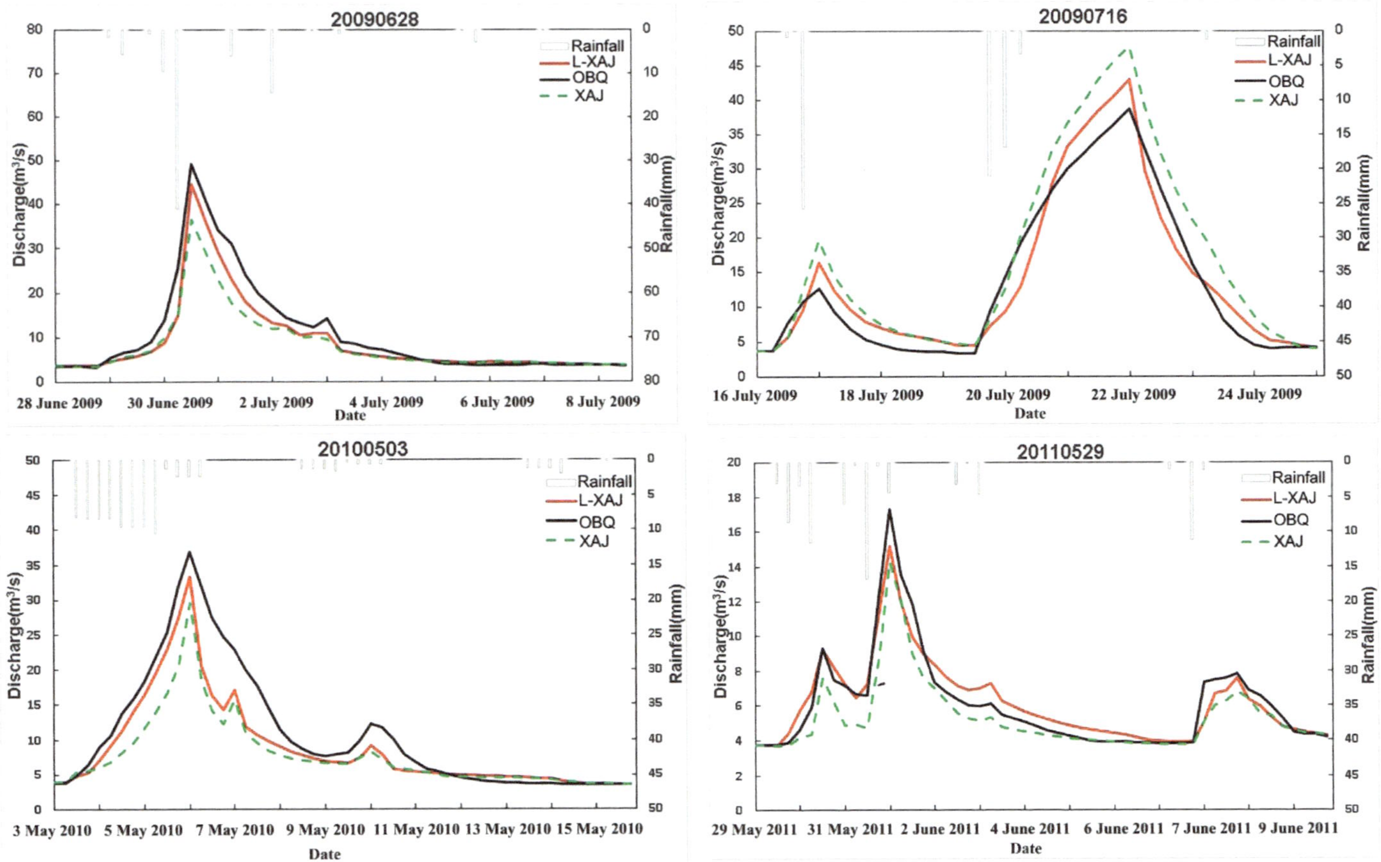

Figure 11. Simulated and observed hydrographs.

On the other hand, a rainfall–runoff model ignoring the underlying surface would perform differently from a model based on land-use form. This difference would be changed by the yielding area and can be shown in Figure 12. As the urban area takes up about 20% of the basin area and crops take up above 70%, the total yielding area (Figure 12a) can be separated easily from each other, such as (0, 0.2) (Figure 12b), [0.2, 0.7] (Figure 12c) and (0.7, 1] (Figure 12d). The larger the yielding area, the smaller difference of the NSE between the XAJ model and L-XAJ model (Figure 12a): when the yielding area was small (Figure 12b), the basin was relatively dry and the impact of 20% of the urban area on the runoff generation was reflected in the L-XAJ model, which can make up for the artificially intercepted rainfall, so the NSE is higher; when the yielding area increases (Figure 12c), the basin was relatively humid and the level of 70% crops was reflected. At this stage, although the NSE of L-XAJ has been improved relative to XAJ, the improvement effect is not as obvious as the previous stage (dry stage). This is because the urban and crops worked together on the runoff yield and the difference in the runoff calculation between XAJ and L-XAJ is not as large as that in the dry stage (Figures 2a and 6b); when the basin was in a near-saturated state (Figure 12d), there was almost no difference between the two models and both are simulated well. These were expected due to the mechanism of land use on runoff generation [71]. These results explained the effect of different land uses on rainfall redistribution when the basin was in different stages. Hence, correspondence between the yield area (in XAJ) and land-use type (in L-XAJ) is characterized.

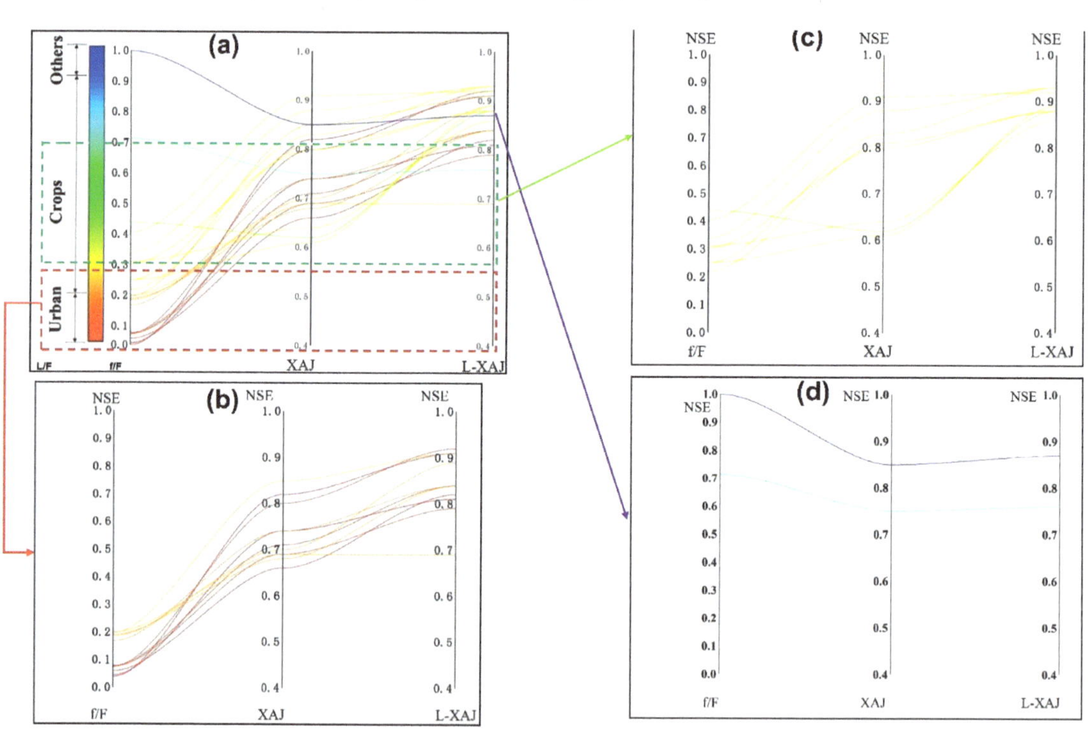

Figure 12. The NSE in partial yielding area ((**a**): all yield area; (**b**): 0 < yield area < 0.2 (**c**) 0.2 ≤ yield area < 0.7; (**d**): yield area ≥ 0.7).

Figure 13 showed the FVE under different yielding areas. Under the control of the urban area, the FVE of the XAJ model was very large and the flood volume was always smaller than observed, while the L-XAJ model not. It was closely related to the influence of the urban area on the runoff yield [72]. While under the control of the crops, the flood volume was always larger than observed, which was because crops had an impact on the runoff yield [73]. The water storage capacity of crops in the Taipingchi basin was relatively large, so the runoff yield in this part was low, but the XAJ model did not consider it.

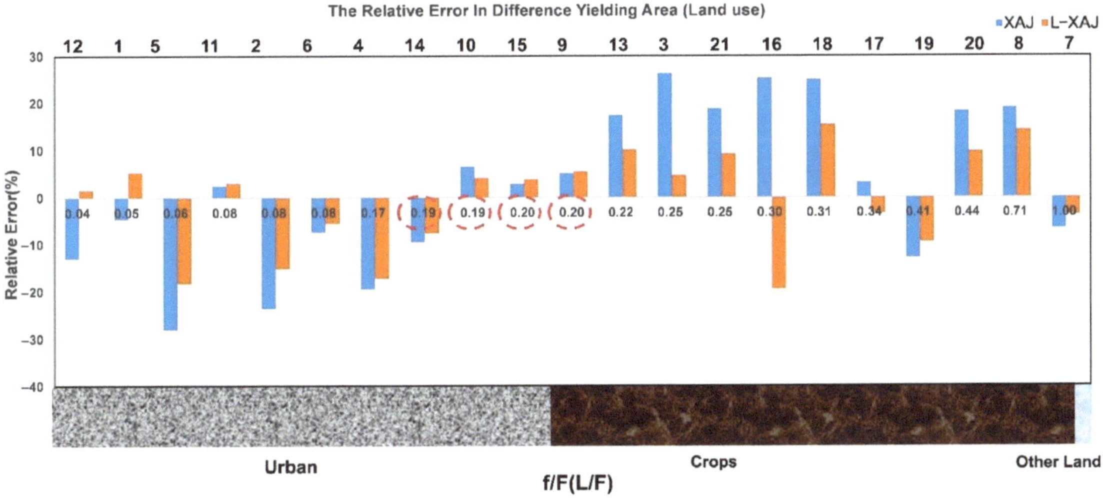

Figure 13. The relative error in different yielding area (different land-use types dominated, the number at the top are flood event ID).

In conclusion, the land-use form influences the runoff process. It could be found that the area of different land types and its storage capacity value correspond to the yielding area of XAJ model. Hence, we verified that f/F corresponds to L/F and that the research objectives that were discussed in the introduction of this paper were met.

3.3. Simulation Results in Different Flood Types

In order to study the sensitivity of the XAJ model and L-XAJ model to the flood magnitude, 21 flood events were divided into three levels: large, medium and small, according to the peak discharge. There were three large flood events, seven medium flood events and eleven small flood events. The NSE and the FVE in different flood levels were shown in Figure 14.

It could be seen from Figure 14a that the NSE of the L-XAJ model was greatly improved compared with the XAJ model in small and medium floods, especially for small floods. The major reason for this was that the runoff of small floods is more easily affected by land use [71,74]. However, the third flood (circled in red in Figure 14a) had not been improved. After analysis, this might be due to the yielding area which was around 0.2 and the difference between the L-XAJ model and XAJ model was not significant.

The FVE of the different flood levels were shown in Figure 14b: for small floods, the FVE was smaller, but compared to the XAJ model, the L-XAJ model had higher accuracy; for medium floods, the FVE of the XAJ model was relatively large. This was because most of the seven medium floods were at the beginning of each year or after the flood season. At this time, for crop growth, many ponds had been artificially established in the basin to store rainfall [75] (Figure 15), which had a significant impact on the runoff yield, however, the runoff yield calculated by L-XAJ is more than that of XAJ at this stage (Figures 2a and 6b), so the store rainfall can be partially offset in L-XAJ. How to consider the rainfall interception in L-XAJ is the main direction of our in-depth research. For big floods, the larger accuracy indicated a better performance of both models, which is in agreement with the relevant literature [71].

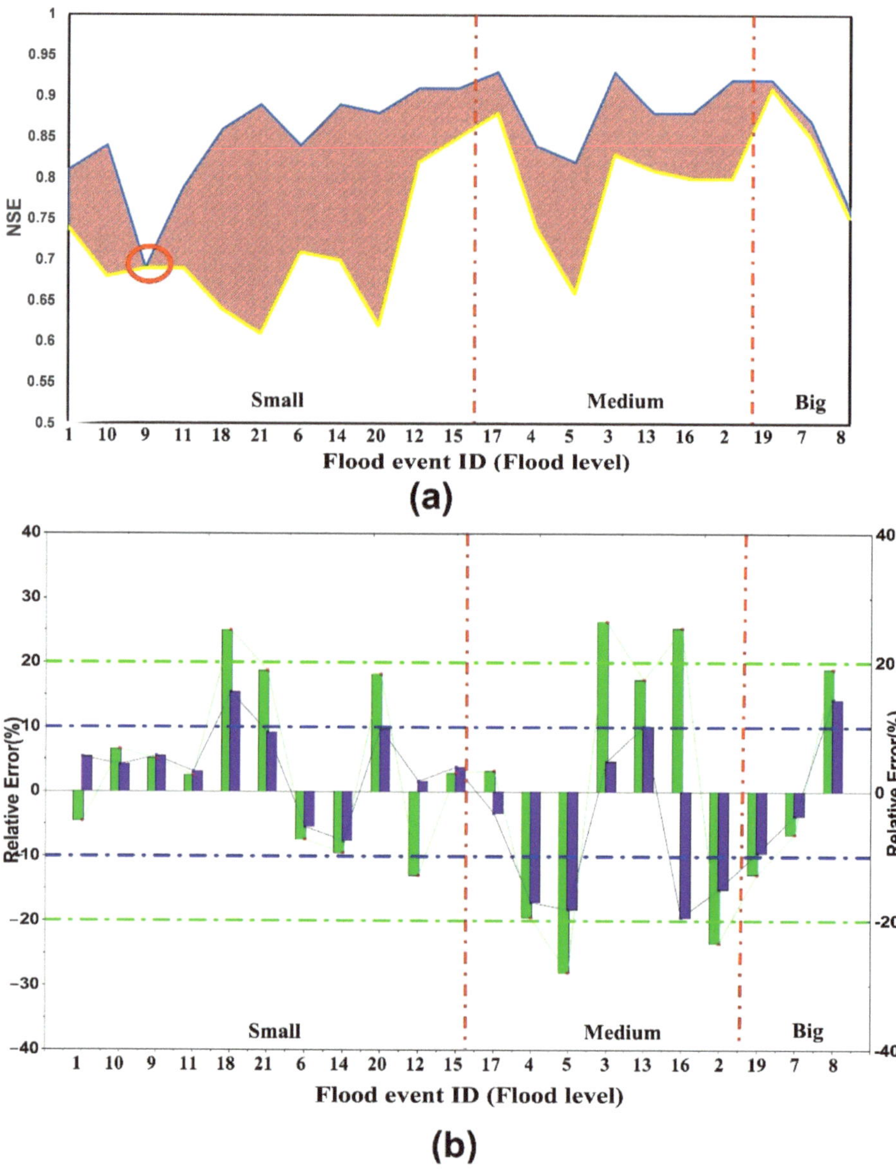

Figure 14. Comparison of NSE and FVE under different flood levels ((**a**): NSE; (**b**) FVE; the green dotted line is the 20% error line and the blue is 10%).

To further analyze the performance of the L-XAJ model compared to XAJ, the improved accuracy of the L-XAJ model relative to the XAJ model was analyzed and the improved results were shown in Figure 16. For the NSE, compared with the XAJ model, the L-XAJ model had a significant improvement of small and medium floods, but almost no improvement for large floods. For the FPE and FVE, in small floods, the FVE increased by 16.57% on average and the FPE by 44.76%; in medium floods, the FVE increased by 38.25% on average and the FPE by 43.42. In large floods, the FVE increased by 31.42% on average and the FPE by 2.66%. Therefore, it was further proved that the L-XAJ model can significantly improve the performance of flood simulation.

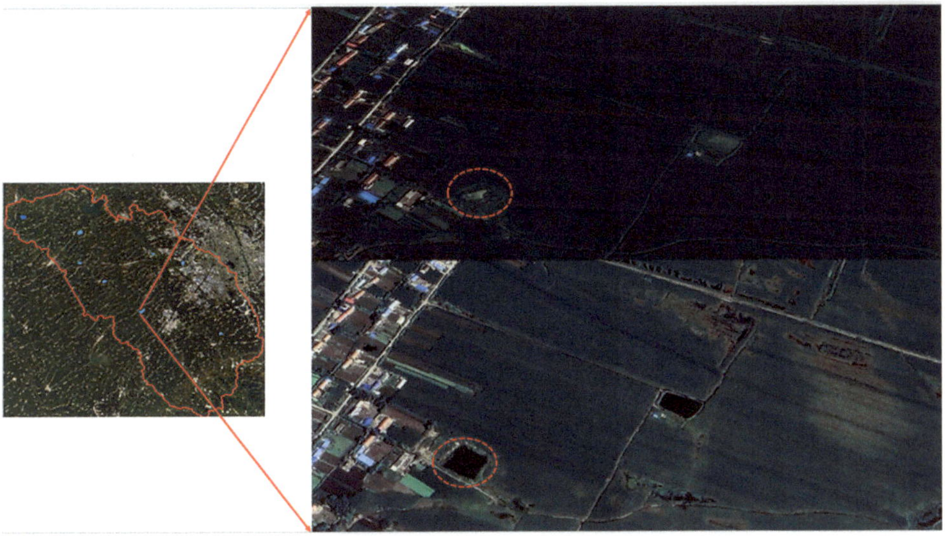

Figure 15. The artificial pond for storing rainfall (the red dotted line is the change of pond area; upper right is before crops growth and lower is after crops growth).

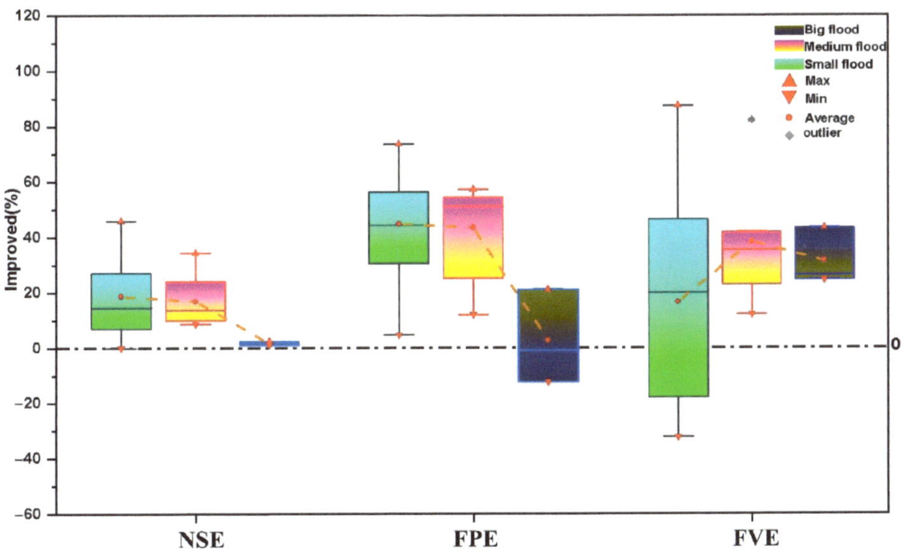

Figure 16. The improvement of L-XAJ model compared to XAJ model (the number of flood events: small: 11, medium: 7, big: 3; NSE: (L-XAJ − XAJ)/XAJ*100%; FVE/FPE: (AV(XAJ) − AV(L-XAJ))/AV(XAJ)*100%, AV: absolute value).

Based on the results in Section 3.2, this change can be easily explained: the runoff yield usually occurs in part of the basin in small floods and medium floods, while the runoff yield of big floods generally occurs in the entire basin.

4. Conclusions

Effective basin water resource management is of significant importance for the basin's sustainable development. The main objective of this research is to study the relationship between the yielding area (f/F) and the area of different land uses (L/F) for better basin flood resource management. The L-XAJ model was constructed by integrating the land-use information into the runoff generation of the XAJ model and the model is shown to improve the performance of the runoff in a typical cultivated–urban binary land-use-type basin: the Taipingchi basin. The major findings of this paper were summarized as follows:

(1) The distribution of the runoff yield can be divided by the land-use form, which is, the areas with the same land-use form are similar in runoff yield, while areas of different land uses are significantly different. In the XAJ model, particularly, that is to say, the yielding area of the XAJ model, (f/F) is determined by the area ratio of different land-use types (L/F) (Section 3.2).

(2) The L-XAJ model can be well used in a rainfall–runoff simulation (Table 3 and Figure 10). It performed better than the XAJ model in a simple land-use-form basin (mean NSE: 0.86 > 0.75, FVE: 8.87% < 14.03%, FPE: 11.10% < 19.06%).

(3) The L-XAJ model can well improve the simulation accuracy of small and medium floods compared to large floods (Figure 16).

Although our preliminary test demonstrated the relationship between the runoff yield and land use, we still know little about the calculation of the runoff yield for specific areas. Ongoing research could focus on analyzing the effect of vegetation type, soil type, topography and other remotely-sensed data on the runoff yield. How to establish an index that integrates various factors to further enhance the physical meaning of the runoff generation in the XAJ model to obtain better results is worthy of further study.

5. Patents

The patent "A method and system for determining runoff yield of artificial watershed" (patent number: CN 202011611825.9) resulted from the work reported in this manuscript.

Author Contributions: Conceptualization, C.X.; Data curation, C.X.; Formal analysis, C.X.; Investigation, C.X. and H.F.; Methodology, C.X.; Project administration, H.F.; Resources, C.X. and Y.W.; Software, C.X. and J.Y.; Supervision, H.F.; Validation, C.X. and L.W.; Visualization, C.X. and Y.W.; Writing—original draft, C.X. and H.F.; Writing review and editing, J.Y., L.W. and Y.W. All authors have read and agreed to the published version of the manuscript.

Funding: This research was funded by Major Science and Technology Program for Water Pollution Control and Treatment, grant number No. 2014ZX07203-008.

Institutional Review Board Statement: Not applicable.

Informed Consent Statement: Not applicable.

Data Availability Statement: Some or all data, models, or codes that support the findings of this study are available from the corresponding author upon reasonable request.

Acknowledgments: The authors would like to acknowledge Xiaoliu Yang of Peking University, China, for his theoretical support and guidance. This research was funded by the Major Science and Technology Program for Water Pollution Control and Treatment (Grant No. 2014ZX07203-008) The authors would like to thank the Jilin Provincial Water Resources Department for providing the hydrological data for this article.

Conflicts of Interest: The authors declare no conflict of interest.

References

1. Huang, F.; Zhang, Y.D.; Zhang, D.R.; Chen, X. Environmental Groundwater Depth for Groundwater-Dependent Terrestrial Ecosystems in Arid/Semiarid Regions: A Review. *Int. J. Environ. Res. Public Health* **2019**, *16*, 763. [CrossRef] [PubMed]
2. Li, P.Y.; Wang, D.; Li, W.Q.; Liu, L.N. Sustainable water resources development and management in large river basins: An introduction. *Environ. Earth Sci.* **2022**, *81*, 179. [CrossRef] [PubMed]

3. Meijer, K.S.; Schasfoort, F.; Bennema, M. Quantitative Modeling of Human Responses to Changes in Water Resources Availability: A Review of Methods and Theories. *Sustainability* **2021**, *13*, 8675. [CrossRef]
4. Li, J.; Liu, C.M.; Wang, Z.G.; Liang, K. Two universal runoff yield models: SCS vs. LCM. *J. Geogr. Sci.* **2015**, *25*, 311–318. [CrossRef]
5. Zhang, J.P.; Zhang, H.; Xiao, H.L.; Fang, H.Y.; Han, Y.P.; Yu, L. Effects of rainfall and runoff-yield conditions on runoff. *Ain Shams Eng. J.* **2021**, *12*, 2111–2116. [CrossRef]
6. Gupta, S.K.; Tyagi, J.; Sharma, G.; Jethoo, A.S.; Singh, P.K. An Event-Based Sediment Yield and Runoff Modeling Using Soil Moisture Balance/Budgeting (SMB) Method. *Water Resour. Manag.* **2019**, *33*, 3721–3741. [CrossRef]
7. Dunne, T.; Black, R.D. Partial Area Contributions to Storm Runoff in a Small New-England Watershed. *Water Resour. Res.* **1970**, *6*, 1296–1311. [CrossRef]
8. Horton; Robert, E. The role of infiltration in the hydrologic cycle. *Trans. Am. Geophys. Union* **1933**, *14*, 446–460. [CrossRef]
9. Bennett, J.C.; Wang, Q.; Li, M.; Robertson, D.; Schepen, A. Reliable long-range ensemble streamflow forecasts by combining dynamical climate forecasts: A conceptual runoff model and a staged error model. *Water Resour. Res.* **2016**, *52*, 8238–8259. [CrossRef]
10. Kan, G.; He, X.; Ding, L.; Li, J.; Liang, K.; Hong, Y. Study on applicability of conceptual hydrological models for flood forecasting in humid, semi-humid semi-arid and arid basins in China. *Water* **2017**, *9*, 719. [CrossRef]
11. Zhang, L.; Yang, X. Applying a Multi-Model Ensemble Method for Long-Term Runoff Prediction under Climate Change Scenarios for the Yellow River Basin, China. *Water* **2018**, *10*, 301. [CrossRef]
12. De Barros, C.; Minella, J.; Schlesner, A.; Ramon, R.; Copetti, A. Impact of data sources to DEM construction and application to runoff and sediment yield modelling using LISEM model. *J. Earth Syst. Sci.* **2021**, *130*, 53. [CrossRef]
13. Nazari-Sharabian, M.; Taheriyoun, M.; Karakouzian, M. Sensitivity analysis of the DEM resolution and effective parameters of runoff yield in the SWAT model: A case study. *J. Water Supply Res. Technol.* **2020**, *69*, 39–54. [CrossRef]
14. Zhao, R.J. The Xinanjiang Model Applied in China. *J. Hydrol.* **1992**, *135*, 371–381. [CrossRef]
15. Tran, Q.Q.; De Niel, J.; Willems, P. Spatially Distributed Conceptual Hydrological Model Building: A Genetic top-Down Approach Starting from Lumped Models. *Water Resour. Res.* **2018**, *54*, 8064–8085. [CrossRef]
16. Zhang, R.; Cuartas, L.A.; Carvalho, L.V.D.; Leal, K.R.D.; Mendiondo, E.M.; Abe, N.; Birkinshaw, S.; Mohor, G.S.; Seluchi, M.E.; Nobre, C.A. Season-based rainfall-runoff modelling using the probability-distributed model (PDM) for large basins in southeastern Brazil. *Hydrol. Process.* **2018**, *32*, 2217–2230. [CrossRef]
17. Gong, J.F.; Yao, C.; Li, Z.J.; Chen, Y.F.; Huang, Y.C.; Tong, B.X. Improving the flood forecasting capability of the Xinanjiang model for small- and medium-sized ungauged catchments in South China. *Nat. Hazards* **2021**, *106*, 2077–2109. [CrossRef]
18. Wang, J.; Bao, W.M.; Gao, Q.Y.; Si, W.; Sun, Y.Q. Coupling Xinanjiang model and wavelet-based random forests method for improved daily streamflow simulation. *J. Hydroinform.* **2021**, *23*, 589–604. [CrossRef]
19. Xu, C.W.; Han, Z.Y.; Fu, H. Remote Sensing and Hydrologic-Hydrodynamic Modeling Integrated Approach for Rainfall-Runoff Simulation in Farm Dam Dominated Basin. *Front. Environ. Sci.* **2022**, *9*, 672. [CrossRef]
20. Zuo, D.P.; Xu, Z.X.; Yao, W.Y.; Jin, S.Y.; Xiao, P.Q.; Ran, D.C. Assessing the effects of changes in land use and climate on runoff and sediment yields from a watershed in the Loess Plateau of China. *Sci. Total Environ.* **2016**, *544*, 238–250. [CrossRef]
21. Paule-Mercado, M.A.; Lee, B.Y.; Memon, S.A.; Umer, S.R.; Salim, I.; Lee, C.H. Influence of land development on stormwater runoff from a mixed land use and land cover catchment. *Sci. Total Environ.* **2017**, *599*, 2142–2155. [CrossRef]
22. Meng, C.Q.; Zhou, J.Z.; Zhong, D.Y.; Wang, C.; Guo, J. An Improved Grid-Xinanjiang Model and Its Application in the Jinshajiang Basin, China. *Water* **2018**, *10*, 1265. [CrossRef]
23. Yao, C.; Zhang, K.; Yu, Z.B.; Li, Z.J.; Li, Q.L. Improving the flood prediction capability of the Xinanjiang model in ungauged nested catchments by coupling it with the geomorphologic instantaneous unit hydrograph. *J. Hydrol.* **2014**, *517*, 1035–1048. [CrossRef]
24. Xu, Y.Y.; Cheng, X.; Gun, Z. What Drove Regional Changes in the Number and Surface Area of Lakes across the Yangtze River Basin during 2000–2019: Human or Climatic Factors? *Water Resour. Res.* **2022**, *58*, e2021WR030616. [CrossRef]
25. Yan, X.L.; Bao, Z.X.; Zhang, J.Y.; Wang, G.Q.; He, R.M.; Liu, C.S. Quantifying contributions of climate change and local human activities to runoff decline in the upper reaches of the Luanhe River basin. *J. Hydro-Environ. Res.* **2020**, *28*, 67–74. [CrossRef]
26. Wang, G.; Yinglan, A.; Xu, Z.; Zhang, S. The influence of land use patterns on water quality at multiple spatial scales in a river system. *Hydrol. Process.* **2015**, *28*, 5259–5272. [CrossRef]
27. Wagener, T. Can we model the hydrological impacts of environmental change? *Hydrol. Process.* **2007**, *21*, 3233–3236. [CrossRef]
28. Zhu, W.B.; Jia, S.F.; Lall, U.; Cao, Q.; Mahmood, R. Relative contribution of climate variability and human activities on the water loss of the Chari/Logone River discharge into Lake Chad: A conceptual and statistical approach. *J. Hydrol.* **2019**, *569*, 519–531. [CrossRef]
29. Zheng, H.X.; Zhang, L.; Zhu, R.R.; Liu, C.M.; Sato, Y.; Fukushima, Y. Responses of streamflow to climate and land surface change in the headwaters of the Yellow River Basin. *Water Resour. Res.* **2009**, *45*. [CrossRef]
30. Gao, H.K.; Cai, H.Y.; Duan, Z. Understanding the impacts of catchment characteristics on the shape of the storage capacity curve and its influence on flood flows. *Hydrol. Res.* **2018**, *49*, 90–106. [CrossRef]
31. Jayawardena, A.; Zhou, M. A modified spatial soil moisture storage capacity distribution curve for the Xinanjiang model. *J. Hydrol.* **2000**, *227*, 93–113. [CrossRef]

32. Shi, P.F.; Yang, T.; Xu, C.Y.; Yong, B.; Huang, C.S.; Li, Z.Y.; Qin, Y.W.; Wang, X.Y.; Zhou, X.D.; Li, S. Rainfall-Runoff Processes and Modelling in Regions Characterized by Deficiency in Soil Water Storage. *Water* **2019**, *11*, 1858. [CrossRef]
33. Chang, B.X.; Wherley, B.; Aitkenhead-Peterson, J.A.; McInnes, K.J. Effects of urban residential landscape composition on surface runoff generation. *Sci. Total Environ.* **2021**, *783*, 146977. [CrossRef]
34. Feng, X.M.; Sun, G.; Fu, B.J.; Su, C.H.; Liu, Y.; Lamparski, H. Regional effects of vegetation restoration on water yield across the Loess Plateau, China. *Hydrol. Earth Syst. Sci.* **2012**, *16*, 2617–2628. [CrossRef]
35. Xu, Y.Y.; Li, J.; Wang, J.D.; Chen, J.L.; Liu, Y.B.; Ni, S.N.; Zhang, Z.Z.; Ke, C.Q. Assessing water storage changes of Lake Poyang from multi-mission satellite data and hydrological models. *J. Hydrol.* **2020**, *590*, 125229. [CrossRef]
36. Sajikumar, N.; Remya, R.S. Impact of land cover and land use change on runoff characteristics. *J. Environ. Manag.* **2015**, *161*, 460–468. [CrossRef]
37. Astuti, I.S.; Sahoo, K.; Milewski, A.; Mishra, D.R. Impact of land use land cover (LULC) change on surface runoff in an increasingly urbanized tropical watershed. *Water Resour. Manag.* **2019**, *33*, 4087–4103. [CrossRef]
38. Zhou, M.; Qu, S.; Chen, X.; Shi, P.; Xu, S.; Chen, H.; Zhou, H.; Gou, J. Impact Assessments of Rainfall–Runoff Characteristics Response Based on Land Use Change via Hydrological Simulation. *Water* **2019**, *11*, 866. [CrossRef]
39. Rogger, M.; Agnoletti, M.; Alaoui, A.; Bathurst, J.C.; Bodner, G.; Borga, M.; Chaplot, V.; Gallart, F.; Glatzel, G.; Hall, J. Land-use change impacts on floods at the catchment scale: Challenges and opportunities for future research. *Water Resour. Res.* **2017**, *53*, 5209–5219. [CrossRef]
40. Farsi, N.; Mahjouri, N. Evaluating the contribution of the climate change and human activities to runoff change under uncertainty. *J. Hydrol.* **2019**, *574*, 872–891. [CrossRef]
41. Yang, L.; Feng, Q.; Yin, Z.; Wen, X.; Si, J.; Li, C.; Deo, R.C. Identifying separate impacts of climate and land use/cover change on hydrological processes in upper stream of Heihe River, Northwest China. *Hydrol. Process.* **2017**, *31*, 1100–1112. [CrossRef]
42. Walega, A.; Salata, T. Influence of land cover data sources on estimation of direct runoff according to SCS-CN and modified SME methods. *Catena* **2019**, *172*, 232–242. [CrossRef]
43. Yin, J.; He, F.; Xiong, Y.J.; Qiu, G.Y. Effects of land use/land cover and climate changes on surface runoff in a semi-humid and semi-arid transition zone in northwest China. *Hydrol. Earth Syst. Sci.* **2017**, *21*, 183–196. [CrossRef]
44. Chen, Y.; Xu, C.Y.; Chen, X.W.; Xu, Y.P.; Yin, Y.X.; Gao, L.; Liu, M.B. Uncertainty in simulation of land-use change impacts on catchment runoff with multi-timescales based on the comparison of the HSPF and SWAT models. *J. Hydrol.* **2019**, *573*, 486–500. [CrossRef]
45. Guo, M.; Zhang, T.; Li, Z.; Xu, G. Investigation of runoff and sediment yields under different crop and tillage conditions by field artificial rainfall experiments. *Water* **2019**, *11*, 1019. [CrossRef]
46. Li, T.; Dong, J.; Yuan, W. Effects of Precipitation and Vegetation Cover on Annual Runoff and Sediment Yield in Northeast China: A Preliminary Analysis. *Water Resour.* **2020**, *47*, 491–505. [CrossRef]
47. Liu, J.; Gao, G.; Wang, S.; Jiao, L.; Wu, X.; Fu, B. The effects of vegetation on runoff and soil loss: Multidimensional structure analysis and scale characteristics. *J. Geogr. Sci.* **2018**, *28*, 59–78. [CrossRef]
48. Nourani, V.; Fard, A.F.; Gupta, H.V.; Goodrich, D.C.; Niazi, F. Hydrological model parameterization using NDVI values to account for the effects of land cover change on the rainfall–runoff response. *Hydrol. Res.* **2017**, *48*, 1455–1473. [CrossRef]
49. Oudin, L.; Salavati, B.; Furusho-Percot, C.; Ribstein, P.; Saadi, M. Hydrological impacts of urbanization at the catchment scale. *J. Hydrol.* **2018**, *559*, 774–786. [CrossRef]
50. Kim, Y.D.; Kim, J.M.; Kang, B. Projection of runoff and sediment yield under coordinated climate change and urbanization scenarios in Doam dam watershed, Korea. *J. Water Clim. Chang.* **2017**, *8*, 235–253. [CrossRef]
51. Schilling, K.E.; Chan, K.-S.; Liu, H.; Zhang, Y.-K. Quantifying the effect of land use land cover change on increasing discharge in the Upper Mississippi River. *J. Hydrol.* **2010**, *387*, 343–345. [CrossRef]
52. Levavasseur, F.; Bailly, J.S.; Lagacherie, P.; Colin, F.; Rabotin, M. Simulating the effects of spatial configurations of agricultural ditch drainage networks on surface runoff from agricultural catchments. *Hydrol. Process.* **2012**, *26*, 3393–3404. [CrossRef]
53. da Silva, R.M.; Santos, C.A.G.; dos Santos, J.Y.G. Evaluation and modeling of runoff and sediment yield for different land covers under simulated rain in a semiarid region of Brazil. *Int. J. Sediment. Res.* **2018**, *33*, 117–125. [CrossRef]
54. Zhang, S.; Li, Z.; Lin, X.; Zhang, C. Assessment of Climate Change and Associated Vegetation Cover Change on Watershed-Scale Runoff and Sediment Yield. *Water* **2019**, *11*, 1373. [CrossRef]
55. Sinha, R.K.; Eldho, T. Effects of historical and projected land use/cover change on runoff and sediment yield in the Netravati river basin, Western Ghats, India. *Environ. Earth Sci.* **2018**, *77*, 111. [CrossRef]
56. Narimani, R.; Erfanian, M.; Nazarnejad, H.; Mahmodzadeh, A. Evaluating the impact of management scenarios and land use changes on annual surface runoff and sediment yield using the GeoWEPP: A case study from the Lighvanchai watershed, Iran. *Environ. Earth Sci.* **2017**, *76*, 353. [CrossRef]
57. Hu, S.; Fan, Y.; Zhang, T. Assessing the effect of land use change on surface runoff in a rapidly urbanized city: A case study of the central area of Beijing. *Land* **2020**, *9*, 17. [CrossRef]
58. Li, F.; Chen, J.; Liu, Y.; Xu, P.; Sun, H.; Engel, B.A.; Wang, S. Assessment of the impacts of land use/cover change and rainfall change on surface runoff in China. *Sustainability* **2019**, *11*, 3535. [CrossRef]
59. Erena, S.H.; Worku, H. Dynamics of land use land cover and resulting surface runoff management for environmental flood hazard mitigation: The case of Dire Daw city, Ethiopia. *J. Hydrol. Reg. Stud.* **2019**, *22*, 100598. [CrossRef]

60. Guo, X.; Li, T.; He, B.; He, X.; Yao, Y. Effects of land disturbance on runoff and sediment yield after natural rainfall events in southwestern China. *Environ. Sci. Pollut. Res.* **2017**, *24*, 9259–9268. [CrossRef]
61. Aghakhani, M.; Nasrabadi, T.; Vafaeinejad, A. Assessment of the effects of land use scenarios on watershed surface runoff using hydrological modelling. *Appl. Ecol. Environ. Res.* **2018**, *16*, 2369–2389. [CrossRef]
62. Yini, H.; Jianzhi, N.; Zhongbao, X.; Wei, Z.; Tielin, Z.; Xilin, W.; Yousong, Z. Optimization of land use pattern reduces surface runoff and sediment loss in a Hilly-Gully watershed at the Loess Plateau, China. *For. Syst.* **2016**, *25*, 9. [CrossRef]
63. Ma, K.; Huang, X.R.; Liang, C.; Zhao, H.B.; Zhou, X.Y.; Wei, X.Y. Effect of land use/cover changes on runoff in the Min River watershed. *River Res. Appl.* **2020**, *36*, 749–759. [CrossRef]
64. Zhang, H.; Wang, B.; Liu, D.L.; Zhang, M.X.; Leslie, L.M.; Yu, Q. Using an improved SWAT model to simulate hydrological responses to land use change: A case study of a catchment in tropical Australia. *J. Hydrol.* **2020**, *585*, 124822. [CrossRef]
65. Wang, Y.P.; Wang, S.; Wang, C.; Zhao, W.W. Runoff sensitivity increases with land use/cover change contributing to runoff decline across the middle reaches of the Yellow River basin. *J. Hydrol.* **2021**, *600*, 126536. [CrossRef]
66. Zhang, J.; Yu, X.L. Analysis of land use change and its influence on runoff in the Puhe River Basin. *Environ. Sci. Pollut. Res.* **2021**, *28*, 40116–40125. [CrossRef]
67. Bartlett, M.S.; Daly, E.; McDonnell, J.J.; Parolari, A.J.; Porporato, A. Stochastic rainfall-runoff model with explicit soil moisture dynamics. *Proc. R. Soc. A Math. Phys. Eng. Sci.* **2015**, *471*, 20150389. [CrossRef]
68. Chen, X.; Chen, Y.D.; Xu, C.Y. A distributed monthly hydrological model for integrating spatial variations of basin topography and rainfall. *Hydrol. Process.* **2007**, *21*, 242–252. [CrossRef]
69. Goegebeur, M.; Pauwels, V.R.N. Improvement of the PEST parameter estimation algorithm through Extended Kalman Filtering. *J. Hydrol.* **2007**, *337*, 436–451. [CrossRef]
70. Beven, K.J.; Kirkby, M.J. A physically based, variable contributing area model of basin hydrology/Un modèle à base physique de zone d'appel variable de l'hydrologie du bassin versant. *Hydrol. Sci. J.* **1979**, *24*, 43–69. [CrossRef]
71. Zhang, Y.; Zhao, Y.; Wang, Q.M.; Wang, J.H.; Li, H.H.; Zhai, J.Q.; Zhu, Y.N.; Li, J.Z. Impact of Land Use on Frequency of Floods in Yongding River Basin, China. *Water* **2016**, *8*, 401. [CrossRef]
72. Zhang, L.M.; Meng, X.Y.; Wang, H.; Yang, M.X. Simulated Runoff and Sediment Yield Responses to Land-Use Change Using the SWAT Model in Northeast China. *Water* **2019**, *11*, 915. [CrossRef]
73. Zumr, D.; Dostal, T.; Devaty, J. Identification of prevailing storm runoff generation mechanisms in an intensively cultivated catchment. *J. Hydrol. Hydromech.* **2015**, *63*, 246–254. [CrossRef]
74. Bian, G.D.; Du, J.K.; Song, M.M.; Zhang, X.L.; Zhang, X.Q.; Li, R.J.; Wu, S.Y.; Duan, Z.; Xu, C.Y. Detection and attribution of flood responses to precipitation change and urbanization: A case study in Qinhuai River Basin, Southeast China. *Hydrol. Res.* **2020**, *51*, 351–365. [CrossRef]
75. Yu, X.Y.; He, X.Y.; Zheng, H.F.; Guo, R.C.; Ren, Z.B.; Zhang, D.; Lin, J.X. Spatial and temporal analysis of drought risk during the crop-growing season over northeast China. *Nat. Hazards* **2014**, *71*, 275–289. [CrossRef]

sustainability

MDPI

Article

A Morphing-Based Future Scenario Generation Method for Stochastic Power System Analysis

Yanna Gao [1], Hong Dong [1], Liujun Hu [1], Zihan Lin [1], Fanhong Zeng [1], Cantao Ye [2,*] and Jixiang Zhang [2]

[1] Guangzhou Power Supply Bureau, Guangdong Power Grid Co., Ltd., Guangzhou 510620, China; 13512778095@163.com (Y.G.); donghong792@163.com (H.D.); 15116330584@163.com (L.H.); lzh7860@163.com (Z.L.); zfh19940214@163.com (F.Z.)

[2] Guangzhou Institute of Energy Conversion, Chinese Academy of Sciences, Guangzhou 510640, China; zhangjx2@ms.giec.ac.cn

* Correspondence: yect@ms.giec.ac.cn; Tel.: +86-20-8705-7791

Abstract: As multiple wind and solar photovoltaic farms are integrated into power systems, precise scenario generation becomes challenging due to the interdependence of power generation and future climate change. Future climate data derived from obsolete climate models, featuring diminished accuracy, less-refined spatial resolution, and a limited range of climate scenarios compared to more recent models, are still in use. In this paper, a morphing-based approach is proposed for generating future scenarios, incorporating the interdependence of power generation among multiple wind and photovoltaic farms using copula theory. The K-means method was employed for scenario generation The results of our study indicate that the average annual variations in dry-bulb temperature (DBT), global horizontal irradiance (GHI), and wind speed (WS) are projected to increase by approximately 0.4 to 1.9 °C, 7.5 to 20.4 W/m^2, and 0.3 to 1.7 m/s, respectively, in the forthcoming scenarios of the four considered Shared Socioeconomic Pathways (SSP1-2.6, SSP2-4.5, SSP3-7.0, and SSP5-8.5). It seems that accumulated maximum wind electricity output (WEO) and solar electricity output (SEO) will increase from 0.9% to 7.3% and 1.1% to 6.8%, respectively, in 2050.

Keywords: future scenario; weather morphing; climate change; cluster analysis; uncertainties

Citation: Gao, Y.; Dong, H.; Hu, L.; Lin, Z.; Zeng, F.; Ye, C.; Zhang, J. A Morphing-Based Future Scenario Generation Method for Stochastic Power System Analysis. *Sustainability* **2024**, *16*, 2762. https://doi.org/10.3390/su16072762

Academic Editor: Bowen Zhou

Received: 2 February 2024
Revised: 18 March 2024
Accepted: 21 March 2024
Published: 27 March 2024

1. Introduction

The last few years have witnessed a swift expansion in renewable energy, with wind and photovoltaic (PV) power emerging as highly promising sources and undergoing rapid development [1–4]. Nevertheless, as the integration of renewable power grows, especially with the escalating impact of global climate change, the inherent randomness of power systems is on the rise. This scenario endangers the stability and dependability of power grids that integrate wind and PV farms. Hence, performing stochastic power system analysis is of great importance to ensure the safety and reliability of power systems.

Employing mathematical transformations, the morphing technique modifies existing weather conditions so that they conform to the anticipated parameters of a climate variability context, as indicated by a general circulation model representing atmospheric, oceanic, cryospheric, and land-surface physical processes [5]. Presupposing the perpetuity of prevailing weather patterns in forthcoming periods, the morphing process preserves indigenous climatic attributes through the metamorphosis of contemporary records. To safeguard the precision of this methodology, it is imperative to synchronize the temporal extent encompassed by contemporary records with the reference period for the envisaged alterations [6]. Significantly, the morphing method minimizes the risk of developing poorly designed power systems for specific locations, thus safeguarding a nation's ability to achieve its carbon neutrality targets [7].

In the face of uncertainties inherent in model predictions, worldwide and localized climate simulations can furnish the requisite meteorological parameters for computations related to electricity generation in current as well as prospective scenarios [8]. This proven methodology is optimal for assessing renewable energy resources and studying projections of renewable energy in future scenarios. Nevertheless, only a limited number of studies have examined the impact of climate change on renewable electricity production, with even fewer utilizing the new CMIP6 data. Based on a pertinent evaluation conducted by [9] within the context of the SSP5-8.5 framework, a 4% fluctuation in the mean annual wind speed was observed. This alteration resulted in a diminished wind power capacity in Northern China, accompanied by a corresponding augmentation of approximately 2% in the southern region. An investigation into the ramifications of these emerging scenarios for the interplay between wind power and solar photovoltaics (PV) in North America revealed that SSP2-4.5 exhibits a marginal advantage in both wind and PV potential when juxtaposed with SSP5-8.5 [10]. Delving into the realm of solar energy, we anticipate a discernible shift in global solar PV potential, with fluctuations expected to fall within the $\pm 10\%$ spectrum. This forecast hinges on specific scenarios outlined in the SSP framework, taking into account diverse regional influences. An exhaustive analysis has unequivocally determined that the foreseen rise in cloud coverage is poised to curtail the availability of solar radiation across the landscapes of Asia and Africa [11]. This aligns seamlessly with empirical observations of diminishing solar exposure. Conversely, a surge in maximum temperatures is poised to catalyze an amplification in solar PV output across the territories of Europe and the eastern seaboard of America [12].

Furthermore, stochastic programming is emerging as a potentially powerful technique for addressing uncertainties related to wind power. However, a key challenge in its implementation lies in the selection of a well-weighted set of scenarios to effectively represent the space of uncertainty. Typically, these methodologies involve fitting forecasted wind power or forecast errors to specific distributions, and scenarios are subsequently generated through the sampling of these derived distributions [13]. The forecast errors, characterized using empirical distributions, are subjected to the inverse transformation method to derive a comprehensive set of scenarios [14]. To enhance accuracy, a generalized Gaussian mixture model was devised to fit forecast errors originating from a multitude of wind farms, and the resulting distribution was then utilized to sample scenarios for probabilistic wind ramp forecasting [15].

Extensively applied and recognized for its efficacy, the scenario generation method plays a pivotal role in optimizing the operation of power systems involving stochastic variables. By scrutinizing historical data linked to these unpredictable factors, this method extrapolates archetypal scenarios. These representative scenarios form the basis for conducting research on the optimal operation of a power system. Integral to this methodology is the extraction of a discrete probability distribution closely mirroring the probability distribution of the primary stochastic variable. This method's effectiveness hinges on the disparity level between the archetypal scenario and the original dataset.

An increasing number of studies have highlighted the importance of spatio-temporal correlation in scenario generation. Typically, this correlation is represented through the use of multivariate joint distributions. In numerous recent studies, the Multivariate Gaussian distribution has been employed to capture correlations among wind power forecasts made at different lead times [16]. However, modeling high-dimensional multivariate non-Gaussian distributions can be challenging, and a commonly adopted approach involves the use of copulas [17]. By applying marginal cumulative distribution functions to stochastic variables, the original variables are transferred from their original space to a common uniform domain. In this domain, correlations among the original variables can be further characterized using copulas. The modeling of spatio-temporal correlations among clustered wind farms using a copula approach has been used to develop a scenario generation method [18].

Multiple renewable power plants are typically integrated, yet the potential impact of climate change on future renewable electricity production is often underestimated in contemporary power systems. Therefore, this paper puts forth an innovative method for generating future scenarios, taking into account the spatio-temporal correlations among multiple renewable farms. Employing weather morphing, copula, and cluster analysis, the innovative approach delineated herein begins by morphing the monthly alterations in EC-Earth3 utilized within the CMIP6 project [19]. Subsequently, the generation of future weather scenarios for each farm is carried out using C-vine copula methods. A k-means method is then employed to cluster hourly profiles of weather data into reduced-number clusters, and renewable power predictions are based on the most similar cluster using a power generation model.

The remainder of this manuscript is structured as follows: Section 2 provides a comprehensive explanation of the newly developed morphing-based future scenario generation method, encompassing cluster analysis and the copula method, elucidating the procedural intricacies of the envisaged methodology for generating future scenarios. In Section 3, the clustered scenario results detailing variations in wind speed, temperature, and incident solar irradiance are presented, and then prognostications for the forthcoming power output from wind and solar photovoltaic sources are delineated. In Section 4, it is revealed that both morphing and scenario generation modeling approaches, along with K-means clustering analysis of multiple scenarios, are deemed necessary to quantify the projected range in the future. Lastly, Section 5 delves into the implications of the primary findings and offers a summary of this study's conclusions.

2. Materials and Methods

2.1. Weahter Morphing

Utilizing the EC-Earth3 general circulation models (GCMs), this methodology involves the use of environmental variables to transform current local weather data into future scenarios and timeframes. Notably, EC-Earth3 distinguishes itself from CMIP6 GCMs by offering comprehensive data for all variables across the four shared socioeconomic pathways (SSP1-2.6, SSP2-4.5, SSP3-7.0, and SSP5-8.5) as compared to alternative models. Distinguished by heightened spatial precision, this model features an increased abundance of spatial elements. It exhibits persistent alignment and minimal deviation in daily air temperatures, encompassing both the extremes and averages.

The morphing procedure follows the general approach outlined in reference [20], which is, in essence, based on the principles presented in Ref. [5]. Utilizing various GCM variables, this process involves the 'shifting' and 'stretching' of multiple fields, such as dry-bulb temperature, global horizontal irradiance, and wind speed. The morphing of future dry-bulb temperature begins with the calculation of a scaling factor for each month, determined through the following equations:

$$\alpha t_m = \frac{\Delta T_m^{\max} - \Delta T_m^{\min}}{\dot{t}_m^{\max} - \dot{t}_m^{\min}} \tag{1}$$

$$t = \dot{t} + \Delta T_m + \alpha t_m \left(\dot{t} - \dot{t}_m\right) \tag{2}$$

where $\Delta T_m^{\max}$ is the GCM change in the average daily maximum dry-bulb temperature, $\Delta T_m^{\min}$ is the GCM change in the average daily minimum dry-bulb temperature, $\dot{t}_m^{\max}$ is the average daily maximum dry-bulb temperature, $\dot{t}_m^{\min}$ is the average daily minimum dry-bulb temperature, $\dot{t}$ is the present dry-bulb temperature, ΔT_m is the GCM change in the mean dry-bulb temperature, $\dot{t}_m$ is the mean of the present dry-bulb temperature, and αt_m is the scaling factor for month m.

The morphing process for future global horizontal irradiance begins with the determination of the scaling factor for downward surface shortwave flux. This scaling factor can be calculated using the provided equations:

$$\alpha I_m = 1 + \frac{\Delta R_m}{\dot{I}_m} \tag{3}$$

$$I = \alpha I_m \cdot \dot{I} \quad (4)$$

where ΔR_m is the GCM change in the mean downward surface shortwave flux, $\dot{I}_m$ is the average of present global horizontal irradiance, I is the future global horizontal irradiance, and αI_m is the scaling factor for downward surface shortwave flux for month m.

To generate future wind speed scenarios, the current wind speed is multiplied by the relative mean change in wind speed from the Global Climate Model (GCM) for a specific month. This relationship can be expressed using the following equation:

$$w_s = \alpha w_m \cdot \dot{w}_s \quad (5)$$

where αw_m is the GCM relative mean change in wind speed, $\dot{w}_s$ is the present wind speed, and w_s is the future wind speed for month m.

The morphing procedure draws on three primary statistical transformations: 'shift', 'stretch', and a combination derived from insights in the literature [5]. In the 'shift' operation, the monthly projected change is added to the current variable, while the 'stretch' operation involves scaling the present-day variable by multiplying it by the fraction of the monthly projected change. Combining 'shift' and 'stretch' allows for adjustments to the mean and variance of the present-day variable or exclusively to the variance. Specifically, the dry-bulb temperature undergoes both 'stretch' and 'shift', global horizontal irradiance undergoes 'shift', and wind speed undergoes 'stretch'.

Employing insights from the literature, the morphing methodology utilizes three fundamental statistical transformations: 'shift', 'stretch', and a hybrid approach [5]. Within the 'shift' procedure, the current variable assimilates the monthly predicted adjustment. Conversely, the 'stretch' process entails adjusting the contemporary factor by multiplying it by a fraction of the anticipated monthly shift. The amalgamation of 'shift' and 'stretch' enables adjustments to either the mean and variance of the present-day variable or exclusively to the variance. Specifically, modifications include both 'stretch' and 'shift' for the dry-bulb temperature, 'shift' for global horizontal irradiance, and 'stretch' for wind speed.

2.2. Copula Approach

Copulas function as connectors, establishing links between univariate marginal distributions within multivariate distribution functions. This fundamental statistical principle, elucidating interdependence and detailed in [21], establishes a critical connection between copulas and the joint distribution of multiple random variables. As delineated in Sklar's theorem, assume X = $[x_1, x_2, \ldots, x_n]$ denotes the random variables with margins F_1 (x_1), F_2 (x_2), ..., F_n (x_n). The joint distribution $F(x_1, x_2, \ldots, x_n)$ can be articulated through a suitable n-dimensional C-copula function, as follows:

$$F(x_1, x_2, \ldots, x_n) = C(F_1(x_1), F_2(x_2), \ldots, F_n(x_n)) \quad (6)$$

If F_1 (x_1), F_2 (x_2), ..., F_n (x_n) are continuous, then the C-copula function is unique.

The practical utility of Sklar's theorem lies in its ability to conveniently disentangle the dependence modeling of a set of random variables by individually considering their marginal distributions and the copula. In the extensive body of work [22], various families of copulas are explored, with prominent examples including the Normal Gaussian copula, the Frank copula, and the Student-t copula.

In the realm of bivariate scenarios, numerous precise copula functions are at our disposal. However, when extending to arbitrary dimensions, the options for suitable copula families become significantly more limited. Traditional multivariate copulas such as the multivariate Gaussian or Student-t, along with interchangeable Archimedean copulas, lack the adaptability necessary for precisely modeling dependence among a larger set of variables. The vine, serving as a versatile graphical model for depicting multivariate copulas through a series of bivariate copulas, emerges as a distinctive alternative. Its resilience to the aforementioned constraints positions it as a potent instrument for capturing multivariate dependence. This article provides a concise examination of the C-vine, exemplifying a prevalent form of regular vines.

In the structure of the C-vine tree, the relationships linked to a particular factor, denoted as the primary root node, are represented through bivariate copulas for each pair. Extending this to another variable, known as the secondary root node, pairwise associations are also modeled. Typically, a root node is designated in each tree, and all pairwise connections concerning this node are modeled, considering all preceding root nodes. This structural characteristic results in C-vine trees exhibiting a star configuration [23]. Building upon the C-vine framework, the decomposition of a multivariate density $f(x_1, \cdots, x_n)$ can be expressed as follows:

$$f(x_1, \cdots, x_n) = c(f(x_1), f(x_2), \cdots, f(x_n)) \times \prod_{i=1}^{n} f_i(x_i) \tag{7}$$

where $f_i(x_i)$ is the density function of the marginal distribution of variables, and $c(f(x_1), f(x_2)$ $\cdots, f(x_n))$ is the copula density function, which can be calculated as follows:

$$c(f(x_1), f(x_2), \cdots, f(x_n)) = \frac{\partial^n C(F(x_1), F(x_2), \cdots, F(x_n))}{\partial F((x_1)\partial F(x_2) \cdots \partial F(x_n))} \tag{8}$$

2.3. K-Means Clustering

K-means is as an extensively employed method for general clustering [24]. Within the framework of K-means, clusters find representation through the centers of mass of their constituent members. The clustering process entails iteratively assigning cluster affiliations to each data vector based on proximity to the cluster center. Simultaneously, each cluster's center is computed as the centroid of its constituent data vectors. Despite the extensive use of K-means clustering, a significant drawback lies in the requirement to anticipate the cluster count based on previous experience. In overcoming this hurdle and determining the best parameter for K-means across different scenarios, this study presents a density-centric metric outlined in [25].

Assume that there is a data matrix $P = \{x_1, x_2, \ldots, x_n\}$, in which each data vector is p-dimensional. c_i is the center of cluster K_i. $N(K_i)$ is the quantity of cluster K_i, and $d(x_i, x_j)$ is the Euclidean distance between x_i and x_j. The optimal parameter can be determined as follows:

$$d(x_i, x_j) = \sqrt{\sum_{k=1}^{n} (x_i^k - x_j^k)^2} \tag{9}$$

$$r_i = \frac{1}{N(K_i)} \sum_{x \in K_i} d(x, c_i) \tag{10}$$

$$c_{ij} = \frac{c_i + c_j \times r_i / r_j}{1 + r_i / r_j} \tag{11}$$

$$C(i) = N(K_i' = \{x | d(x, c_i) \le r_i\}) \tag{12}$$

$$B(i) = \frac{1}{K-1} \sum_{j=1, j \ne i}^{K} N\Big(K_j' = \{x | d(x, c_{ij}) \le [(r_i + r_j)/2]\}\Big) \tag{13}$$

$$DBI(i) = \sum_{i=1}^{K} C(i) / \sum_{i=1}^{K} B(i) \tag{14}$$

$$K_b = Max\{DBI(i)\} \tag{15}$$

where r_i is the cluster radius, K_i is the element, $N(K_i)$ is the quantity of cluster K_i, $d(x, c_i)$ is the Euclidean distance from datum x to the cluster center, c_i is the center of cluster K_i, c_{ij} is the midpoint between clusters, $C(i)$ is the cluster center density, $B(i)$ is the cluster margins density, DBI is the ratio of cluster center density to cluster margin density, and K_b is the optimal number of clusters.

Imagine an ideal clustering scenario where data vectors within a cluster are closely grouped, while the space between two clusters exhibits significant dispersion. In other words, the optimal clustering scheme in K-means aligns with maximizing the Davies–Bouldin Index (DBI). Consequently, determining the optimal parameter within specified parameter ranges involves a comparison of DBI values.

2.4. Energy Potential Calculations

In determining the electrical generation from each sustainable energy source, we performed computations utilizing the latest technological advancements, securing precise assessments of power generation potential for each renewable source. This approach emerges as the optimal method for scrutinizing the impact of varying climate change scenarios on the potential for electricity generation, offering heightened precision in forecasting electrical output. Furthermore, it is noteworthy that renewable wind farms typically operate for 20 to 25 years, while solar farms typically have lifespans exceeding 25 years. This implies that newly installed renewable energy systems will remain operational through 2050. Concerning wind energy generation, onshore wind turbines typically have an average installed capacity of approximately 3.5 MW [26]. Utilizing a realistic power curve representative of onshore turbines, such as Vestas V126-3.45 [MW] shown in Figure 1, ensures accuracy in estimating power output. Wind speed data were obtained from the model at a standard hub height of 100 m, a widely accepted parameter in wind resource assessments [27–31].

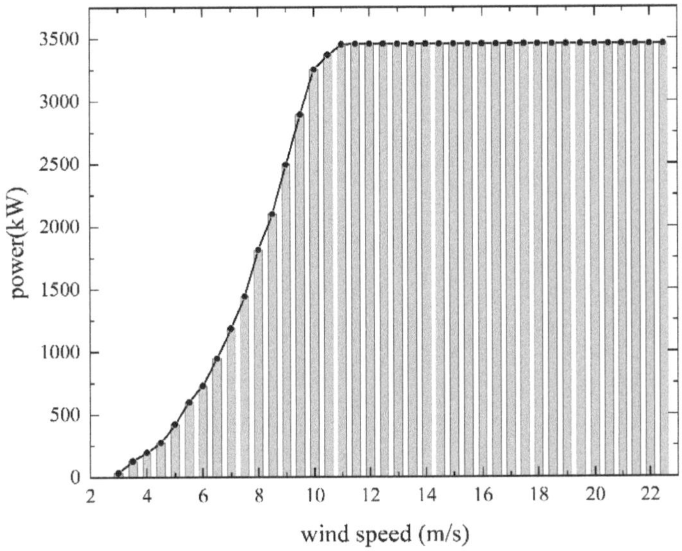

Figure 1. Power curves for the onshore Vestas V126-3.45 wind turbines.

Solar PV panels exhibit a complex reaction to diverse environmental factors, including DB, GHI, and WS. Their performance is notably impacted by distinct panel specifications, where heightened incident irradiance does not uniformly result in amplified power output. This discrepancy is attributed to factors such as temperature response coefficient, capacity factor, and cell temperature. Precise determination of achievable maximum photovoltaic power output (PVO) requires meticulous attention to specific attributes. Calculating cell temperature involves considering parameters and coefficients reflecting the thermal response, derived from an advanced monocrystalline silicon solar panel. Subsequent PVO calculation incorporates the prevailing market standard for average efficiency, which is approximately 17% [32]. The values for solar irradiance, ambient temperature, and surface wind speed are provided as follows:

$$T_{cell} = c_1 + c_2 T_a + c_3 G + c_4 w_s \tag{16}$$

$$P_{pv} = G\eta_p[1 + \mu(T_{cell} - T_r)] \tag{17}$$

where T_{cell} is PV cell temperature; T_a is the ambient temperature; G is solar irradiance; c_1, c_2, c_3, and c_4 are the distinctive attributes inherent to a monocrystalline silicon solar panel, i.e., 4.3 °C, 0.943, 0.028 °C m^2/W, and −1.528 °C s/m, respectively; η_p is monocrystalline silicon solar panel efficiency; μ is the thermal efficiency factor associated with temperature changes; T_r is the reference temperature; and P_{pv} is the solar PV power output.

2.5. Scenario Generation

The fundamental stages in the suggested methodology for scenario generation can be described as follows.

Firstly, accurately project future changes in climate data. Subsequently, apply the morphing method to three future weather elements—dry-bulb temperature (DBT), global horizontal irradiance (GHI), and wind speed (WS)—for SSP1, SSP2, SSP3, and SSP5 in the GCM EC-Earth3, using the 'shift' and 'stretch' approaches to align with the median year of the 2050 timeframe.

Secondly, leverage the maximum likelihood estimate technique [33] to pinpoint the most advantageous category and configurations for each pair-copula, progressing through the following steps:

1. Transform the arbitrary continuous random variable into a uniform distribution using a cumulative distribution function (CDF) transformation, resulting in a uniform distribution in the interval [0, 1].

2. Identify the optimal pair-copula function by selecting the potential copula associated with the minimum Euclidean distance. Quantify the Euclidean gap between the CDF of the observed copula and each potential copula through the following calculation:

$$D_e = \sum_{i=1}^{n} \left| C_n(u_i, v_i) - C_p(u_i, v_i) \right|^2 \tag{18}$$

where D_e is the Euclidean distance, C_n is the CDF of the observed copula, and C_p is the CDF of potential copula.

Thirdly, generate future weather scenarios for DBT, GHI, and WS to calculate renewable energy in terms of electricity power output. Apply the C-vine copula technique employing the most effective pair-copula for generating simulated data. Assume $w_1, \ldots, w_n$ are independent and uniform in the interval [0, 1]; the sample of $x_1, \ldots, x_n$ can be expressed as follows:

$$\begin{cases} x_1 = w_1 \\ x_2 = F^{-1}(w_2|x_1) \\ x_3 = F^{-1}(w_3|x_1, x_2) \\ \vdots \\ x_n = F^{-1}(w_n|x_1, \cdots, x_{n-1}) \end{cases} \tag{19}$$

where $F(w_n|x_1, \cdots, x_{n-1})$ is the distribution function under specified conditions, calculable through Equation (6).

In the fourth step, determine the optimal parameter for clustering data vectors by comparing the DBI values across various parameters. Following this, classify data vectors using K-means clustering with the identified optimal parameter.

Finally, compute the electricity production output for renewable energy under future weather scenarios.

3. Results

To exemplify this methodology's applications, a simulation was executed for three adjacent wind and solar PV facilities located in Guangdong, China. Among the three renewable energy power plants, the first one is a wind and solar power generation site with a rated wind power generation capacity of 50 MW and a rated solar power generation capacity of 70 MW. The second one is also a wind and solar power generation site, with a rated wind power generation capacity of 60 MW and a rated solar power generation capacity of 40 MW. The third one is a photovoltaic power generation site with a rated solar power generation capacity of 40 MW. The simulation spans the current scenario and envisions the future conditions in 2050, taking into account the approximate lifespan of wind turbines and solar PV panels, ranging from 20 to 25 years. The objective was to comprehend the variations in renewable energy electricity production output amidst future climate changes in southern China. Conducted in alignment with the year 2050 for the GCM EC-Earth3, the simulations encompass diverse scenarios, including SSP1, SSP2, SSP3, and SSP5. Illustrated in Figure 2 is a visual representation that displays the average annual values of chosen contemporary environmental factors and the corresponding fluctuations during the 2050 timeframe, effectively highlighting the transformative output. Aligned with global patterns, the outcomes of the morphing process for Guangdong province unveil a progression in temperatures, wind speed, and solar irradiance in prospective scenarios, surpassing the intensity observed in current climate conditions.

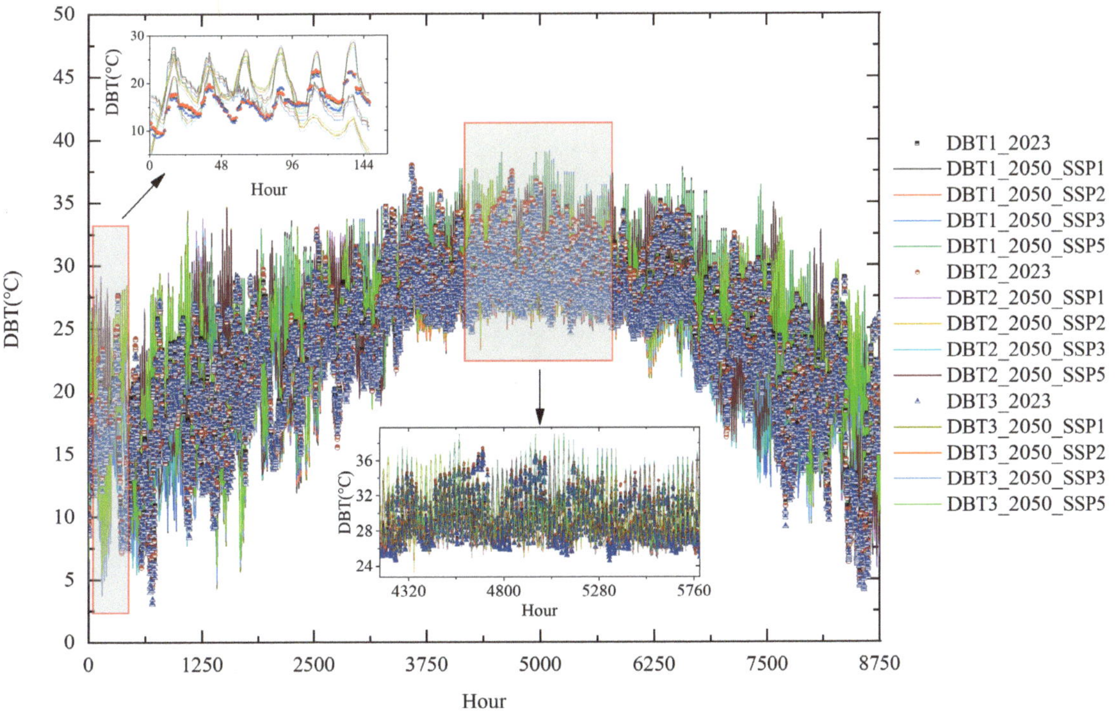

Figure 2. *Cont.*

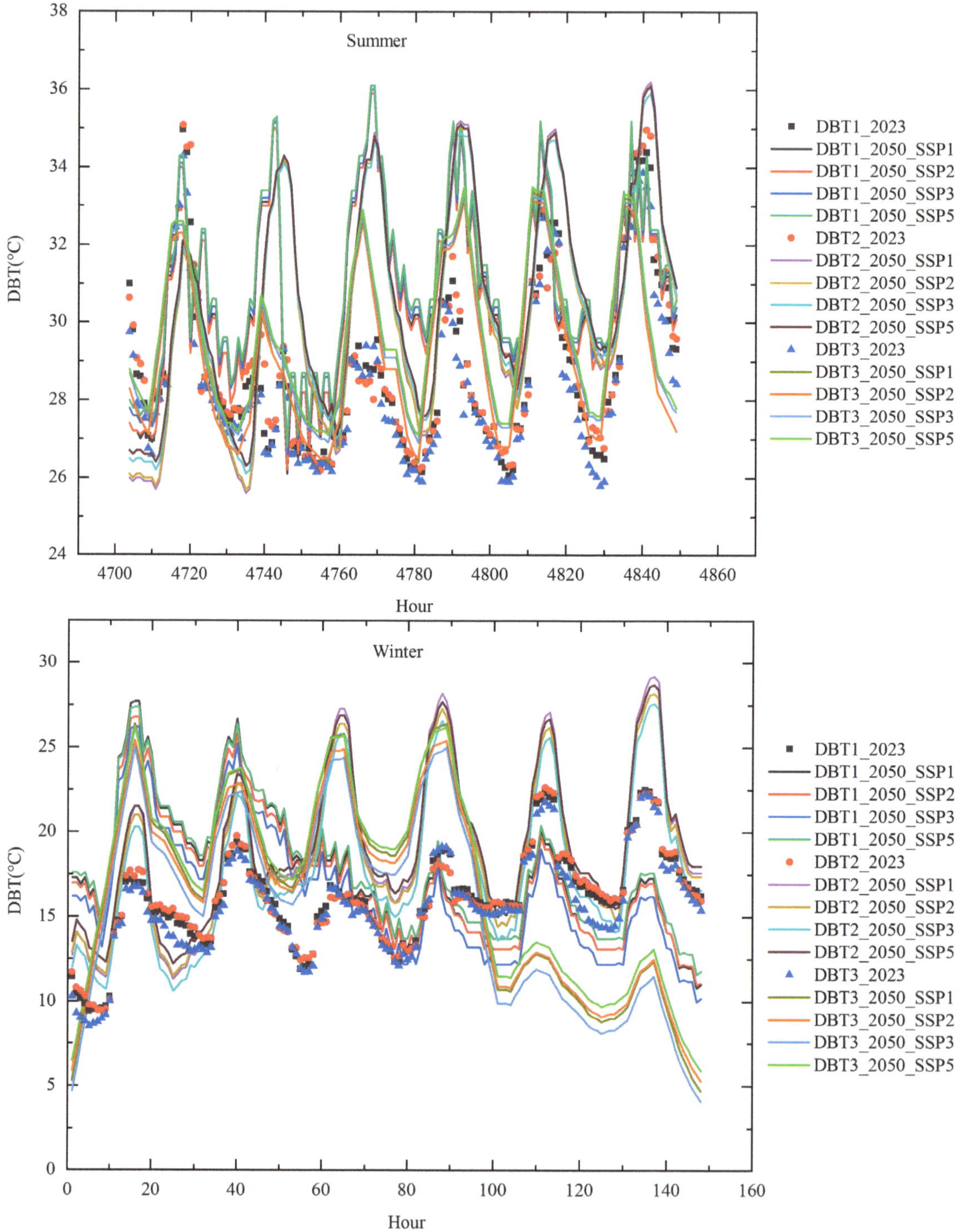

(**a**)

Figure 2. *Cont.*

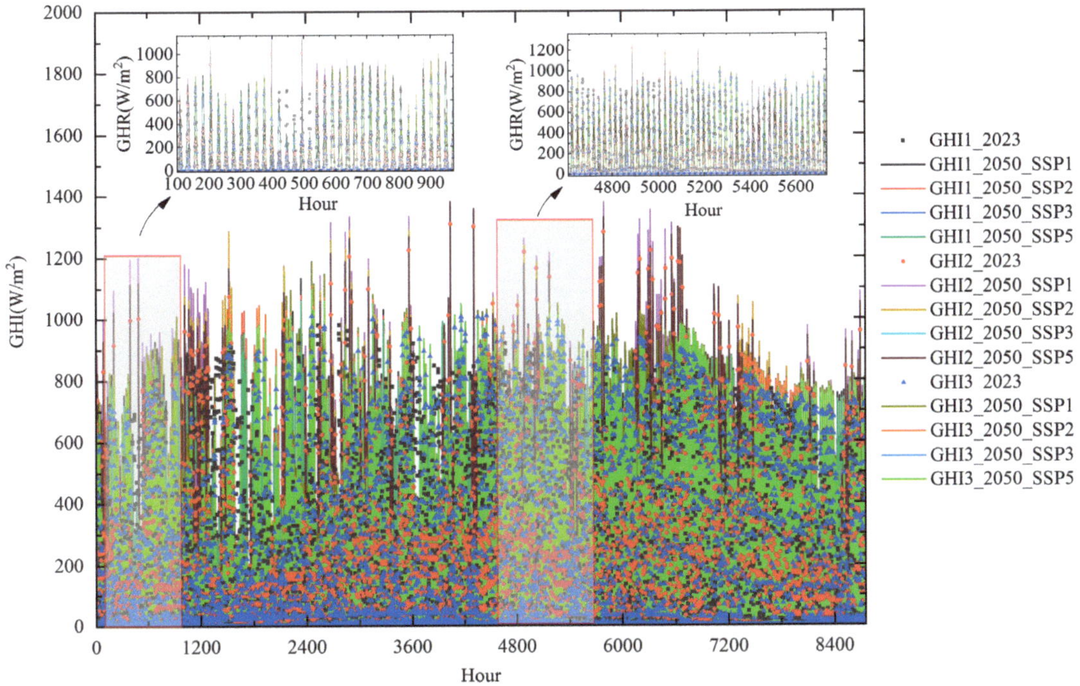

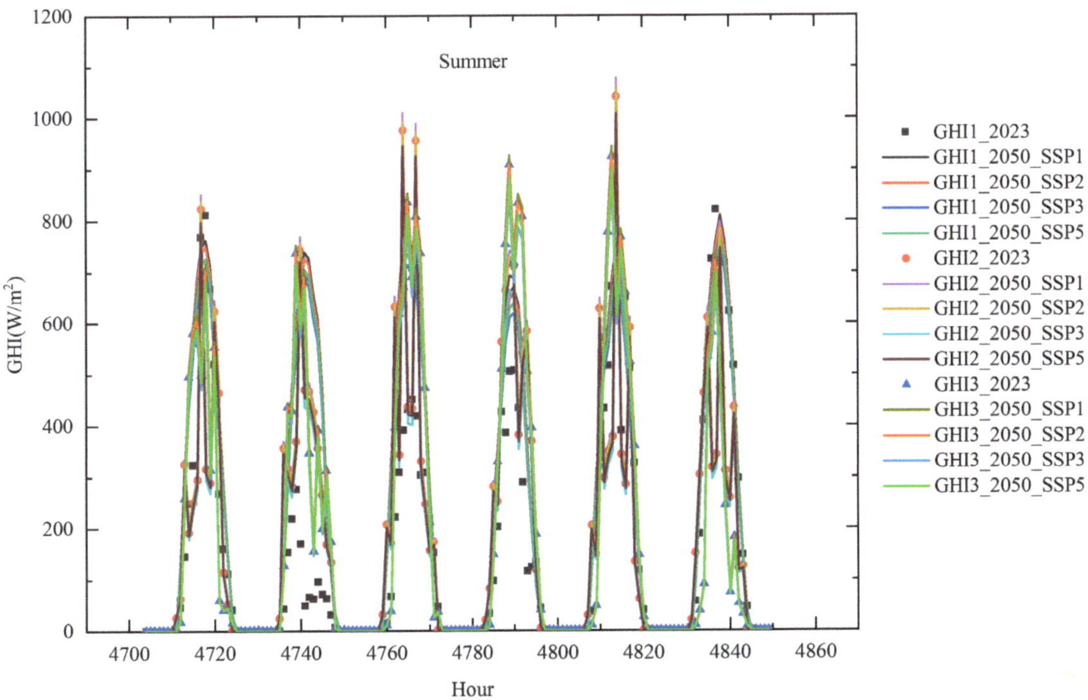

Figure 2. *Cont.*

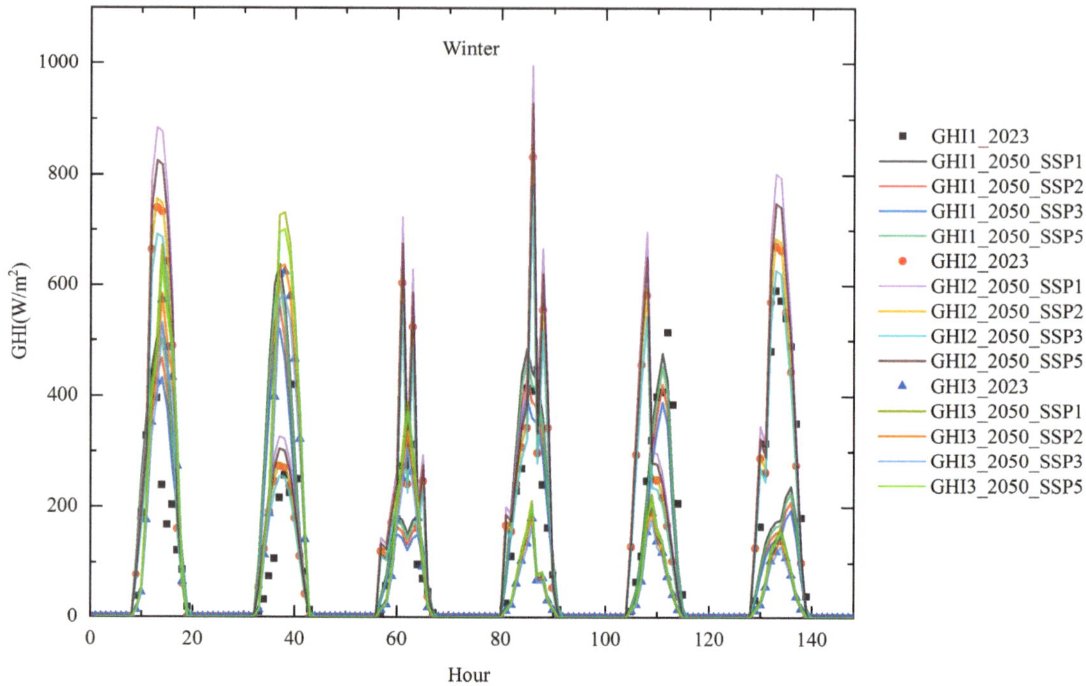

(**b**)

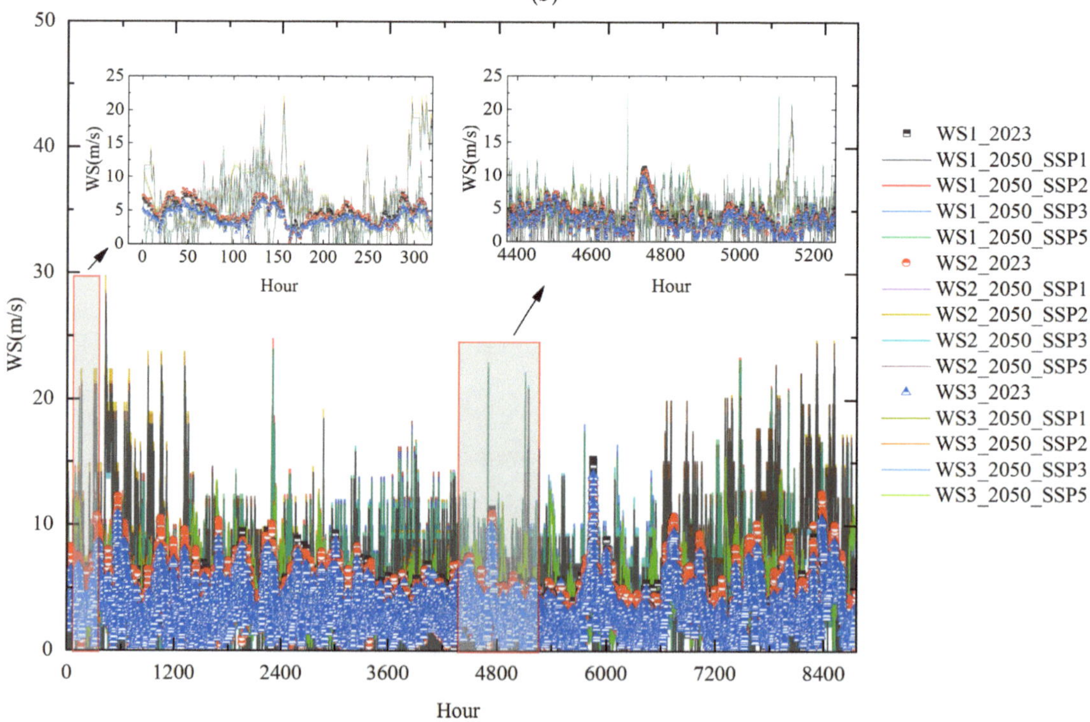

Figure 2. *Cont.*

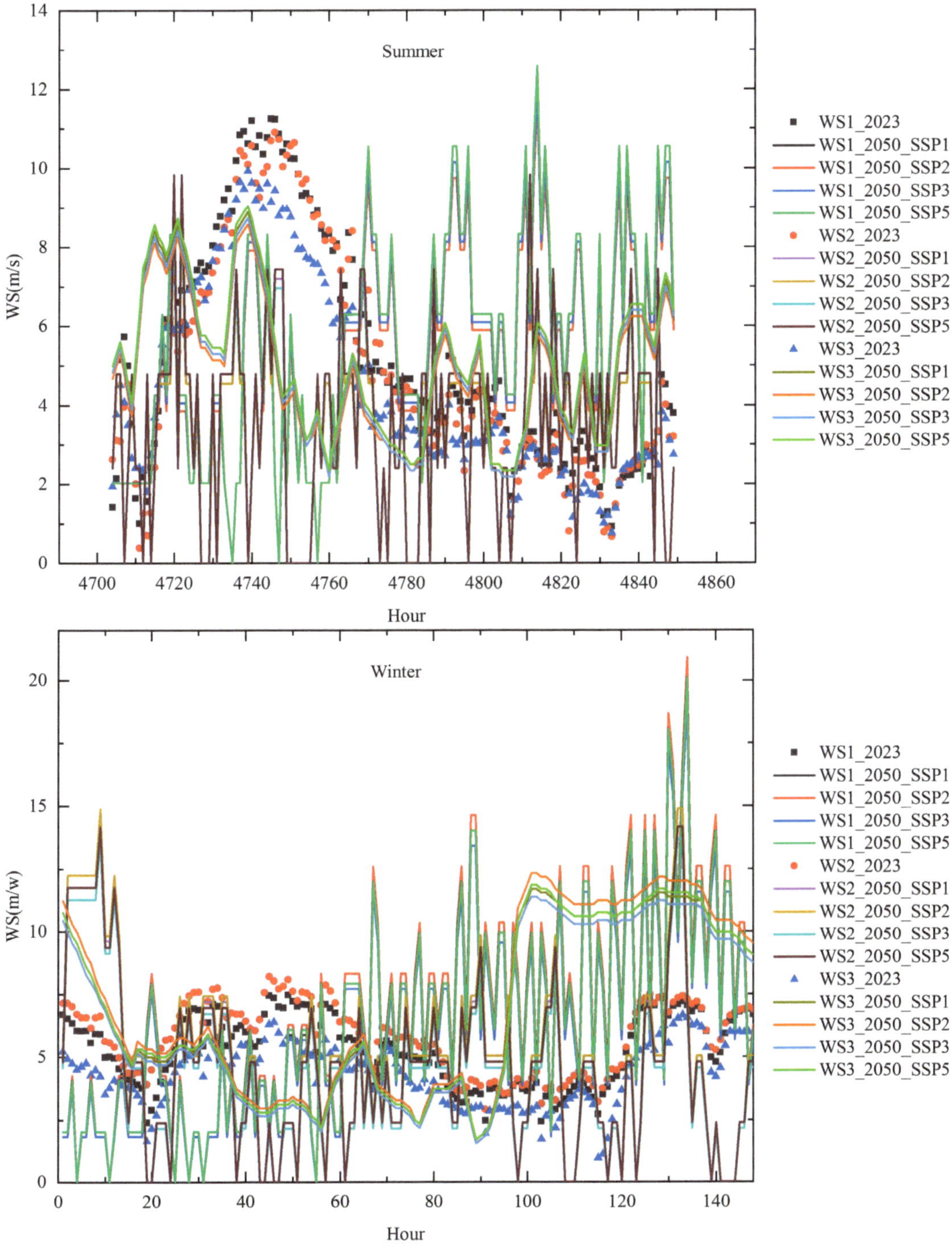

(c)

Figure 2. Mean annual weather elements for 2023 and 2050 across four distinct SSPs: (**a**) DBT, (**b**) GHI, and (**c**) WS.

In Figure 2, DBT1, DBT2, and DBT3 represent the respective dry-bulb temperatures of three solar photovoltaic fields, while GHI1, GHI2, and GHI3 correspond to their individual solar irradiance levels. Additionally, WS1, WS2, and WS3 represent the respective wind speeds at each site. SSP1-2.6, SSP2-4.5, SSP3-7.0, and SSP5-8.5 represent various Shared Socioeconomic Pathways (SSPs) coupled with different radiative forcing levels, measured in Watts per square meter (W/m^2). These abbreviations correspond to scenarios used in the Intergovernmental Panel on Climate Change (IPCC)'s Fifth Assessment Report to depict different trajectories of societal development and greenhouse gas emissions. SSP1-2.6 represents a sustainable development pathway with low greenhouse gas emissions (with the radiative forcing being equal to 2.6 W/m^2). It is an optimistic scenario indicating significant global emission reduction measures. SSP2-4.5 illustrates a moderate greenhouse gas emission pathway (with the radiative forcing being equal to 4.5 W/m^2). This represents a scenario with intermediate levels of greenhouse gas reduction. SSP3-7.0 depicts an unsustainable development pathway with high greenhouse gas emissions (in which the radiative forcing is 7.0 W/m^2). This is a pessimistic scenario, suggesting a lack of effective global emission reduction measures. SSP5-8.5 represents a high-emission pathway with very high greenhouse gas emissions (with the radiative forcing equaling 8.5 W/m^2). This extreme scenario signifies a failure to mitigate greenhouse gas emissions effectively in the coming decades. These scenarios are utilized for studying possible trajectories of climate change and global warming, providing distinct future paths for societal and economic development.

Figure 2 illustrates the distribution of three meteorological elements in different temporal and spatial scenarios. In the forthcoming scenarios of the four considered Shared Socioeconomic Pathways (SSP1-2.6, SSP2-4.5, SSP3-7.0, and SSP5-8.5), the average annual variations in dry-bulb temperature (DBT), global horizontal irradiance (GHI), and wind speed (WS) are projected to increase by approximately 0.4 to 1.9 °C, 7.5 to 20.4 W/m^2, and 0.3 to 1.7 m/s, respectively.

The variation in DBT is depicted in Figure 2a, and the results show the following: In 2023, the annual average is 23.3 °C, the summer average is 28.7 °C, the winter average is 15.9 °C, the maximum for the year is 38 °C, and the minimum is 3 °C. By 2050, the annual average is projected to range between 23.7 and 25.2 °C, the summer average will range between 29.2 and 31.1 °C, and the winter average will range between 15.6 and 19.2 °C, with the maximum for the year reaching 39.1 °C and the minimum being 3.7 °C. Across various scenarios, there is an approximate increase in the annual average temperature of 0.4–1.9 °C, with growth rates ranging from approximately 1.5% to 8.3%. The summer average temperature is expected to rise by about 0.9–2.3 °C, with growth rates of around 1.5–8.3%. The winter average temperature is projected to increase by about 0.1–3.1 °C, with growth rates ranging from approximately 0.6% to 19.5%. In the SSP5 scenario, the maximum increases in annual average and summer average temperatures are observed, reaching 1.9 °C and 2.3 °C, respectively. The magnitude of winter temperature rise is larger than that of summer, and the number of days with high temperatures in summer is gradually increasing.

The GHI variation is illustrated in Figure 2b, and the results indicate the following: In 2023, the annual average is 234 W/m^2, the summer average is 296.5 W/m^2, and the winter average is 212.7 W/m^2, with the annual maximum reaching 1308 W/m^2. By 2050, the annual average is projected to range between 241.5 and 254.4 W/m^2, the summer average will range between 351.3 and 490.1 W/m^2, and the winter average will range between 140.2 and 324.9 W/m^2, with the annual maximum reaching 1380 W/m^2 in the summer Across various scenarios, there is an approximate annual increase of 7.5–20.4 W/m^2, with an average growth rate of about 6%. The summer average increase is approximately 124.2 W/m^2, with a growth rate of around 42%, while the winter average increase is about 20 W/m^2, with a growth rate of approximately 9.4%. In the SSP1 scenario, the maximum increase in the summer average occurs, reaching 124.2 W/m^2, with a larger magnitude of increase in the summer compared to that in the winter, and the peak value occurs in

August. The maximum cumulative increase in the summer is approximately 11.2 kWh/m^2, with a maximum growth rate of about 48%, while the maximum cumulative increase in the winter is approximately 3.5 kW/m^2, with a maximum growth rate of about 19.7%.

The WS variation is shown in Figure 2c, and the results show the following: In 2023, the annual average is 3.7 m/s, the summer average is 4.9 m/s, and the winter average is 5.1 m/s, with the annual maximum reaching 15.2 m/s. By 2050, the annual average is projected to range between 4 and 5.4 m/s, the summer average will range between 3.4 and 5.5 m/s, and the winter average will range between 4.4 and 7 m/s, with the annual maximum reaching 29.8 m/s. Across various scenarios, there is an approximate annual increase of 0.3–1.7 m/s, with an average growth rate exceeding 8%. The maximum increase in the summer is approximately 0.6 m/s, with a maximum growth rate of about 13.1%, while the maximum increase in the winter is approximately 2.7 m/s, with an average growth rate of no less than 53%. In the SSP2 scenario, the maximum increases in the annual average and winter average occur, reaching 1.7 m/s and 2.7 m/s, respectively. The magnitude of the winter increase is larger than that of the summer, and the number of days with strong winds in the summer is gradually increasing.

These climate data fluctuations will directly impact the efficiency of renewable energy power generation in future scenarios and, consequently, their annual power generation output.

The optimal copula parameters have been determined for each future scenario of the three adjacent wind and solar PV farms, resulting in the generation of 600 clusters of random scenarios under the four future climate scenarios, as depicted in Figure 3.

As shown in Figure 3, GHI_2050_AVG, DBT_2050_AVG, and WS_2050_AVG represent the annual average hourly meteorological elements GHI, DBT, and WS under four SSP scenarios in the year 2050. The light-blue area represents the annual average hourly standard deviation of the three meteorological elements for 600 random scenario clusters under each SSP scenario. The results indicate the following ranges: SSP1-2.6 scenario—1.9–3.3 °C for DBT, 0–141.9 W/m^2 for GHI, and 0.2–1.3 m/s for WS; SSP2-4.5 scenario—1.0–3.3 °C for DBT, 0–144.2 W/m^2 for GHI, and 0.2–1.1 m/s for WS; SSP3-7.0 scenario—1.3–3.0 °C for DBT, 0–127.3 W/m^2 for GHI, and 0.2–1.2 m/s for WS; and SSP5-8.5 scenario—1.8–3.4 °C for DBT, 0–131.2 W/m^2 for GHI, and 0.3–1.1 m/s for WS.

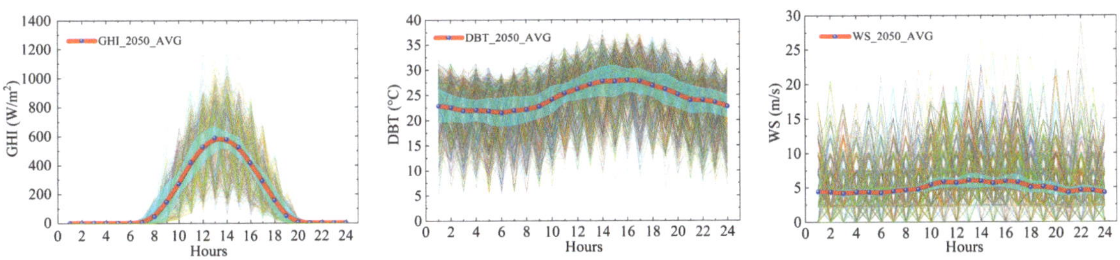

(**a**)

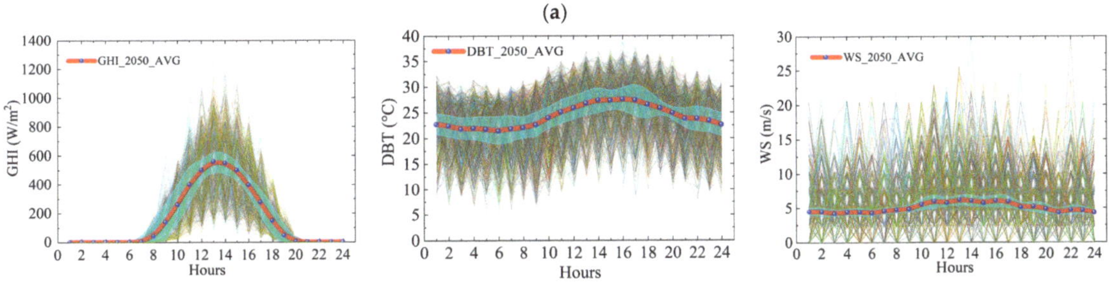

(**b**)

Figure 3. *Cont.*

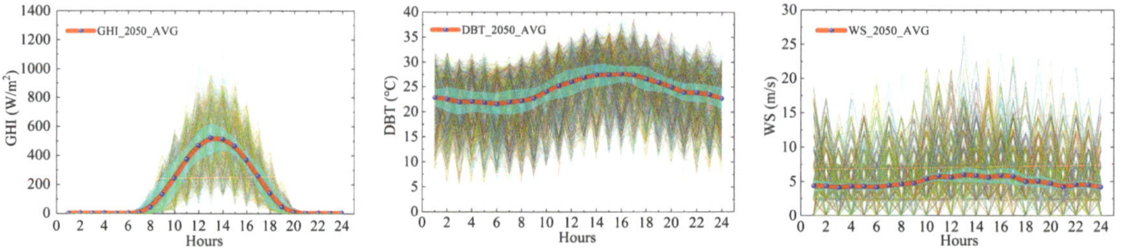

(c)

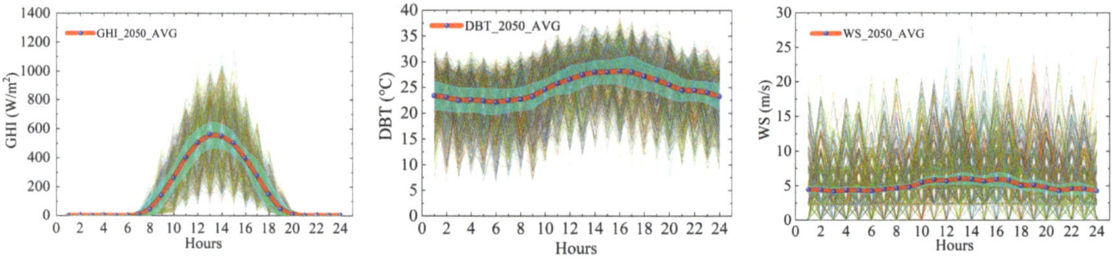

(d)

Figure 3. The 600 clusters of random scenarios under four future climate scenarios for each three weather element in terms of GHI, DBT, and WS: (**a**) SSP1-2.6, (**b**) SSP2-4.5, (**c**) SSP3-7.0, and (**d**) SSP5-8.5.

The meteorological element scenario characteristic curves under the SSP1-2.6 scenario are shown in Figure 3a. The results indicate that GHI ranges from 0 to 1337.4 W/m^2, with peaks in the range of 88.8–1337.4 W/m^2, reached at around 1 p.m. DBT fluctuates within the range of 4.8–39.0 °C, with a peak occurring at around 2 p.m. and the valley occurring around midnight at 00:00. WS fluctuates between 0 and 29.3 m/s, with a peak occurring around 10 p.m., consistent with the future meteorological prediction model's range and characteristics.

Under the SSP2-4.5 scenario, the meteorological element scenario characteristic curves, presented in Figure 3b, indicate the following: GHI spans from 0 to 1261.2 W/m^2, with peaks within the range of 62.2–1261.2 W/m^2, occurring at around noon; DBT fluctuates between 5.1 and 38.8 °C, with peak moments at around 4 p.m. and troughs at around 5 a.m.; WS fluctuates between 0 and 30.3 m/s, with peak moments around 2 p.m. These results align with the projected range and variation features of future meteorological prediction models.

As for the SSP3-7.0 scenario, the meteorological element scenario characteristic curves, depicted in Figure 3c, reveal the following: GHI ranges from 0 to 1316.8 W/m^2, with peaks within the range of 87–1316.8 W/m^2, occurring between 2 and 3 p.m.; DBT fluctuates between 5.2 and 3 8.9° C, with peak moments at around 4 p.m. and troughs at around 3 a.m.; WS fluctuates between 0 and 26.9 m/s, with peak moments at around 1 p.m. These results align with the expected range and variation features of future meteorological prediction models.

The meteorological element scenario characteristic curves under the SSP5-8.5 scenario, as depicted in Figure 3d, reveal the following: GHI ranges from 0 to 1308 W/m^2, with peaks between 82.6 and 1308 W/m^2, occurring at around 1 p.m.; DBT fluctuates between 5.9 and 38.9 °C, with peaks at around 4 p.m. and valleys at around 6 a.m.; WS fluctuates between 0 and 28.1 m/s, with peaks at around 2 p.m. These results are in accordance with the range and variation characteristics of future meteorological prediction models.

According to Equations (9)–(15), within the range of 2 to 10 for K clusters in K-means clustering, the maximum DBI values for the corresponding number of K classifications at 24 typical daily time points were calculated. The maximum DBI values for each hourly interval are highlighted with grey boxes. The results, shown in Figure 4a, indicate that the maximum K classification is 4 at 4 a.m. and 4 p.m.; in Figure 4b, it is 6 at 5 p.m.; in Figure 4c, it is 8 at 10 a.m.; and in Figure 4d, it is 7 at 11 a.m. Thus, among the 24 sets of hourly DBI values, the corresponding maximum K classification is 8.

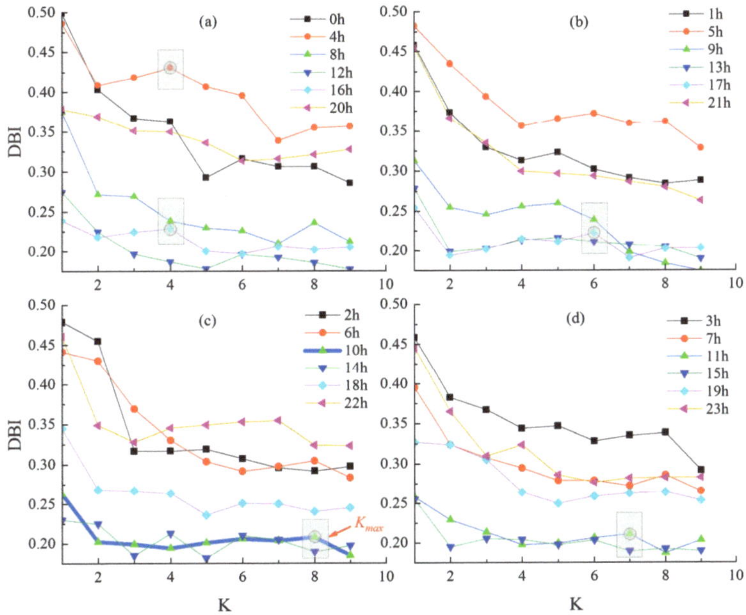

Figure 4. Correlation between the clustering parameter k and the scores derived from the DBI: (**a**) The change in DBI at time points 0, 4, 8, 12, 16, and 20 with respect to K, (**b**) The change in DBI at time points 1, 5, 9, 13, 17, and 21 with respect to K, (**c**) The change in DBI at time points 2, 6, 10, 14, 18, and 22 with respect to K, and (**d**) The change in DBI at time points 3, 7, 11, 15, 19, and 23 with respect to K.

In the quest to pinpoint the optimal parameter for K-means clustering, the parameter range for clustering was established within the interval of 2 to 10. The determination of the optimal parameter was achieved through a thorough comparison of DBI values. Presented in Figure 4 are the simulation results that elucidate the correlation between DBI and clustering parameters. The graph in Figure 4 distinctly shows that the DBI attains its peak value of 8 at 10 a.m., while the maximum DBI values of the remaining 23 hours are between 2 and 7, indicating that the most effective parameter for this specific case study is 8 for each of the four hourly meteorological factors of DBT, GHI, and WS. Utilizing this optimal parameter, the measured data should be condensed into eight clusters for each renewable power plant, as shown in Figure 5.

As shown in Figure 5, the annual average hourly standard deviations of the three elements GHI, DBT, and WS for 24 typical scenario clusters after being clustered under the four SSP scenarios are as follows: SSP1-2.6 scenario—1.5–3.5 °C for DBT, 0–137.2 W/m^2 for GHI, and 1.9–3.9 m/s for WS; SSP2-4.5 scenario—1.0–3.5 °C for DBT, 0–138.3 W/m^2 for GHI, and 1.7–3.7 m/s for WS; SSP3-7.0 scenario—1.4–2.7 °C for DBT, 0–134.3 W/m^2 for GHI, and 1.7–4.1 m/s for WS; and SSP5-8.5 scenario—1.7–3.8 °C for DBT, 0–133.0 W/m^2 for GHI, and 2.0–3.9 m/s for WS. The typical scenario clusters after clustering better reflect the hourly random fluctuation characteristics of the GHI, DBT, and WS elements compared to those before clustering.

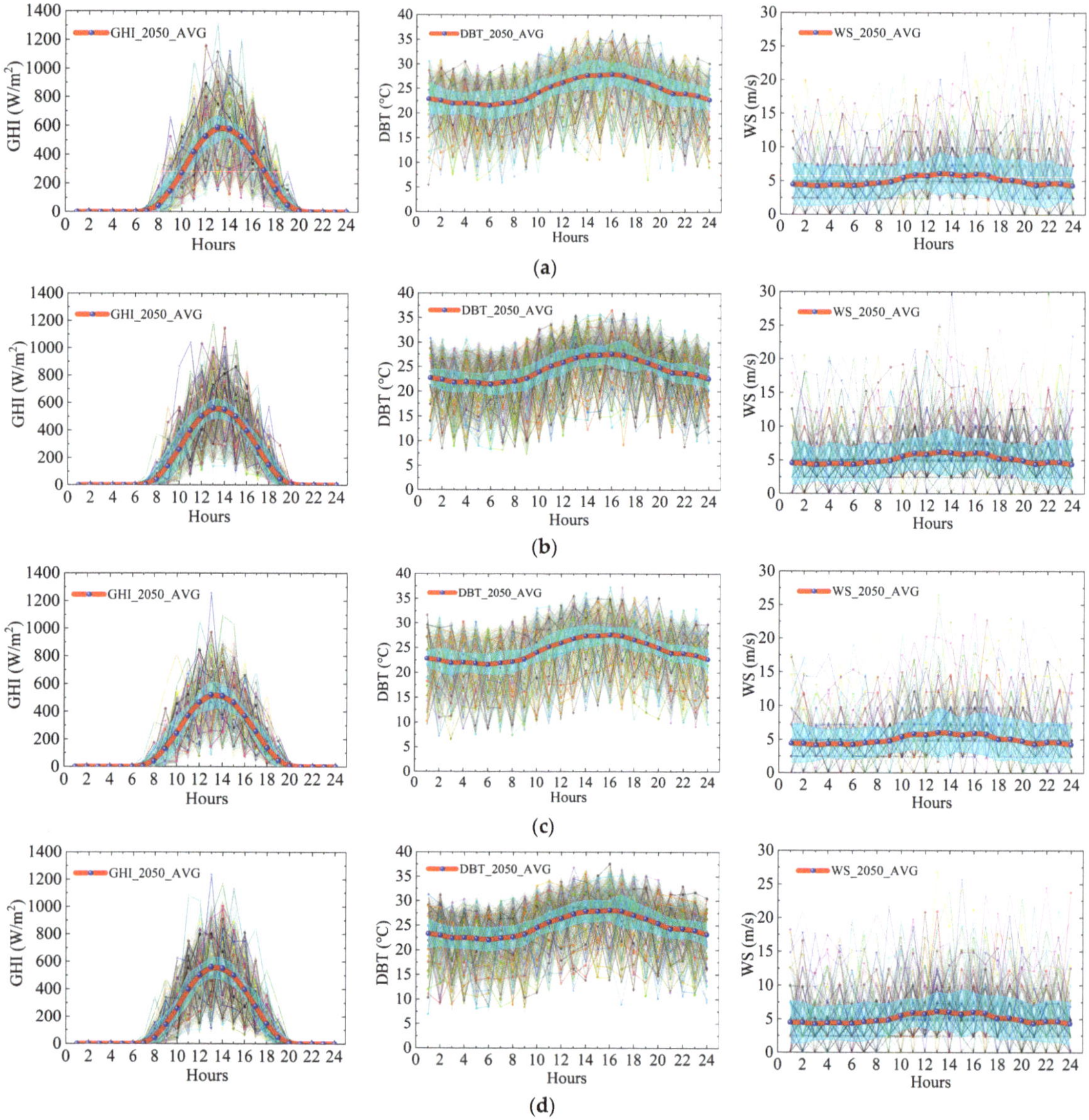

Figure 5. The results of cluster analysis, wherein the parameter K is set to 8, covering four future scenarios: (**a**) SSP1-2.6, (**b**) SSP2-4.5, (**c**) SSP3-7.0, and (**d**) SSP5-8.5.

Under the SSP1-2.6 scenario, the typical feature curves of meteorological elements after reduction are depicted in Figure 5a. The results indicate that GHI ranges from 0 to 1316.6 W/m^2, with peaks occurring between 153.9 and 1316.6 W/m^2, reaching a maximum at around 1 p.m. DBT fluctuates within the range of 5.5–36.7 °C, with peaks at around 4:00 PM and valleys at around 1 a.m. WS fluctuates between 0 and 29.1 m/s, with peaks at around 10 p.m. For the SSP2-4.5 scenario, the typical feature curves of meteorological elements after reduction are shown in Figure 5b. GHI ranges from 0 to 1190.3 W/m^2, with peaks between 142 and 1190.3 W/m^2, occurring between 1 p.m. and 2 p.m. DBT fluctuates between 8.2 and 36.5 °C, with peaks at around 3–4 p.m. and valleys at around

4 a.m. WS fluctuates between 0 and 30.3 m/s, with peaks at around 2 p.m. Under the SSP3-7.0 scenario, the typical feature curves of meteorological elements after reduction are illustrated in Figure 5c. GHI ranges from 0 to 1256.1 W/m^2, with peaks between 144.6 and 1256.1 W/m^2, occurring between 1 p.m. and 2 p.m. DBT fluctuates between 6.6 and 37.2 °C, with peaks at around 4 p.m. and valleys at around 3 a.m. WS fluctuates between 0 and 26.5 m/s, with peaks at around 1 p.m. In the SSP5-8.5 scenario, the typical feature curves of meteorological elements after reduction are presented in Figure 5d. GHI ranges from 0 to 1236.5 W/m^2, with peaks between 143.1 and 1236.5 W/m^2, occurring between 1 p.m. and 2 p.m. DBT fluctuates between 7 and 37.6 °C, with peaks at around 3–4 p.m. and valleys at around 1 a.m. WS fluctuates between 0 and 26.9 m/s, with peaks at around 1 p.m.

GHI exhibits strong regularity, and the reduced typical scenes generally present an "envelope" shape. There are some differences in peak values between typical scenes, but the high peak periods are consistently between 1 and 2 p.m. DBT shows certain regularity, and the overall reduced scenes also exhibit an "envelope" shape. There are some differences in peak values between typical scenes, but the high peak periods are consistently between 3 and 4 p.m. WS demonstrates strong randomness, and the overall reduced scenes also exhibit an "envelope" shape. There are some differences in peak values between typical scenes, and the high peak periods may occur between 1 and 10 p.m.

Therefore, the daily cumulative maximum electricity energy output for renewable energy was calculated across eight representative future scenarios, as detailed in Table 1. A comparative analysis was conducted with the existing standard scenario in 2023.

Table 1. Differences in typical daily accumulated maximum wind electricity output (WEO) and solar electricity output (SEO) between 2023 and the future climate change scenarios (SSP1-2.6, SSP2-4.5, SSP3-7.0, and SSP5-8.5) for the three contiguous renewable energy farms, involving two wind and solar PV farms along with one solar PV farm.

Farm	Electricity Energy	2023	SSP1	SSP2	SSP3	SSP5
1	WEO_1 [kWh]	46,1349.2	473,796.5	478,072.1	495,245.9	491,962.5
	SEO_1 [kWh]	556,597.8	578,999.0	594,298.1	550,116.4	562,506.8
2	WEO_2 [kWh]	750,435.0	785,459.7	796,404.4	775,946.3	756,890.0
	SEO_2 [kWh]	421,665.7	436,951.8	440,305.0	423,951.9	432,780.3
3	SEO_3 [kWh]	349,954.1	368,499.0	354,377.2	334,422.0	342,274.7
Sum [kWh]		2,540,001.9	2,643,706.1	2,663,456.8	2,579,682.5	2,586,414.3

The accumulated daily differences in WEO between current and future scenarios follow a pattern akin to that depicted in Figure 2, with minor modifications in spatial allocation attributed to the non-linear power curves inherent in wind turbines. Remarkably, Wind and Solar PV Farm 1 witness the most substantial increases in WEO, particularly in the SSP3-7.0 and SSP5-8.5 future scenarios, ranging from 7.3% for SSP3-7.0 to over 6.6% for SSP5-8.5. While Farm 2 experiences marginal increases in four of the future scenarios, the most notable increment is 6.1% for SSP2-4.5, accompanied by minor upticks of 0.9% for SSP5-8.5. Both scenarios exhibit variations in comparison to the current state, showcasing significant alterations in their day-to-day variability, with a particular emphasis on offshore locations.

The alterations in accumulated SEO are considerably lower compared to those for WEO, a result primarily attributed to two factors.

In comparison to WEO, the levels of change in accumulated SEO are significantly lower, primarily due to two main reasons. In the first place, GHI exhibits fluctuations of approximately 5% to 10% across the entire domain, and the changes in GHI are not as pronounced as those in WS. Secondly, wind turbines generally exhibit higher efficiency in capturing available resources and converting them into electrical energy. Consequently, even in areas displaying similar percentage changes in incident solar irradiance and wind speed, this variation will lead to a lower change in SEO compared to WEO.

Significant variations endure in the scrutinized scenarios regarding their daily fluctuation and broader trends. SSP1-2.6 and SSP2-4.5 foresee upticks of 3.6% to 5.3% and 1.3% to 6.8%, while SSP3-7.0 indicates a decline ranging from −1.2% to −4.4%. Conversely, in SSP5-8.5, there is a positive prediction for SEO, presenting relatively modest values of 1.1% to 2.6%. The alterations in cumulative SEO parallel the fluctuations in solar irradiance across diverse climate scenarios. The anticipated augmentations in cloud coverage and heightened wind speed notably influence solar PV panel output, leading to diminished output in SSP5-8.5 or slight increases in more advantageous conditions under SSP2-4.5.

4. Discussion

Future climate change could have both positive and negative implications for the electricity production sector. The envisaged rise in wind speed has the capacity to elevate wind power generation, while increased solar irradiance may bolster solar PV power generation. Nevertheless, both encounter obstacles resulting in diminished electricity production. Renewable power plants, notably, could experience deviations in their standard output during peak periods due to anticipated fluctuations in intra-annual resource variability. In the realm of stochastic power system analysis, the holistic assessment of these advantages and drawbacks must align with energy demand. For example, a decline in winter electricity production may not present an issue if energy demand diminishes owing to climate-change-induced reductions in heating requirements or substantial shifts in energy consumption patterns facilitated by diverse tariff systems or scheduling mechanisms. Lastly, significant differences exist between the SSP1-2.6, SSP2-4.5, SSP3-7.0, and SSP5-8.5 scenarios, underscoring the uncertainty in future assessments of renewable power production. To quantify the projected range in WEO and SEO, both morphing and scenario generation modeling approaches, coupled with K-means clustering analysis of various scenarios, are essential. While the SSP3-7.0 scenario favors wind power production, it concurrently reveals greater declines in accumulated SEO. In contrast, the SSP2-4.5 scenario exhibits a less intense but consistently increasing trend, suggesting a potentially safer pathway.

5. Conclusions

This paper introduces a future scenario generation approach utilizing the morphing method and analysis in conjunction with copula and K-means clustering techniques. The future weather elements, including DBT, GHI, and WS, were morphed for the SSP1-2.6, SSP2-4.5, SSP3-7.0, and SSP5-8.5 scenarios in the 2050 timeframe using the GCM EC-Earth3. Existing copula techniques based on the C-vine method were then applied to generate 600 scenarios for the three future weather elements. Subsequently, based on K-means clustering with optimal parameters, future scenarios were condensed into eight typical daily clusters. Finally, the renewable energy electricity production output in future weather scenarios was calculated. The collective outcomes suggest the feasibility of crafting a future scenario that is both intricate and comprehensive, accounting for the interdependence among various wind and solar PVOs. The results of our study indicate that the average annual variations in DBT, GHI, and WS are projected to increase by approximately 0.4 to 1.9 °C, 7.5 to 20.4 W/m^2, and 0.3 to 1.7 m/s, respectively, in the four future scenarios, i.e., SSP1-2.6, SSP2-4.5, SSP3-7.0, and SSP5-8.5. Climate data fluctuations will directly impact the efficiency of renewable energy power generation in future scenarios and, consequently, their annual power generation output. It seems that accumulated maximum WEO and SEO will increase 0.9% to 7.3% and 1.1% to 6.8%, respectively, in 2050.

An inclusive approach to future electricity production under varying climate scenarios can yield a spectrum of potential electricity production options, aiding in the selection of the optimal renewable mix for stochastic power system analysis. This not only fosters confidence in climate change scenario analyses but also ensures a well-considered and effective strategy for sustainable development. Subsequent research should encompass all primary sources of renewable energy production alongside detailed energy consumption projections at a high hourly temporal resolution. This methodology allows for more accurate estimates of supply and demand patterns in climate change scenarios, providing valuable insights with which to mitigate uncertainty in renewable electricity production assessments. Moreover, applying this methodology to multiple climate change scenarios can generate a range of typical future clusters.

Author Contributions: Y.G.: Conceptualization, methodology, validation, formal analysis, and writing—original draft preparation. H.D.: Validation, formal analysis, and data curation. L.H.: Investigation and data curation. Z.L.: Resources and visualization. F.Z.: Resources and data curation. C.Y.: Conceptualization, methodology, software, and writing—review and editing. J.Z.: Resources, supervision, and funding acquisition. All authors have read and agreed to the published version of the manuscript.

Funding: This work was financially supported by the Science and Technology Project of Guangzhou Power Supply Bureau of Guangdong Power Grid [GDKJXM20222457 (037700KK52222011)].

Institutional Review Board Statement: Not applicable.

Informed Consent Statement: Not applicable.

Data Availability Statement: Data are contained within the article.

Acknowledgments: The authors thank the editors and anonymous reviewers for their valuable comments, which greatly improved the quality of this paper.

Conflicts of Interest: The authors declare no conflicts of interest. The funding sponsors had no role in the design of the study; in the collection, analyses, or interpretation of data; in the writing of the manuscript, and in the decision to publish the results.

Nomenclature

DBT	Dry-bulb temperature	CDF	Cumulative distribution function
GHI	Global horizontal irradiance	SSPs	Shared socioeconomic pathways
WS	Wind speed	GCM	General circulation model
PV	Photovoltaic	PVO	Photovoltaic power output
DBI	Davies–Bouldin Index	WEO	Wind electricity output
		SEO	Solar electricity output

References

1. Fang, R.; Shang, R.; Wang, Y.; Guo, X. Identification of vulnerable lines in power grids with wind power integration based on a weighted entropy analysis method. *Int. J. Hydrog. Energy* **2017**, *42*, 20269–20276. [CrossRef]
2. Ehsan, A.; Yang, Q. Optimal integration and planning of renewable distributed generation in the power distribution networks: A review of analytical techniques. *Appl. Energy* **2018**, *210*, 44–59. [CrossRef]
3. Kaplan, O.; Temiz, M. A novel method based on Weibull distribution for short-term wind speed prediction. *Int. J. Hydrog. Energy* **2017**, *42*, 17793–17800. [CrossRef]
4. Yang, Q.; Fang, X. Demand response under real-time pricing for domestic households with renewable DGs and storage. *IET Gener. Transm. Distrib.* **2017**, *11*, 1910–1918. [CrossRef]
5. Belcher, S.E.; Hacker, J.N.; Powell, D.S. Constructing design weather data for future climates. *Build. Serv. Eng. Res. Technol.* **2005**, *26*, 49–61. [CrossRef]
6. Herrera, M.; Natarajan, S.; Coley, D.A.; Kershaw, T.; Ramallo-González, A.P.; Eames, M.; Fosas, D.; Wood, M. A review of current and future weather data for building simulation. *Build. Serv. Eng. Res. Technol.* **2017**, *38*, 602–627. [CrossRef]
7. Urge-Vorsatz, D.; Petrichenko, K.; Staniec, M.; Eom, J. Energy use in buildings in a long-term perspective. *Curr. Opin. Environ. Sustain.* **2013**, *5*, 141–151. [CrossRef]
8. Russo, M.; Carvalho, D.; Martins, N.; Monteiro, A. Forecasting the inevitable: A review on the impacts of climate change on renewable energy resources. *Sustain. Energy Technol. Assess.* **2022**, *52*, 102283. [CrossRef]
9. Zhuo, C.; Guo, J.; Li, W.; Zhang, F.; Xiao, C.; Pan, Z. Changes in wind energy potential over China using a regional climate model ensemble. *Renew. Sustain. Energy Rev.* **2022**, *159*, 112219. [CrossRef]
10. Costoya, X.; Decastro, M.; Carvalho, D.; Gómez-Gesteira, M. Assessing the complementarity of future hybrid wind and solar photovoltaic energy resources for North America. *Renew. Sustain. Energy Rev.* **2023**, *173*, 113101. [CrossRef]
11. Dutta, R.; Chanda, K.; Maity, R. Future of solar energy potential in a changing climate across the world: A CMIP6 multi-model ensemble analysis. *Renew. Energy* **2022**, *188*, 819–829. [CrossRef]

12. Niu, J.; Qin, W.; Wang, L.; Zhang, M.; Wu, J.; Zhang, Y. Climate change impact on photovoltaic power potential in China based on CMIP6 models. *Sci. Total Environ.* **2023**, *858*, 159776. [CrossRef] [PubMed]
13. Pinson, P. Very-short-term probabilistic forecasting of wind power with generalized logit–normal distributions. *J. R. Stat. Soc. Ser. C Appl. Stat.* **2012**, *61*, 555–576. [CrossRef]
14. Ma, X.Y.; Sun, Y.Z.; Fang, H.L. Scenario generation of wind power based on statistical uncertainty and variability. *IEEE Trans. Sustain. Energy* **2013**, *4*, 894–904. [CrossRef]
15. Cui, M.; Zhang, J.; Wang, Q.; Krishnan, V.; Hodge, B.-M. A data-driven methodology for probabilistic wind power ramp forecasting. *IEEE Trans. Smart Grid* **2017**, *10*, 1326–1338. [CrossRef]
16. Pinson, P.; Madsen, H.; Nielsen, H.A.; Papaefthymiou, G.; Klöckl, B. From probabilistic forecasts to statistical scenarios of short-term wind power production. *Wind Energy Int. J. Prog. Appl. Wind Power Convers. Technol.* **2009**, *12*, 51–62. [CrossRef]
17. Papaefthymiou, G.; Kurowicka, D. Using copulas for modeling stochastic dependence in power system uncertainty analysis *IEEE Trans. Power Syst.* **2008**, *24*, 40–49. [CrossRef]
18. Zhang, N.; Kang, C.; Xu, Q.; Jiang, C.; Chen, Z.; Liu, J. Modelling and simulating the spatio-temporal correlations of clustered wind power using copula. *J. Electr. Eng. Technol.* **2013**, *8*, 1615–1625. [CrossRef]
19. Eyring, V.; Bony, S.; Meehl, G.A.; Senior, C.A.; Stevens, B.; Stouffer, R.J.; Taylor, K.E. Overview of the Coupled Model Intercomparison Project Phase 6 (CMIP6) experimental design and organization. *Geosci. Model Dev.* **2016**, *9*, 1937–1958. [CrossRef]
20. Jentsch, M.F.; James, P.A.B.; Bourikas, L.; Bahaj, A.S. Transforming existing weather data for worldwide locations to enable energy and building performance simulation under future climates. *Renew. Energy* **2013**, *55*, 514–524. [CrossRef]
21. Sklar, M. Fonctions de répartition à n dimensions et leurs marges. *Ann. l'ISUP* **1959**, *8*, 229–231.
22. Aas, K.; Czado, C.; Frigessi, A.; Bakken, H. Pair-copula constructions of multiple dependence. *Insur. Math. Econ.* **2009**, *44*, 182–198. [CrossRef]
23. Sun, M.; Konstantelos, I.; Strbac, G. C-vine copula mixture model for clustering of residential electrical load pattern data. *IEEE Trans. Power Syst.* **2016**, *32*, 2382–2393. [CrossRef]
24. Kanungo, T.; Mount, D.M.; Netanyahu, N.S.; Piatko, C.D.; Silverman, R.; Wu, A.Y. An efficient k-means clustering algorithm: Analysis and implementation. *IEEE Trans. Pattern Anal. Mach. Intell.* **2002**, *24*, 881–892. [CrossRef]
25. Tong, W.; Liu, S.; Gao, X.Z. A density-peak-based clustering algorithm of automatically determining the number of clusters *Neurocomputing* **2021**, *458*, 655–666. [CrossRef]
26. Watson, S.; Moro, A.; Reis, V.; Baniotopoulos, C.; Barth, S.; Bartoli, G.; Bauer, F.; Boelman, E.; Bosse, D.; Cherubini, A.; et al. Future emerging technologies in the wind power sector: A European perspective. *Renew. Sustain. Energy Rev.* **2019**, *113*, 109270. [CrossRef]
27. Akinsanola, A.A.; Ogunjobi, K.O.; Abolude, A.T.; Salack, S. Projected changes in wind speed and wind energy potential over West Africa in CMIP6 models. *Environ. Res. Lett.* **2021**, *16*, 044033. [CrossRef]
28. Carvalho, D.; Rocha, A.; Costoya, X.; Decastro, M.; Gómez-Gesteira, M. Wind energy resource over Europe under CMIP6 future climate projections: What changes from CMIP5 to CMIP6. *Renew. Sustain. Energy Rev.* **2021**, *151*, 111594. [CrossRef]
29. Matthew, O.J.; Ohunakin, O.S. Simulating the effects of climate change and afforestation on wind power potential in Nigeria. *Sustain. Energy Technol. Assess.* **2017**, *22*, 41–54. [CrossRef]
30. Moemken, J.; Reyers, M.; Feldmann, H.; Pinto, J.G. Future changes of wind speed and wind energy potentials in EURO-CORDEX ensemble simulations. *J. Geophys. Res. Atmos.* **2018**, *123*, 6373–6389. [CrossRef]
31. Wang, S.; Zhu, J.; Huang, G.; Baetz, B.; Cheng, G.; Zeng, X.; Wang, X. Assessment of climate change impacts on energy capacity planning in Ontario, Canada using high-resolution regional climate model. *J. Clean. Prod.* **2020**, *274*, 123026. [CrossRef]
32. Gernaat, D.E.; de Boer, H.S.; Daioglou, V.; Yalew, S.G.; Müller, C.; van Vuuren, D.P. Climate change impacts on renewable energy supply. *Nat. Clim. Chang.* **2021**, *11*, 119–125. [CrossRef]
33. Wiboonpongse, A.; Liu, J.; Sriboonchitta, S.; Denoeux, T. Modeling dependence between error components of the stochastic frontier model using copula: Application to intercrop coffee production in Northern Thailand. *Int. J. Approx. Reason.* **2015**, *65*, 34–44. [CrossRef]

sustainability

Article

Multi-Attention Network for Sewage Treatment Plant Detection

Yue Shuai [1,2], Jun Xie [3], Kaixuan Lu [1] and Zhengchao Chen [1,*]

[1] Airborne Remote Sensing Center, Aerospace Information Research Institute, Chinese Academy of Sciences, Beijing 100094, China
[2] University of Chinese Academy of Sciences, Beijing 100049, China
[3] College of Geoscience and Surveying Engineering, China University of Mining and Technology, Beijing 100083, China
* Correspondence: chenzc@radi.ac.cn

Abstract: As an important facility for effectively controlling water pollution discharge and recycling waste water resources, accurate sewage treatment plant extraction is very important for protecting quality, function, and sustainable development of the water environment. However, due to the presence of rectangular and circular treatment facilities in sewage treatment plants, the shapes are diverse and the scales are different, resulting in the poor performance of conventional object detection algorithms. This paper proposes a multi-attention network (MANet) for sewage treatment plants using remote sensing images. MANet consists of three major components: a light backbone used to obtain multi-scale features, a channel and spatial attention module that realizes the feature representation of the channel dimension and spatial dimension, and a scale attention module to obtain scale-aware features. The results from the extensive experiments performed on the sewage treatment plant dataset suggest that our proposed MANet exhibits a superior performance compared with other competing methods. Meanwhile, we used a well-trained model to predict the sewage treatment plant from the GF-2 data for the Beijing area. By comparing the results with the data of manually obtained sewage treatment plants, our method can achieve an accuracy of 80.1% while maintaining the recall rate at a high level (90.4%).

Keywords: deep learning; sewage treatment plant detection; Beijing area; attention module; remote sensing images

Citation: Shuai, Y.; Xie, J.; Lu, K.; Chen, Z. Multi-Attention Network for Sewage Treatment Plant Detection. *Sustainability* **2023**, *15*, 5880. https://doi.org/10.3390/su15075880

Academic Editors: Marc A. Rosen and Andrea G. Capodaglio

Received: 15 February 2023
Revised: 24 March 2023
Accepted: 27 March 2023
Published: 28 March 2023

1. Introduction

Wastewater treatment contributes to the achievement of 11 of the 17 Sustainable Development Goals that have currently been adopted globally [1]. As the main carrier of wastewater treatment, sewage treatment plants are important assistants for effectively curbing sewage discharge and recycling wastewater in industrial society, and they are increasingly important for water quality protection and the sustainable development of man and nature [2]. They are especially important in world-class cities such as Beijing, where a large amount of industrial sewage and domestic wastewater are produced every day; if they were directly discharged without being treated by the sewage treatment plant, they would cause a huge disaster to the natural environment and further affect people's lives, which is not conducive to the sustainable development of man and nature [3,4]. Realizing the automatic extraction of large-scale sewage treatment plants will provide basic data support for people to study the details of sewage treatment plants and provide further technical support for the realization of the Sustainable Development Goals.

The contributions made by sewage treatment plants are significant, but there are also negative problems. Due to the need to collect sewage for purification, the surrounding environment has a high level of pollutants, which has an adverse impact on the surrounding ecological environment and on people's lives [5,6]. Under the current conditions, we cannot immediately obtain the distribution information of sewage treatment plants in a certain area—especially on a large scale, such as the provincial or national level—or the sewage treatment plants that are built by some factories. In this study, we attempted to realize the automatic identification of large-scale sewage treatment plants through technical research, obtain their spatial location information and quantity information, and make this information easier for people or managers to obtain. Based on the relevant information of sewage treatment plants, they can provide a reference for the selection of the living location of the relevant population or allow people to further enhance the protection awareness of water resources with better understanding. In addition, the information can also provide a reference for city managers for optimizing the layout of sewage treatment plants and for the scientific construction of cities [7].

The goal of our technical research is to conduct a realization of large-scale sewage treatment plant extraction based on big data and computer vision methods. The convolutional neural network [8], designed by simulating the function of human neurons, has strong feature fitting and learning capabilities for input data through the stacking of network depths and the setting of nonlinear activation functions. Compared with the fully connected neural network, a convolutional neural network that extracts data features by setting the size of the convolution kernel area has a higher computing efficiency for image data; with the assistance of graphics computing hardware, it is the best choice for processing large image data [9]. The deep learning object detection algorithm based on the convolutional neural network has been rapidly developed after recent in-depth research; many classic algorithms have been sequentially proposed, such as Faster RCNN [10], SSD [11], RetinaNet [12], YOLO series [13–16], etc., and successfully applied in many fields of computer vision. A study of the feasibility detection algorithm that is based on the deep learning object detection algorithm and combined with the sewage treatment plant's characteristics will greatly improve the recognition efficiency and automation level of sewage treatment plants.

As a long-distance detection technology, remote sensing has the characteristics of wide monitoring ranges, short periods, and low costs [17]. It can be used as a technical means to objectively obtain sewage treatment plant distributions. In recent years, with the launch of a large number of satellites, it is very convenient to obtain large-scale, high-resolution, and short-period optical remote sensing image data. Using optical remote sensing data to carry out high-precision, high-frequency monitoring of sewage treatment plant times, extraction has become possible [18,19]. Different from natural images, optical remote sensing satellite images that are captured from the top-down view of the Earth contain rich and complicated ground object information. Directly transferring the object detection algorithm applied to natural images to optical remote sensing images will reduce the model's accuracy. As a building facility, sewage treatment plants contain modules such as circular and rectangular purification pools for filtering sewage, and the overall characteristics are consistent. However, there are also local differences in the scale and shape characteristics; furthermore, because there are many ground objects with similar characteristics, the characteristics cause certain challenges in detecting sewage treatment plants. There are a lot of studies on the detection difficulties caused by the multi-scale and large shape differences of objects similar to sewage treatment plants in optical satellite remote sensing images [20–23].

In view of the multi-scale characteristics of remote sensing ground objects, multi-scale information fusion modules are commonly designed for feature extraction. For example, the FPN [24] (feature pyramid network) can account for both deep and shallow features to preserve the multi-scale information of the object [25,26]. On the basis of multi-scale detection, Yan et al. [27] balanced the training weights of differently scaled objects for the loss function and strengthened the robustness of the algorithm to different scales. In addition, the attention mechanism introduced by the transformer model [28] has been proven to have a good effect in multi-scale object detection. Zhu et al. [29] used the transformer model to improve the prediction network of YOLOv5 and combine the self-attention mechanism to achieve multi-scale object detection. In view of the characteristics of the large differences in the shapes of remote sensing ground objects, the main method used adjusts the type and quantity of the anchor frames in the detection stage to adapt to different shapes of the same objects or multiple types of objects [25,30,31]. For example, reset the scale of the anchor box, the aspect ratio parameters, or increase the angle variable, etc., and use a deformable convolutional network [32–34] to adapt to the target shape. The disadvantage of these methods is that the increase in anchor frame parameters and addition of deformable convolution will add a large number of parameters to the network, increase the difficulty of the model training, and lead to unfavorable model convergence. The detection model based on key points can overcome the problem of large changes in the object's shape to a certain extent, but the detection accuracy is basically the same as that of the anchor frame method; there is still large room for improvement and optimization.

However, sewage treatment plant characteristics in remote sensing images are different from other ground objects, and the above work still cannot directly meet the detection needs of sewage treatment plants. In this paper, starting from the detection of sewage treatment plants in remote sensing images in Beijing, a MANet sewage treatment plant detection network is proposed to solve the problems of the large differences in the shape and scale of sewage treatment plants, as well as their inconsistent local features. MANet integrates the channel and spatial attention in the feature extraction module and innovates a scale attention algorithm for network feature optimization, which better solves for the detection difficulties of sewage treatment plants and greatly improves the interpretation of sewage treatment plant target precision. The main contributions of this paper are as follows:

(1) We introduced a lightweight channel and spatial attention module (CSAM) to improve the feature expression ability of the extracted target in the spatial and channel dimensions;
(2) We innovated a novel scale attention modeule (SAM) algorithm to improve the feature learning ability of the network at different levels for targets with large-scale changes;
(3) We added the above two attention modules based on RetinaNet, proposed a MANet sewage treatment plant detection network, and achieved better results in the dataset test. In the actual scene, based on GF-2 remote sensing images, the sewage treatment plant detection in the Beijing area was realized. The results show that our method can achieve an accuracy of 80.1% while maintaining the recall rate at a high level (90.4%).

2. Methodology

In this section, we introduce the architecture of our proposed approach, MANet (Section 2.1), the backbone for the feature extraction (Section 2.2), the channel and spatial attention module (Section 2.3), and the scale attention module (Section 2.4).

2.1. Model Overview

The attention mechanism in deep learning approaches imitates the human visual system. When a human being observes an object, they first quickly scan the entire area, select the target from the area, and invest more visual resources to obtain more detailed information; however, the neural network needs to scan each pixel when scanning an image. The attention module is used to ensure that the CNN learns and pays more attention to key features instead of learning useless background information. In the object detection task, the useful information refers to the target's location and category information on the image, which essentially uses the $C \times H \times W$ feature map as an input and provides $1 \times H \times W$ as the output attention map. This attention map is then element-wise multiplied with the input feature map to obtain a more refined and salient output. In general, the attention mechanism is mainly applied to the spatial dimension or the channel dimension and is integrated in the residual structure of the network.

Sewage treatment plants have different shapes, including rectangular and circular treatment devices. Targets usually appear in completely different shapes, rotations, and positions, and the spatial variation of targets needs to be considered. At the same time, sewage treatment plants have different scales, ranging from large to small. The scale variation of the targets needs to be considered. Due to the complexity of sewage treatment plants' characteristics, the traditional deep learning object detection model has missed detection and falsely detected plants in sewage treatment plant recognition [35]. By analyzing the characteristics of sewage treatment plants using high-resolution remote sensing images and aiming to resolve the detection difficulties of sewage treatment plants, we designed MANet. Its overall architecture is shown in Figure 1. It mainly includes three parts: (1) the feature extraction part, which contains a backbone for obtaining multi-scale feature maps; (2) CSAM, which includes the spatial attention module and channel attention module, which learns the best features of the target from the two dimensions of the space and channel; and (3) SAM, which is only processed in the feature layer dimension, which learns the relative importance of multiple semantic layers and enhances features at the appropriate level according to the scale of the object.

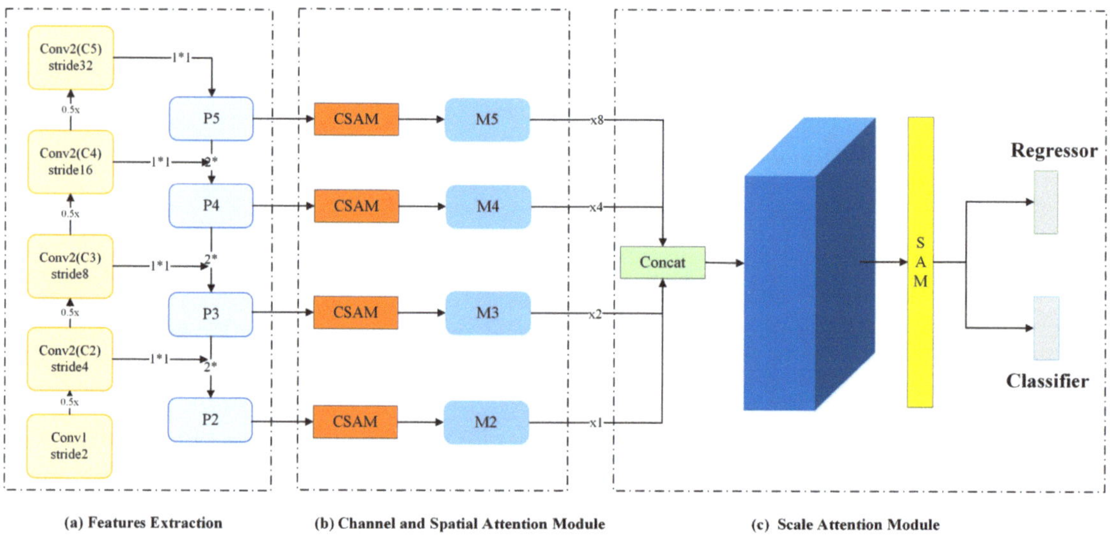

Figure 1. The framework of our proposed approach.

2.2. Feature Extraction

This part uses the structure of the ResNet+Feature Pyramid Network (FPN) to extract low-resolution features from images (input RGB images, size $H \times W \times 3$) and obtain multi-scale feature maps ($\frac{H}{S} \times \frac{W}{S} \times C$) through different stage steps (S4, 8, 16, and 32). In order to reduce the weight, this paper chooses a relatively simple ResNet-50+FPN structure The basic structural unit of ResNet-50 is the residual structure. As shown in Figure 1, the entire network is divided into five blocks, namely conv1, conv2-x, conv3-x, conv4-x, and conv5-x. The convolution kernel size of conv1 is set to 7×7, the step size is set to 2, and the expansion is set to 3; then, the maximum pooling is performed. The pooled convolution kernel size is set to 3×3, the step size is set to 2, and the expansion setting is 0. There are three convolution blocks in the conv2-x part, where the convolution kernel sizes are set to 1×1, 3×3, and 1×1, respectively. The three parts of conv3-x, conv4-x, and conv5-x are similar to the structure of conv2-x; the difference is that the number of convolution blocks is different. The conv3-x part has four convolution blocks, the conv4-x part has six convolution blocks, and the conv5-x part has three convolution blocks. Finally, the multi-scale features are obtained after the blocks calculate the image data.

The structure of a light FPN typically consists of the following components: (1) top-down pathway: this pathway starts from the high-level semantic feature maps obtained from the backbone network and passes them through up-sampling operations to obtain feature maps at lower scales; (2) bottom-up pathway: this pathway starts from the low-level feature maps obtained from the backbone network and passes them through up-sampling operations to obtain feature maps at higher scales; (3) fusion layer: the feature maps from the top-down and bottom-up pathways are combined using element-wise summation or concatenation to obtain the final feature maps at each scale. By combining the features from multiple scales, the FPN can capture both the fine-grained details and the high-level context of the input image, leading to improved performance in object detection tasks.

2.3. Channel and Spatial Attention Module

In addition to the variable scale of sewage treatment plants mentioned in this paper, the characteristics of different shapes and colors make it difficult for the network to distinguish between them. Therefore, we introduce a CSAM to further optimize the features and cause these feature pairs to be more distinguishable. It is a very lightweight module that does not incur excessive memory and computational overhead. As shown in Figure 2, CSAM consists of two sub-modules, a channel attention module, and a spatial attention module to help strengthen the useful information in extracted features.

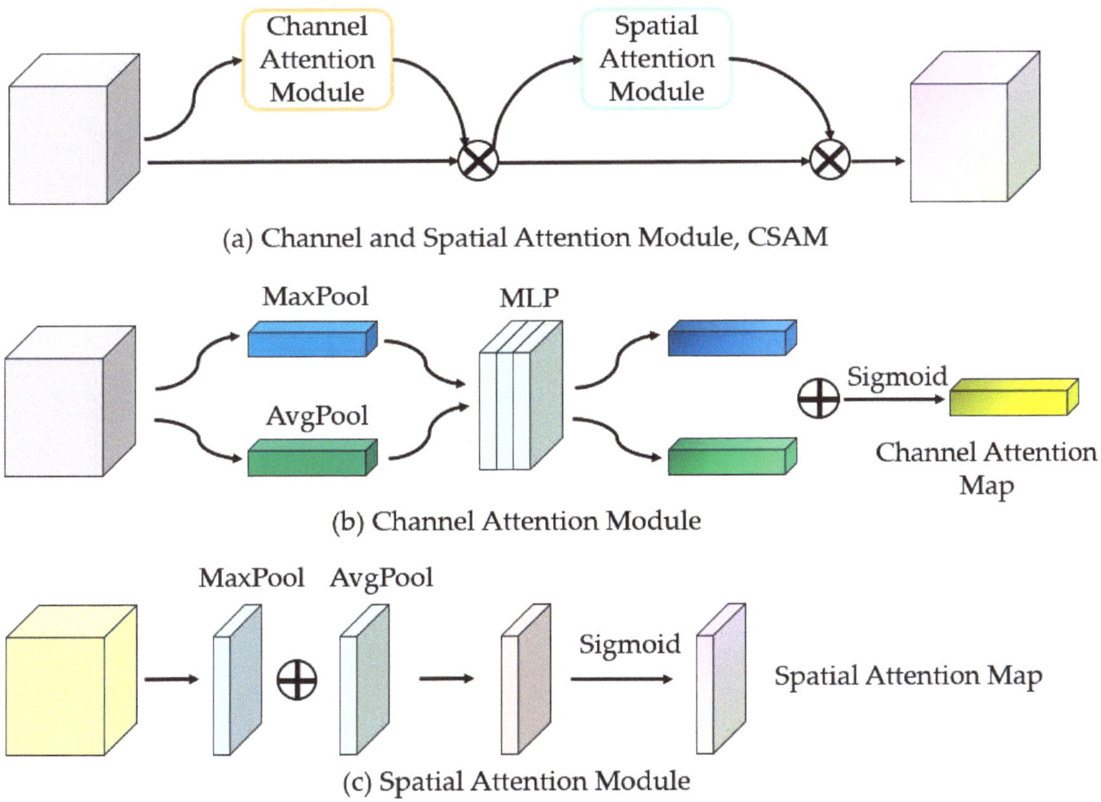

Figure 2. The structure of the channel and spatial attention module, CSAM. (**a**) Modules included in the CSAM. (**b**) Structural details of the channel attention module. (**c**) Structural details of the spatial attention module.

The channel attention module is a channel-based attention module in the convolutional neural networks that aims to capture the long-term contextual information of channel directions through channel attention maps. To efficiently compute channel attention, we aggregate the spatial information of the feature maps using two pooling operations (average pooling and max pooling) to generate two 2D feature maps $F_1 \in R^c$ and $F_2 \in R^c$ These two feature maps represent the average pooled features and max pooled features in the channel, respectively. The channel attention module can be calculated using the following formula:

$$M_c(F) = \sigma(MLP(AvgPool(F)) + MLP(MaxPool(F))) \quad (1)$$

where σ denotes the sigmoid function.

The spatial attention module is a module that is applied to spatial attention in the convolutional neural network. It uses the spatial relationship of features to generate a spatial attention map and concentrates on mining target location information. To compute the spatial attention, we aggregate the channel information of feature maps through two pooling operations (average pooling and max pooling) to generate two 2D feature maps $F_1 \in R^{H\times W}$ and $F_2 \in R^{H\times W}$. These two feature maps represent the average pooled feature and maximum pooled feature in the channel, respectively, and they are concatenated and convolved by a standard convolutional layer to generate a two-dimensional spatial attention map $Attn_s \in R^{H\times W}$. This attention map shows how much the model pays attention to the position. The spatial attention module can be calculated using the following formula:

$$M_s(F) = \sigma(f([AvgPool(F)]; MaxPool(F))) \quad (2)$$

where σ denotes the sigmoid function and f represents a convolution operation.

2.4. Scale Attention Module

The object scale difference is related to the features of different levels. Improving the representation learning ability of different feature levels is conducive to improving the detection accuracy of target detection. However, the features at different levels are usually extracted from different depths of the network, which results in an obvious semantic gap, and it is not optimal to directly fuse feature layers at different levels. To solve this problem, we introduce a SAM to dynamically fuse the features of different scales based on semantics. The structure diagram of SAM is shown in Figure 3.

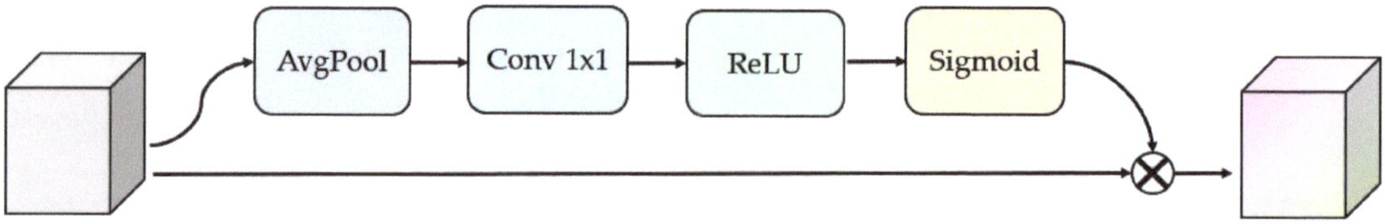

Figure 3. Structural details of the scale attention module, SAM.

We first sample features at different scales to $\frac{H}{4} \times \frac{W}{4}$ and connect them together to form $F_{level} \in R^{L\times C\times \frac{H}{4}\times \frac{W}{4}}$, where L is 4 and C is 256. Next, we use the scale attention module to obtain the scale attention feature map. Then, the scale attention module is composed of average pooling, 1×1 convolution, and relu. Finally, we use a sigmoid normalization to obtain the final scale attention feature map $Attn_{level} \in R^{L\times 1\times 1\times 1}$. The scale attention calculation formula is as follows:

$$M(F) = \sigma(f(AvgPool(\sum F))) \quad (3)$$

where σ denotes the sigmoid function and f represents a 1×1 convolution operation.

3. Experimental Results and Discussion

We performed related experiments to evaluate the proposed MANet architecture's effectiveness. In this section, we introduce the selected study area; the used experimental data, experimental setting, and evaluation metrics; the comparative experiment used to verify the performance of MANet; the ablation experiment to compare the effects of multiple attention modules; and, finally, the detection results of MANet in the actual sewage treatment plant scene.

3.1. Study Area and Experimental Data

We chose Beijing as the study area. Beijing is located in the northern part of the North China Plain, which is adjacent to Tianjin. It is located at 115.7°–117.4° east longitude and 39.4°–41.6° north latitude with a total area of 16,410.54 square kilometers. The climate is a typical northern temperate semi-humid continental monsoon climate. As a world-class city, Beijing has a dense population and a large number of factories engaged in production; thus, it requires sewage treatment plants with better operation layouts. Realizing the extraction of sewage treatment plants in this area, in addition to obtaining the number and location information of sewage treatment plants, will provide a reference for the subsequent development and construction planning of emerging cities. Figure 4 is a regional image of Beijing.

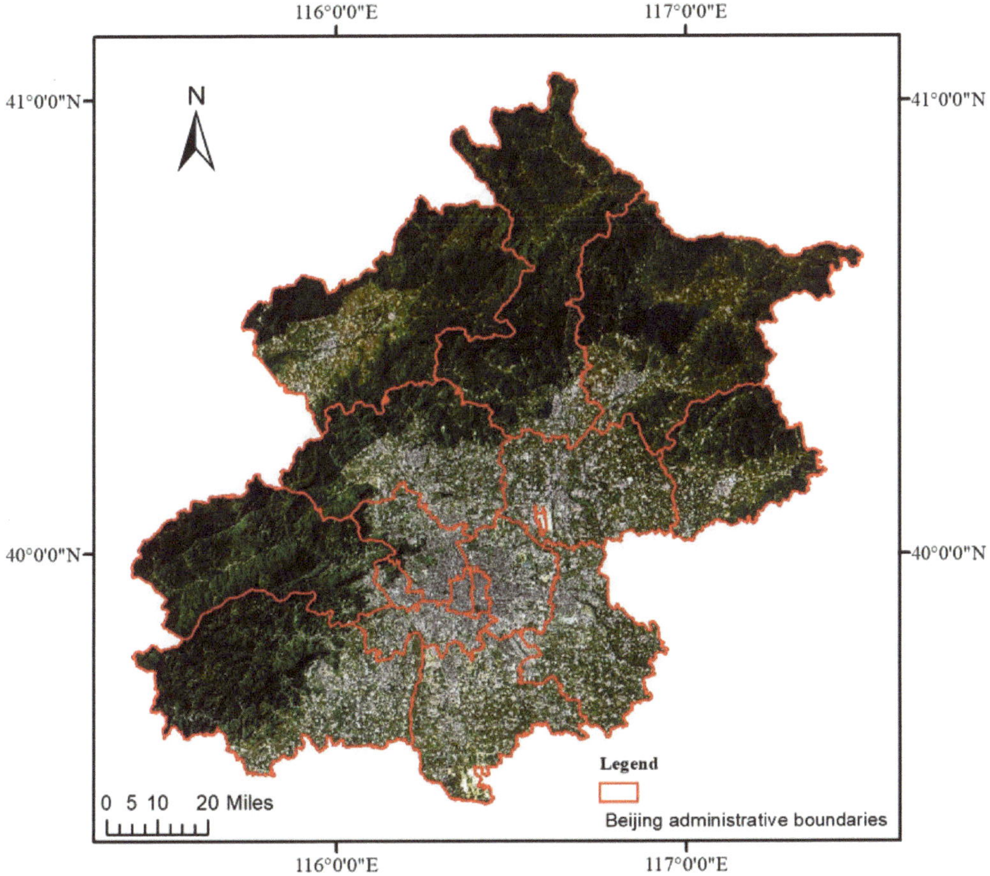

Figure 4. Beijing area.

In order to ensure the clarity of the sewage treatment plant in the remote sensing image, we use the 2 m resolution GF-2 satellite image data domestically produced in China to create a sample dataset of the sewage treatment plant. As shown in Figure 5, we use a sewage treatment plant containing circular and rectangular sedimentation tank structures as detection targets for the sample labeling. Considering the size of the sewage treatment plant target in the 2 m remote sensing image, we use a resolution of 1536 × 1536 for slice production. After manual labeling, 3000 samples of sewage treatment plants were obtained, and the dataset was divided into training and validation sets according to the ratio of 10:1

Figure 5. Sewage treatment plant in remote sensing images.

3.2. Experiment Setting and Evaluation Metrics

The method proposed in this paper and the related experiments were all run on the Ubuntu 16.04.7 LTS operating system using an NVIDIA GeForce RTX 3090 GPU with a 24GB memory size. All of the algorithm model experiments were carried out on the PyTorch deep learning framework, and the relevant parameters of the model were kept consistent during the training and testing processes. The input size and batch size were set to $1024 \times 1024 \times 3$ and 8, respectively. The total number of iterations for all experiments was 12 epochs, and all backbones were pre-trained on the ImageNet-1K dataset. The experiment used the stochastic gradient descent (SGD) optimizer, and the original learning rate parameter of network training was set to 0.01. The momentum parameter used to accelerate and stabilize the optimal solution of the function was set to 0.9. The weight decay parameter, which is conducive to the network convergence and fitting data, was set to 0.0001.

Regarding the evaluation metrics of the experiments, we adopt a confusion matrix, which is often used to evaluate object detection results. The confusion matrix's composition is shown in Table 1. Among them, TP indicates that the target is a sewage treatment plant and is correctly predicted; TN indicates that the target is not a sewage treatment plant and is correctly predicted; FP indicates that the target is not a sewage treatment plant but is predicted to be one; and FN indicates that the target is a sewage treatment plant but is predicted to not be one.

Table 1. Confusion matrix.

		Ground Truth	
		True	**False**
Predicted Label	True	TP (True Positive)	FP (False Positive)
	False	FN (False Negative)	TN (True Negative)

Based on the confusion matrix, we further use precision AP (average precision) and recall AR (average recall) to evaluate the detection results. The relevant calculation formulae are as follows:

$$AP = \frac{TP}{TP + FP} \tag{4}$$

$$AR = \frac{TP}{TP + FN} \tag{5}$$

where AP indicates the proportion of correctly predicted sewage treatment plants in the detection results and AR indicates the proportion of correctly predicted sewage treatment plants in the validation set. When its threshold is set to 0.5, AP50 indicates the proportion of correctly predicted sewage treatment plants in the detection results.

3.3. Experimental Results

MANet uses RetinaNet as the baseline network and integrates CSAM and SAM modules based on the ResNet-50+FPN structure. We selected some advanced and representative object detection algorithms to conduct comparative experiments using MANet and used AP and AP50 to evaluate the experimental results. Then, the effectiveness of CSAM and SAM was verified through ablation experiments, and the effects of different modules on MANet performance were analyzed.

3.3.1. Comparison of Model Performance

To evaluate the performance of MANet, we selected four of the most advanced and mature object detection methods (RetinaNet, Cascade RCNN [36], ATSS [37], and Faster RCNN) to conduct the experiments and compare the results in the same environment and settings. The experimental settings are in Section 3.2. The *AP* and *AP*50 curves obtained from the experimental results of the five networks are shown in Figure 6.

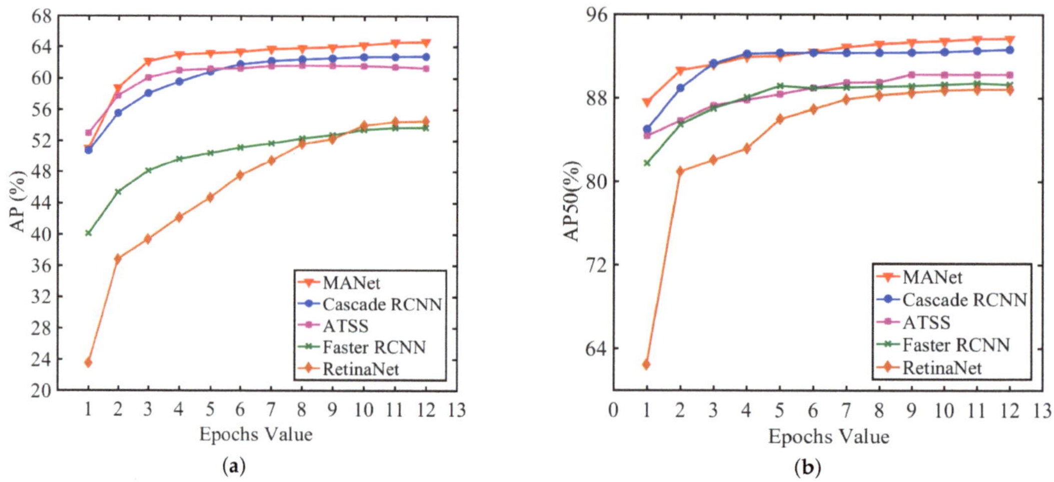

Figure 6. Comparative experimental results of MANet and related networks; (**a**) AP results; (**b**) AP50 results.

As shown in Figure 6, the proposed MANet method achieved higher precision than the other four methods. MANet can obtain the highest AP value of 64.6%, which is significantly higher than the baseline network RetinaNet's value of 54.48%. The AP values of Cascade RCNN and ATSS are closer to MANet but are still below the AP curve of MANet. When the threshold is 0.5, the AP50 value of MANet is still the highest at up to 93.62%, and the AP50 value is basically ahead of the other four networks during the training process. It can be seen that MANet has obvious advantages in the task of detecting sewage treatment plant targets, can more effectively learn the remote sensing image features of sewage treatment plants, and can achieve higher recognition capabilities.

3.3.2. Ablation Studies

In order to verify the effectiveness of CSAM and SAM for MANet to identify sewage treatment plant targets, we conducted ablation experiments on CSAM and SAM. The experiment used RetinaNet as the baseline network, which is based on the ResNet-50+FPN structure, and used the control variable method to experiment with CSAM or SAM. CSAM was disassembled into a channel attention module (Channel-AM) and a spatial attention module (Spatial-AM) for the experiments. The experimental results were evaluated using the AP and AP50 values, and the training time was also involved in the comparison.

Table 2 shows the results of the ablation experiments. Analyzing the experimental results reveals that both of the proposed CSAM and SAM models can improve the network's performance and that the AP values are increased by 4.02% and 6.01%, respectively, compared with the baseline network. When the two models work together, the AP value increases by 10.12%, the AP value reaches 64.6%, and the AP50 can reach 93.62%. These results show that both the CSAM and SAM modules can efficiently extract the sewage treatment plant's features and have strong robustness regarding the shape and scale changes of the sewage treatment plant. In particular, SAM's processing of differently scaled features at different levels contributes more to the performance of the model. Compared with the baseline, the channel attention module and spatial attention module of CSAM have improved by 1.8% and 2.22%, respectively, indicating that CSAM has improved the ability to learn the features of sewage treatment plants in the channel and space dimensions.

Table 2. Ablation experiment results of CSAM and SAM.

Baseline-RetinaNet	Channel-AM	Spatial-AM	Scale-AM	AP (%)	AP50 (%)	Time/h
✓				54.48	88.78	7.97
✓	✓			56.28	89.65	8.14
✓		✓		56.70	89.85	8.18
✓			✓	60.49	91.68	8.54
✓	✓	✓		58.50	90.72	8.35
✓	✓	✓	✓	**64.60**	**93.62**	**8.92**

3.4. Extraction Results of Beijing Sewage Treatment Plant

We used the model obtained by training MANet to detect the sewage treatment plant on the 2 m GF-2 remote sensing image of Beijing, compared the detection results with the actual number of the manual statistics, and used the confusion matrix to evaluate the detection results. The model is set at thresholds of 0.5, 0.6, 0.7, 0.8, and 0.9 when detecting sewage treatment plants, and the results are shown in Table 3.

Table 3. Assessment of detection results of sewage treatment plant in the Beijing area.

Threshold	Actual Amount	Predicted Amount	TP	FP	FN	AP (%)	AR (%)
0.5		210	**149**	61	**2**	70.9	**97.7**
0.6		203	146	57	5	71.6	95.4
0.7	151	180	140	40	11	77.3	92.4
0.8		170	137	33	14	80.1	90.4
0.9		160	130	**30**	21	**81.5**	85.4

The detection results in Table 3 highlight that the AP of the detection results increases with the increase in the threshold, while the AR decreases. When the threshold is 0.5, the number of detections of sewage treatment plants is the largest, reaching 149; however, the false positive detections are serious. Additionally, the AP is the lowest at 70.9%, and the AR can reach 97.7%. When the threshold is 0.9, the predicted number of the sewage treatment plant is the lowest, but the AP is the highest at a value of 81.5%; the AR is 85.4%. Overall, the network has a good detection effect.

According to the sewage treatment plant detection results, we obtained the location information of the sewage treatment plants and created a distribution map of them for the Beijing area in a TIF image. As shown in Figure 7, the red dots represent the detected targets of the sewage treatment plants. From the picture, we can learn more about the sewage treatment plant distribution in Beijing, which can provide a reference for the site selection of sewage treatment plants and help future urban construction.

To further demonstrate the detection effect, we selected some sewage treatment plant targets from the detection results to assist in the description, as shown in Figure 8. It can be seen that MANet can more effectively overcome the problems of the varying shapes and scales of sewage treatment plants and inconsistent local features, has good generalization performance, and can accurately realize the detection of sewage treatment plants.

3.5. Discussion

In this study, we designed a multi-attention network MANet containing multiple modules for the characteristics of large scale changes, large shape differences, and complex semantic information of sewage treatment plants as derived from remote sensing images. The CSAM of the network improves its ability to extract target features from the spatial and channel dimensions, and the SAM processes feature maps of different scales from the scale dimension to reduce the impact of target scale changes. Based on this, MANet was constructed and a deep learning model was trained. The performance of the model was verified in experiments. Finally, the extraction of sewage treatment plants in Beijing was realized based on 2 m GF-2 satellite remote sensing images. The accurate and fast extraction of sewage treatment plants in a large area fully demonstrates the advantages of deep learning methods in the interpretation of remote sensing image objects.

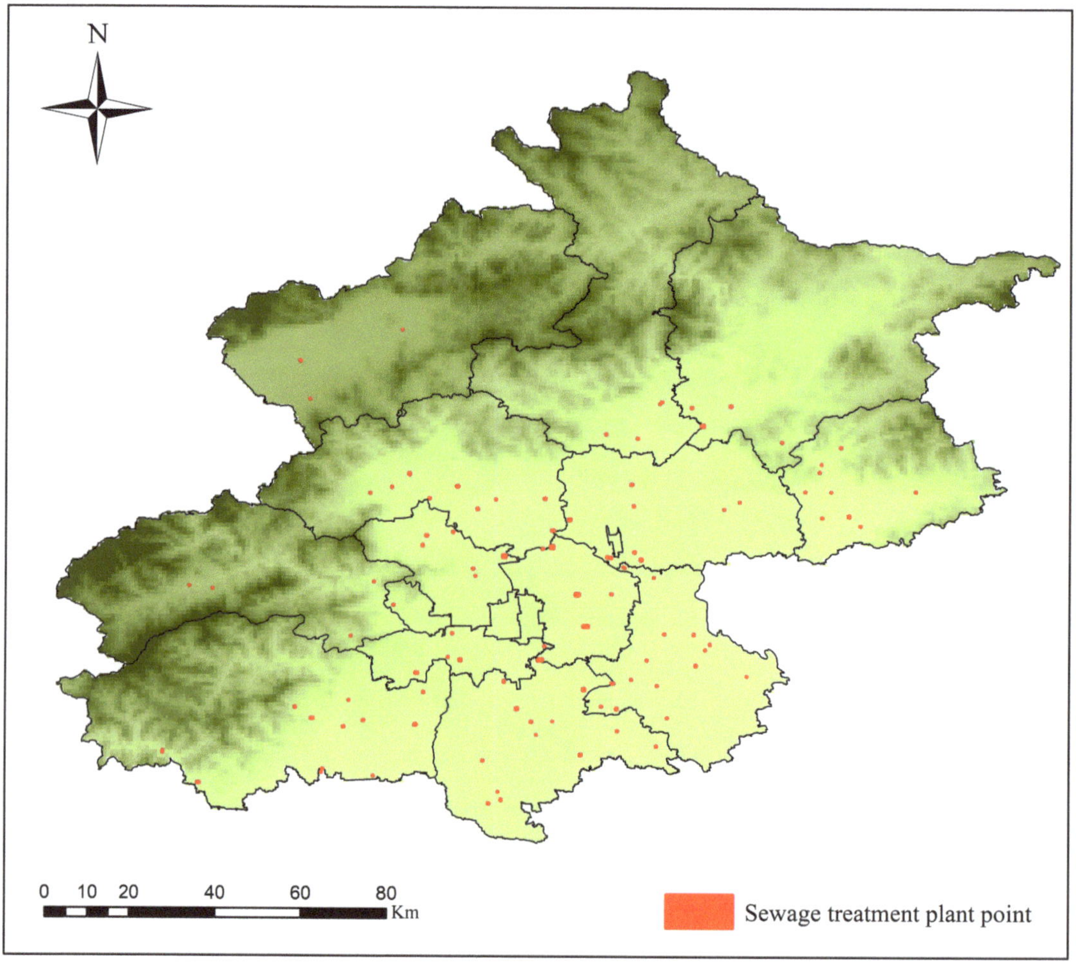

Figure 7. Distribution map of the detection results of Beijing sewage treatment plants.

During the research, we found that the number of samples restricted the accuracy of the model to a certain extent; however, the most fundamental problem was that there were not many sewage treatment plants in reality. Therefore, under the premise of ensuring the generalization performance of the model, we have performed data enhancement on the samples to increase the number of samples. Based on the characteristics of the circular and rectangular sedimentation tank structures contained in the sewage treatment plant, we first focused on the method of sample labeling. On the basis of the original labeling, mark the modules inside the sewage treatment plants and establish a mathematical model for the spatial distance between the modules to judge the target of the sewage treatment plants. In addition, based on the extraction results of the sewage treatment plants, how to fully mine its information according to the key structure of the object, such as further estimating the sewage treatment capacity of the sewage treatment plant based on the target recognition network, may require the designing of a more powerful network.

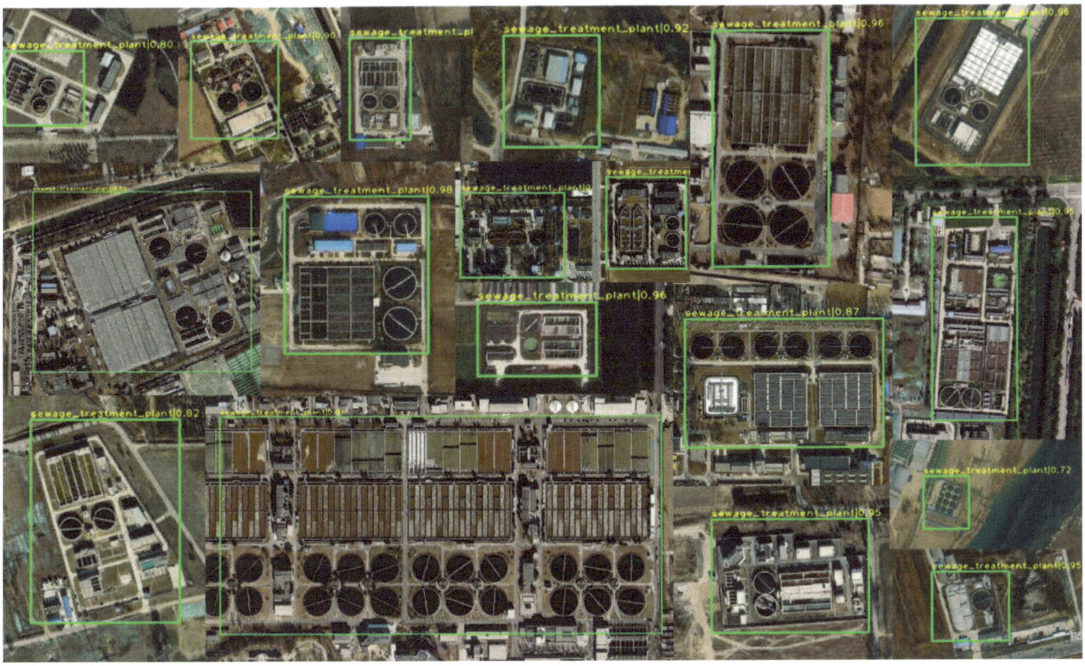

Figure 8. Detection results of sewage treatment plants.

4. Conclusions

Sewage treatment plants in remote sensing images have the characteristics of varying shapes and scales and inconsistent local features. It is difficult to detect sewage treatment plants using traditional deep learning object detection algorithms. In this study, we proposed a novel and effective sewage treatment plant detection network, MANet, which has obvious advantages compared with other advanced object detection algorithms. Based on the model trained by MANet, we realized the detection of the sewage treatment plant on the 2 m resolution GF-2 satellite remote sensing image in the Beijing area and obtained a distribution location map. The conclusions are drawn as follows:

(1) We introduced a lightweight CSAM using channel attention and spatial attention, which can efficiently improve the feature learning ability of MANet in spatial and channel dimensions;
(2) A novel SAM was proposed, which can improve the feature learning ability of MANet at different levels when extracting sewage treatment plant targets with large-scale changes;
(3) Based on the addition of CSAM and SAM to the RetinaNet model, a sewage treatment plant detection network called MANet was proposed, and better results were achieved in the dataset experiments. In the actual scene, based on GF-2 remote sensing images, sewage treatment plant detection for the Beijing area was realized. The results show that our method can achieve an accuracy of 80.1% while maintaining the recall rate at a high level (90.4%).

Author Contributions: Conceptualization, Y.S.; Methodology, Y.S. and K.L.; Data Curation, J.X.; Investigation, Y.S.; Resources, J.X.; Visualization, Y.S.; Writing—Original Draft, Y.S. and K.L.; Reviewing and Editing, Y.S., Z.C. and J.X. All authors have read and agreed to the published version of the manuscript.

Funding: This work is supported by the National Key Research and Development Program of China (Grant No. 2021YFB3901202).

Institutional Review Board Statement: Not applicable.

Informed Consent Statement: Not applicable.

Data Availability Statement: Not applicable.

Acknowledgments: Thanks to Haoyang Wang for his support in making the distribution map.

Conflicts of Interest: The authors declare no conflict of interest.

References

1. Obaideen, K.; Shehata, N.; Sayed, E.T.; Abdelkareem, M.A.; Mahmoud, M.S.; Olabi, A. The role of wastewater treatment in achieving sustainable development goals (SDGs) and sustainability guideline. *Energy Nexus* **2022**, *7*, 100112. [CrossRef]
2. Shanmugam, K.; Gadhamshetty, V.; Tysklind, M.; Bhattacharyya, D.; Upadhyayula, V.K. A sustainable performance assessment framework for circular management of municipal wastewater treatment plants. *J. Clean. Prod.* **2022**, *339*, 130657. [CrossRef]
3. Gautam, S.K.; Sharma, D.; Tripathi, J.K.; Ahirwar, S.; Singh, S.K. A study of the effectiveness of sewage treatment plants in Delhi region. *Appl. Water Sci.* **2013**, *3*, 57–65. [CrossRef]
4. Jin, L.; Zhang, G.; Tian, H. Current state of sewage treatment in China. *Water Res.* **2014**, *66*, 85–98. [CrossRef] [PubMed]
5. Bao, R.; Wang, Z.; Qi, H.; Mehmood, T.; Cai, M.; Zhang, Y.; Yang, R.; Peng, L.; Liu, F. Occurrence and distribution of microplastics in wastewater treatment plant in a tropical region of China. *J. Clean. Prod.* **2022**, *349*, 131454. [CrossRef]
6. Xie, J.; Jin, L.; Wu, D.; Pruden, A.; Li, X. Inhalable antibiotic resistome from wastewater treatment plants to urban areas: Bacterial hosts, dissemination risks, and source contributions. *Environ. Sci. Technol.* **2022**, *56*, 7040–7051. [CrossRef]
7. Liu, B.; Tang, J.; Qu, Y.; Yang, Y.; Lyu, H.; Dai, Y.; Li, Z. A GIS-based method for identification of blindness in former site selection of sewage treatment plants and exploration of optimal siting areas: A case study in Liao River basin. *Water* **2022**, *14*, 1092 [CrossRef]
8. Hubel, D.H.; Wiesel, T.N. Receptive fields, binocular interaction and functional architecture in the cat's visual cortex. *J. Physiol.* **1962**, *160*, 106. [CrossRef]
9. Raina, R.; Madhavan, A.; Ng, A.Y. Large-scale deep unsupervised learning using graphics processors. In Proceedings of the 26th Annual International Conference on Machine Learning, Montreal, QC, Canada, 14–18 June 2009; pp. 873–880.
10. Ren, S.; He, K.; Girshick, R.; Sun, J. Faster R-CNN: Towards Real-Time Object Detection with Region Proposal Networks *IEEE Trans. Pattern Anal. Mach. Intell.* **2016**, *39*, 1137–1149. [CrossRef]
11. Liu, W.; Anguelov, D.; Erhan, D.; Szegedy, C.; Reed, S.; Fu, C.Y.; Berg, A.C. Ssd: Single shot multibox detector. In Proceedings of the Computer Vision–ECCV 2016: 14th European Conference, Amsterdam, The Netherlands, 11–14 October 2016; Part I 14 Springer: Berlin/Heidelberg, Germany, 2016; pp. 21–37.
12. Lin, T.Y.; Goyal, P.; Girshick, R.; He, K.; Dollár, P. Focal Loss for Dense Object Detection. *arXiv* **2017**, arXiv:1708.02002.
13. Redmon, J.; Divvala, S.; Girshick, R.; Farhadi, A. You Only Look Once: Unified, Real-Time Object Detection. *arXiv* **2015**, arXiv:1506.02640.
14. Redmon, J.; Farhadi, A. YOLO9000: Better, Faster, Stronger. *arXiv* **2016**, arXiv:1612.08242.
15. Redmon, J.; Farhadi, A. Yolov3: An incremental improvement. *arXiv* **2018**, arXiv:1804.02767.
16. Bochkovskiy, A.; Wang, C.Y.; Liao, H.Y.M. Yolov4: Optimal speed and accuracy of object detection. *arXiv* **2020**, arXiv:2004.10934
17. Rogan, J.; Chen, D. Remote sensing technology for mapping and monitoring land-cover and land-use change. *Prog. Plan.* **2004**, *61*, 301–325. [CrossRef]
18. Zhao, S.; Wang, Q.; Li, Y.; Liu, S.; Wang, Z.; Zhu, L.; Wang, Z. An overview of satellite remote sensing technology used in China's environmental protection. *Earth Sci. Inform.* **2017**, *10*, 137–148. [CrossRef]
19. Toth, C.; Jóźków, G. Remote sensing platforms and sensors: A survey. *ISPRS J. Photogramm. Remote. Sens.* **2016**, *115*, 22–36 [CrossRef]
20. Ren, Y.; Zhu, C.; Xiao, S. Small object detection in optical remote sensing images via modified faster R-CNN. *Appl. Sci.* **2018**, *8*, 813. [CrossRef]
21. Chen, J.; Wan, L.; Zhu, J.; Xu, G.; Deng, M. Multi-scale spatial and channel-wise attention for improving object detection in remote sensing imagery. *IEEE Geosci. Remote. Sens. Lett.* **2019**, *17*, 681–685. [CrossRef]
22. Liu, W.; Ma, L.; Chen, H. Arbitrary-oriented ship detection framework in optical remote-sensing images. *IEEE Geosci. Remote Sens. Lett.* **2018**, *15*, 937–941. [CrossRef]
23. Deng, Z.; Sun, H.; Zhou, S.; Zhao, J.; Lei, L.; Zou, H. Multi-scale object detection in remote sensing imagery with convolutional neural networks. *ISPRS J. Photogramm. Remote. Sens.* **2018**, *145*, 3–22. [CrossRef]
24. Lin, T.Y.; Dollár, P.; Girshick, R.; He, K.; Hariharan, B.; Belongie, S. Feature Pyramid Networks for Object Detection. *arXiv* **2016**, arXiv:1612.03144.

25. Yang, X.; Sun, H.; Fu, K.; Yang, J.; Sun, X.; Yan, M.; Guo, Z.J. Automatic ship detection in remote sensing images from google earth of complex scenes based on multiscale rotation dense feature pyramid networks. *Remote Sens.* **2018**, *10*, 132. [CrossRef]
26. Li, Y.; Huang, Q.; Pei, X.; Jiao, L.; Shang, R.J. RADet: Refine feature pyramid network and multi-layer attention network for arbitrary-oriented object detection of remote sensing images. *Remote Sens.* **2020**, *12*, 389. [CrossRef]
27. Yan, J.; Wang, H.; Yan, M.; Diao, W.; Sun, X.; Li, H.J. IoU-adaptive deformable R-CNN: Make full use of IoU for multi-class object detection in remote sensing imagery. *Remote Sens.* **2019**, *11*, 286. [CrossRef]
28. Vaswani, A.; Shazeer, N.; Parmar, N.; Uszkoreit, J.; Jones, L.; Gomez, A.N.; Kaiser, L.; Polosukhin, I. Attention Is All You Need. *arXiv* **2017**, arXiv:1706.03762.
29. Zhu, X.; Lyu, S.; Wang, X.; Zhao, Q. TPH-YOLOv5: Improved YOLOv5 Based on Transformer Prediction Head for Object Detection on Drone-captured Scenarios. *arXiv* **2021**, arXiv:2108.11539.
30. Long, H.; Chung, Y.; Liu, Z.; Bu, S. Object detection in aerial images using feature fusion deep networks. *IEEE Access* **2019**, *7*, 30980–30990. [CrossRef]
31. Yang, X.; Yang, J.; Yan, J.; Zhang, Y.; Zhang, T.; Guo, Z.; Xian, S.; Fu, K. SCRDet: Towards More Robust Detection for Small, Cluttered and Rotated Objects. *arXiv* **2018**, arXiv:1811.07126.
32. Dai, J.; Qi, H.; Xiong, Y.; Li, Y.; Zhang, G.; Hu, H.; Wei, Y. Deformable Convolutional Networks. *arXiv* **2017**, arXiv:1703.06211.
33. Zhu, J.; Fang, L.; Ghamisi, P. Deformable convolutional neural networks for hyperspectral image classification. *IEEE Geosci. Remote Sens. Lett.* **2018**, *15*, 1254–1258. [CrossRef]
34. Guo, H.; Bai, H.; Yuan, Y.; Qin, W. Fully deformable convolutional network for ship detection in remote sensing imagery. *Remote Sens.* **2022**, *14*, 1850. [CrossRef]
35. Li, H.; Zech, J.; Hong, D.; Ghamisi, P.; Schultz, M.; Zipf, A. Leveraging openstreetmap and multimodal remote sensing data with joint deep learning for wastewater treatment plants detection. *Int. J. Appl. Earth Obs. Geoinf.* **2022**, *110*, 102804. [CrossRef] [PubMed]
36. Cai, Z.; Vasconcelos, N. Cascade R-CNN: Delving into High Quality Object Detection. *arXiv* **2017**, arXiv:1712.00726.
37. Zhang, S.; Chi, C.; Yao, Y.; Lei, Z.; Li, S.Z. Bridging the Gap Between Anchor-based and Anchor-free Detection via Adaptive Training Sample Selection. *arXiv* **2019**, arXiv:1912.02424.

sustainability

MDPI

Article

A Convolutional Neural Network for Coastal Aquaculture Extraction from High-Resolution Remote Sensing Imagery

Jinpu Deng [1,2], Yongqing Bai [1], Zhengchao Chen [1], Ting Shen [1], Cong Li [1] and Xuan Yang [1,*]

[1] Airborne Remote Sensing Center, Aerospace Information Research Institute, Chinese Academy of Sciences, Beijing 100094, China
[2] University of Chinese Academy of Sciences, Beijing 100049, China
* Correspondence: yangxuan@radi.ac.cn

Abstract: Aquaculture has important economic and environmental benefits. With the development of remote sensing and deep learning technology, coastline aquaculture extraction has achieved rapid, automated, and high-precision production. However, some problems still exist in extracting large-scale aquaculture based on high-resolution remote sensing images: (1) the generalization of large-scale models caused by the diversity of remote sensing in breeding areas; (2) the confusion of breeding target identification caused by the complex background interference of land and sea; (3) the boundary of the breeding area is difficult to extract accurately. In this paper, we built a comprehensive sample database based on the spatial distribution of aquaculture, and expanded the sample database by using confusing land objects as negative samples. A multi-scale-fusion superpixel segmentation optimization module is designed to solve the problem of inaccurate boundaries, and a coastal aquaculture network is proposed. Based on the coastline aquaculture dataset that we labelled and produced ourselves, we extracted cage culture areas and raft culture areas near the coastline of mainland China based on high-resolution remote sensing images. The overall accuracy reached 94.64% and achieved a state-of-the-art performance.

Keywords: deep learning; negative samples; superpixel optimization; Gaofen-2; semantic segmentation

Citation: Deng, J.; Bai, Y.; Chen, Z.; Shen, T.; Li, C.; Yang, X. A Convolutional Neural Network for Coastal Aquaculture Extraction from High-Resolution Remote Sensing Imagery. *Sustainability* **2023**, *15*, 5332. https://doi.org/10.3390/su15065332

Academic Editor: George P. Kraemer

Received: 25 February 2023
Revised: 15 March 2023
Accepted: 15 March 2023
Published: 17 March 2023

1. Introduction

Seafood is an essential source of nutrition for humans. According to statistics from the Food and Agriculture Organization of the United Nations, the total global seafood production in 2018 was about 179 million tons around the world, of which about 156 million tons were for direct human consumption [1]. China has a vast territory, and its long coastline covers tropical, subtropical, and temperate climate zones [2]. It has unique natural and geographical conditions for aquaculture. The development of aquaculture has created many jobs for people in coastal areas. However, while creating economic benefits, it also brings a series of environmental problems, such as water pollution caused by a large amount of breeding bait and chemical fertilizers [3], eutrophication caused by cultured crop metabolites [4], antibiotic abuse [5], and the impact of the uncontrolled expansion of breeding areas on coastal wetland systems and marine traffic [6]. Therefore, the aquaculture industry's enormous economic value and environmental benefits make it necessary to conduct periodic and rapid supervision and statistics on the aquaculture industry. Limited by the growth cycle of marine organisms [7] and the ecological complexity of water areas [8] traditional field surveys have great difficulties and limitations. The monitoring work of aquaculture must be completed on time and needs to be repeated year on year. Therefore, improving the monitoring capability of aquaculture is of great significance for improving the quality and efficiency of fisheries, improving the supply capacity of aquatic products, and promoting the green development of offshore ecology.

The satellite remote sensing technology developed in the 20th century has provided a large amount of satellite image data to assist in monitoring ground objects due to its wide imaging area and fast data collection cycle [9]. Compared with manual field surveys, it has provided huge benefits. Early studies mainly used related methods such as visual interpretation and geographical information systems to explore breeding areas' spatial and temporal changes from a macro perspective [10–13]. Methods such as the water body index [14–16] and object-oriented classification [17–20] use prior knowledge to separate water from land, and use the similarity characteristics of different regions to improve the interpretation accuracy from a higher level.

The deep-seated value in massive remote sensing data has also been exploited thanks to improved computing power and developed deep learning algorithms [21]. Deep learning algorithms are currently the most widely used automatic information extraction method. Automatic information extraction refers to the automatic extraction of feature information in image data by machines using deep learning methods. Automatic information extraction by deep learning algorithms has gradually replaced manual visual interpretation methods. The deep learning method automatically encodes and extracts image features by stacking many convolutional neural networks without manually designing feature extractors based on expert knowledge [22]. It has better feature extraction capabilities and large-scale generalization capabilities. Semantic segmentation technology in computer vision corresponds to coastal aquaculture information extraction. Semantic segmentation algorithms based on deep learning are mainly divided into four types: encoder–decoder, backbone, hybrid, and transformer. The encoder–decoder network extracts features of different scales through multiple pooling and upsampling [23,24]. The backbone network extracts features of different scales through parallel feature modules [25,26]. The hybrid network fully combines the advantages of the encoder–decoder and backbone types to improve the ability to extract features of different scales [27]. The transformer network uses the self-attention mechanism to suggest the global correlation of features [28]. It avoids the problem of limited receptive fields of local features at different scales.

The convolutional neural network is the most commonly used method in deep learning. The following discusses some studies that used deep learning methods to extract aquaculture features: Liu et al. [29] used Landsat 8 images to extract the main coastal aquaculture areas within 40 km of the shore. Cheng et al. [30] used UNet and dilated convolution to expand the receptive field, solved the problem that the breeding area was easily gridded, and optimized the issue of misidentifying floating objects and sediments on the water surface. Su et al. [31] used the RaftNet to optimize the extraction accuracy of turbid seawater and made the model adaptable to different scales of raft culture areas. However, the extraction of breeding areas often has blurred and glued edges, usually caused by the complexity of the ocean background. The edge enhancement method has a certain optimization effect for extracting aquaculture areas. Ottinger et al. [32] proposed that a fine edge detection method using enhanced linear structure can improve the accuracy of breeding area monitoring. Cui et al. [33] used the UPSNet with the PSE structure to adapt to multi-scale feature maps to extract complex environments and improve the edge blurring and adhesion that often occur in raft culture extraction. Fu et al. [34] extracted aquaculture areas with an automatic labelling method based on convolutional neural networks. They used multi-layer cascaded networks to aggregate multi-scale information captured by dilated convolutions. Furthermore, the channel attention mechanism and spatial attention mechanism modules are used to refine the feature layer to obtain better extraction results, but the network model extraction speed is slow. Cui et al. [35] extracted the raft aquaculture in the Lianyungang based on a fully convolutional neural network. Still, their network could only identify a single raft aquaculture area under a simple seawater background. Lin et al. [36] proposed a method for extracting ocean, land, and ships based on a fully convolutional network. A multi-scale convolutional neural network adapts to the differences in scales of land and sea ships. The idea is also applicable to extraction in aquaculture areas. Feng et al. [37] designed a homogeneous convolutional neural network to extract small-scale aquaculture rafts in the image, sea and land were separated while extracting and obtaining better results on the Gaofen-1 image. Ferriby et al. [38] used a Laplacian

5×5 convolution filter to improve the edge of fish pond extraction results, but it made many pixels appear grey. Zhang et al. [39] proposed a segmentation network NSCT method combined with non-subsampling contour transformation, which can enhance the main contour features of raft culture in the ocean. The SE2Net [40] embedded the self-attention mechanism module in the network and introduced the Laplacian operator to enhance edge information. The simple edge enhancement algorithm has a certain effect on the edge extraction of the breeding area. Still, it cannot solve the problem of a large number of broken edges in a high-detail area. As a result, the breeding area is discontinuous and broken due to the influence of ocean currents, clouds, etc. The extracted results are challenging to form a completely large area from the blurred and disconnected areas.

For aquaculture extraction using remote sensing images, it is necessary to overcome the environmental complexity and species diversity of target objects and adapt to the characteristics of the spatial and temporal differences in the remote sensing images in the study area. Extracting large-scale aquaculture based on high-resolution remote sensing images still has the following problems.

- **Remote sensing diversity of breeding areas.** There are differences in the types of aquaculture in different regions. Cage and raft cultures are different in size, colour, shape, and scale. As a result, the model's generalization ability faces significant challenges in large-scale research areas, and the spatial distribution of samples is an essential research factor.
- **The complex background interference of land and sea.** Although the background of aquaculture is relatively simple in the ocean, there will be complex sea–land interlacing in offshore aquaculture areas. In addition, cages and rafts will also appear in tidal flats and ponds on land. The diversity and comprehensiveness of samples is also a key research factor to avoid aquaculture sea–land interference.
- **The boundaries of breeding areas are difficult to accurately extract.** Because seawater may randomly submerge the edges of cage and raft cultures, the boundaries are not completely straight, and irregular edges will appear. Therefore, affected by complex imaging factors, it is not easy to extract the precise boundaries of breeding areas.

In response to the above problems, we propose a coastal aquaculture network (CANet) to effectively extract large-scale aquaculture areas. The main contributions of this paper are as follows:

- We constructed the sample database from the perspective of the balance of spatial distribution. Considering the differences in the size, colour, and shape of aquaculture areas in diverse regions, representative samples covering each region are labelled. In this way, the model has a good large-scale generalization ability in all areas.
- We expanded the sample database by taking confused land objects as negative samples. For the complex background conditions of land, the target of the land prone to misdetection by the model is labelled as the negative sample. Then, the interference of confusing land objects with aquaculture extraction from land areas is solved.
- We designed a multi-scale-fusion superpixel segmentation optimization module. Considering the problem of inaccurate boundaries of extraction results, we take full advantage of the sensitivity of superpixel segmentation to edge features and the abstraction of features by semantic segmentation networks. In this way, the network effectively optimized the accuracy of boundary identification and improved the overall accuracy of aquaculture extraction.

- Based on 640 scenes of Gaofen-2 satellite images, we extracted cage and raft culture areas near the coastline in mainland China, covering a range of 30 km outward from the coastline. The overall accuracy was satisfactory, and it can support the breeding area and quantity statistics. Compared with other mainstream methods, our proposed CANet achieved state-of-the-art performance.

2. Materials and Methods

In this section, we first propose an overall framework for the deep learning-based aquaculture extraction workflow. The overall workflow includes data acquisition, preprocessing, sample production and iteration, model training and iteration, and product production. Among these, sample production and negative sample iteration are key contributors to the overall workflow. Then, for model training, based on the DeepLabV3+ network, we design a multi-scale feature superpixel optimization method and build a CANet model. CANet is a key contribution to the method design of deep learning models.

2.1. Study Area

The seashore in eastern China stretches from the mouth of the Yalu River in the north to the mouth of the Kunlun River in the south, with a total coastline of more than 18,000 km [41]. Hangzhou Bay roughly bounds the coast of China. To the north of Hangzhou Bay, the coastline passes through several uplift and subsidence zones, presenting an interlaced pattern of rising mountainous harbour coasts and descending plain coasts. To the south of Hangzhou Bay, the coastline continues the same uplift zone, and has relatively consistent characteristics. The average elevation of the coastline is below 500 m, with a temperate marine monsoon climate, a subtropical marine monsoon climate, and a tropical monsoon climate [42,43]. Most coasts are sea erosion coasts, characterized by steep twists and turns and dangerous terrain. For millions of years, complex geological structures, ocean currents, biological effects, and climate conditions have formed many coastal landforms, such as bays and estuaries, providing a sufficient breeding ground for aquaculture [44].

There are more than 200 bays of various sizes in China, with more than 150 bays with an area larger than 10 km^2. Affected by factors such as regional structure, the bays are distributed along the coastal sections with relatively solid dynamics and zigzagging coastlines. Among them, the Zhejiang and Fujian coastal bays have the largest number of distributions, followed by Shandong and Guangdong. Next are Liaoning, Guangxi, and Hainan. The remaining coastal bays of Hebei, Tianjin, Jiangsu, and Shanghai are relatively small in size. We selected the coastlines of the coastal provinces in mainland China as the research area. The selected coastal bays were Liaoning, Hebei, Tianjin, Shandong, Jiangsu, Shanghai, Zhejiang, Fujian, Guangdong, Guangxi, and Hainan from north to south.

2.2. Data Source

We extracted aquaculture based on Gaofen-2 satellite images. The data was sourced from the China Centre for Resources Satellite Data and Application (https://data.cresda.cn/#/2dMap, accessed on 15 March 2023). The Gaofen-2 satellite is equipped with a panchromatic camera with a resolution of 0.8 m and a multispectral camera with a resolution of 3.2 m, with an imaging width of 45 km. The study selected 640 scenes of Gaofen-2 images from 2019 covering the waters outside the coastline and about 30 km in the estuary. The selection of images was determined by the growth cycle of farmed crops and the difficulty of obtaining images. The distribution of marine animal cages does not change with the seasons. However, the optimum temperature for marine plant growth is usually below 19 °C, so the date of the selected image should be near the crops vigorous growth period. In addition, due to the influence of the downdraft in the coastal zone, the coastal area is often covered by thick clouds throughout the year, especially the southern area during summer, resulting in fewer available images in summer. The temporal resolution of Gaofen-2 is limited, making it very difficult to obtain high-quality, low-cloud images that completely cover the entire coastline in a short period. Therefore, the proportion of selected images distributed from October to March is relatively large, while the number from June to September is relatively small. The seawater temperature used for mariculture varies very little throughout the year, and aquaculture can be carried out all year round. There is no significant change in the aquaculture area between summer and winter, so using images from other months is acceptable. Even so, some areas are still completely covered by clouds These area were supplemented or replaced with images from the same period in 2018 and 2020. Furthermore, there is an overlap between the images as clouds cover obscured many images. However, these unobstructed images are still usable.

2.3. Overall Framework

We first selected the high-resolution image data according to the growth cycle of marine animals and plants, image quality, and cloud cover to obtain the original image set. In the data preprocessing stage, the original image underwent three steps of de-hazing/colour restoration, fusion, and registration to obtain a high-quality image set Then, a small number of representative images were selected from the high-quality image set, and combined with prior knowledge and feature analysis, visual interpretation was performed to obtain the primary samples. The next step was to build a computer vision-based network model and use the primary samples to train a primary model and evaluate its accuracy. Next, the primary model was used to predict images of the surrounding area of the primary sample, and obtain corresponding prediction results. We performed human visual judgment on the prediction results. The correct prediction results were added to the sample set. The results with certain flaws were added to the sample set after being manually labelled and modified. The results of the wrong predictions were added to the sample set as a negative sample to expand the sample set. We used the new sample set to retrain the model and again predict images of the surrounding area of the sample set, to realize multiple rounds of sample and model iteration. Finally, we obtained a larger sample set and a more stable advanced model. We used the advanced models to predict images of the whole study area and evaluate the accuracy of the prediction results. Since the accuracy of the model can be improved with multiple iterations of the sample set, we can set an accuracy expectation value according to the task requirements, computing resources, and time conditions. If the accuracy evaluation does not meet expectations, we can optimize the model structure and training strategy to retrain the sample set. If the accuracy meets expectations, we can obtain three major achievements: the cage and raft culture dataset, coastline aquaculture area extraction products, and the coastline aquaculture area automatic extraction model. The overall framework of this study is shown in Figure 1.

2.4. Dataset

2.4.1. Data Processing

The raw Gaofen-2 images must undergo a processing series before they can be used Seawater showed low radiation intensity in the image. The image's original information needs to be enhanced, especially to identify breeding areas in marine areas. The images covered a large span of time and space. Data processing ensured that the processed data had decent consistency.

First, the raw images generally have a bluish cast. Second, due to the weather, mist exists in many images. In addition, the blue band of some images is missing or damaged, which distorts image colour. Before the images can be input into the neural network, we must precisely register the raw images between multiple periods in the same area between different scenes. It is also necessary to ensure that the images are rich in colour, moderate in saturation, and have good visibility to facilitate visual interpretation and machine learning. The image is then de-hazed, and its colour restored in the form of batch processing, and the resolution of its multispectral bands is improved by image fusion from the panchromatic band.

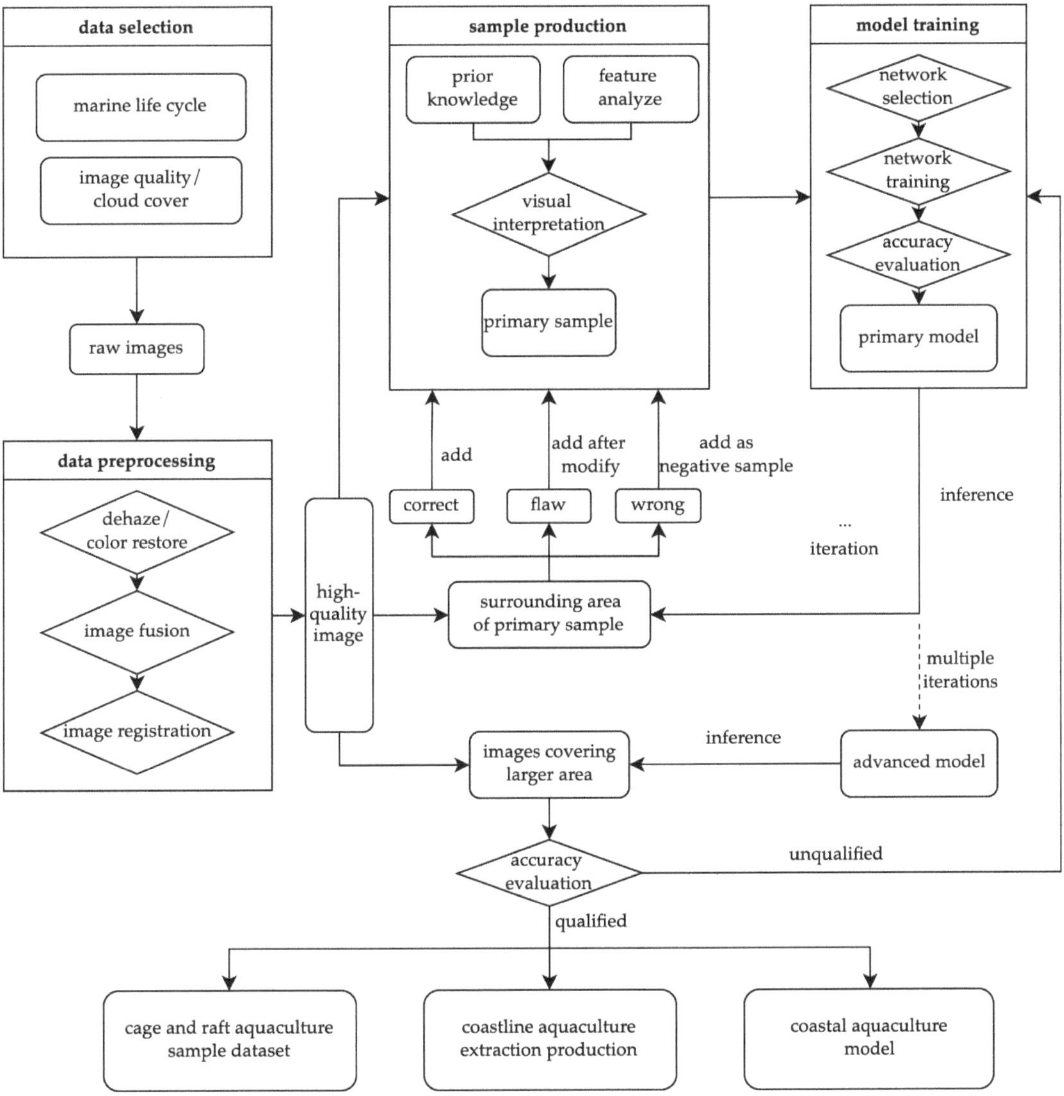

Figure 1. The overall framework of coastal aquaculture extraction.

2.4.2. Samples

The production of samples needs to consider the remote sensing diversity of the breeding area and the complex background interference of the land and sea. The species of cultured organisms located in different regions are different for aquaculture cages. The scale and materials of the breeding cages vary in scale, size, colour, and shape, and various forms and distribution patterns appeared in the remote sensing images. Marine plants in the images will show their unique seasonal growth cycles for aquaculture rafts. This makes breeding areas in different phases in the same area display periodic changes in colour, texture, and structure, and the original biological characteristics will disappear after harvesting. Therefore, we need to consider the representativeness and comprehensiveness of the sample from the distribution of geographical factors.

We selected sea areas with an average water depth of 2∼15 m and an average annual seawater velocity of less than 3 m/s, including tropical, subtropical, and temperate climate zones in the range of 18° N ∼ 45° N latitude. This was combined with the fishery statistics in each province's aquaculture production and large-scale marine pasture areas, after removing important marine transportation hubs such as ports. The original primary samples were labelled by human visual interpretation. All samples consisted of a remote sensing image and corresponding ground truth label. As shown in Table 1, most of the selected areas were bays, where the seawater depth is moderate, the flow rate is slow, and the area is well-lit, which is suitable for aquaculture. These areas represent the differences in breeding areas between the north and south, and cover multiple dimensional and climate zones, ensuring the comprehensiveness and diversity of the samples. In addition, assuming sufficient images for the same area, we tried to obtain multi-phase images for samples where possible. In this way, the differences caused by the remote sensing data in terms of time and irradiation conditions can be resolved. Therefore, we also considered the differences in the images caused by different lighting and seasonal factors and images of different quality when selecting samples.

Table 1. The latitude, average water depth, climate, and average annual sunshine hours of the bays used for sample labelling.

Area	Latitude	Average Water Depth	Climate	Average Annual Sunshine Hours
Jinzhou Bay in Liaoning	39° N	3∼9 m	temperate monsoon climate	2200∼3000
Sanggou Bay & Allen Bay in Shandong	37° N	5∼10 m	temperate marine monsoon climate	2200∼3000
Haizhou Bay in Jiangsu	34° N	5∼10 m	temperate monsoon climate	2200∼3000
Sansha Bay in Fujian	26° N	>10 m	subtropical monsoon climate	2200∼3000
Zhenhai Bay in Guangdong	21° N	5∼7 m	subtropical monsoon climate	2200∼3000
Leizhou Bay in Guangdong	20° N	5∼7 m	tropical monsoon climate	1400∼2000
Qinzhou Bay in Guangxi	20° N	2∼18 m	subtropical marine monsoon climate	2400∼2600
South Bay in Hainan	18° N	2∼10 m	tropical marine monsoon climate	2400∼2600

Although the marine environment will change due to waves, sea winds, ships, marine life, etc., it can still be considered a relatively simple background compared to the complex surface environment on land. Ground objects present the phenomena of same object with a different spectrum and different object with the same spectrum in remote sensing data There will be a large number of images containing ground objects that are easily confused with aquaculture targets. However, the terrestrial environment cannot be completely discarded when extracting the aquaculture area because many aquaculture cages and rafts are distributed in tidal flats and ponds on land. Therefore, in the land area and complex background area of the coastline, we labelled some ground features with similar characteristics or as easily confused features. We used these as negative samples to assist in model training, thereby reducing the model's false detection rate and improving model accuracy. Figure 2 shows some examples of the sample database.

(a) Cage Culture (b) Raft Culture (c) Background

Figure 2. Sample database visualization for (**a**) cage culture, (**b**) raft culture, and (**c**) background.

2.5. Coastal Aquaculture Network

2.5.1. Baseline

To maximize the advantages of the encoder–decoder and backbone networks simultaneously, we choose the hybrid network DeepLabV3+ [27] as the baseline network architecture. The DeepLabV3+ network adopts the encoder–decoder structure to fuse the multi-scale information of the network to meet the feature extraction requirements of breeding areas of different sizes. However, encoder–decoder networks usually use pooling operations to increase receptive fields and aggregate features. This leads to the problem of a decrease in the resolution of the features. Although the upsampling is restored to the decoder's original size, the features' details is still irreversibly lost, resulting in decreased accuracy. DeepLabV3+ adopts a dilated convolution operation commonly used in backbone networks to increase the receptive field while avoiding the reduction in feature resolution and the loss of feature information. Keeping the feature resolution unchanged causes difficulties in feature aggregation. To solve this problem, DeepLabV3+ adopts a multi-scale feature pyramid structure, encoding the features with dilated convolutions with different dilation rates, and fuses the multi-scale information of the features. In addition, the encoder adopts a depth-wise separable convolution to improve the speed of network feature extraction.

2.5.2. Superpixel Optimization

In the multi-scale feature pyramid structure of DeepLabV3+, five parallel feature aggregation encoding operations are performed on the feature map. These operations include 1×1 convolution, dilated convolutions at rates of 6, 12, and 18, and global pooling operations. The encoded features, in addition to fusion and upsampling according to the original DeepLabV3+ network, also construct five feature copies. After the five feature copies are fused, the simple linear iterative clustering (SLIC) algorithm [45] is used for the unsupervised segmentation of high-level features.

In the superpixel segmentation module of the network, the feature maps are converted into approximate, compact, and uniform superpixel blocks for aggregation. The specific implementation process of the SLIC algorithm is as follows:

After dimensionality reduction, the feature map is treated as an RGB colour space image and converted to a CIELab colour space. The LAB colour space consists of three components. L represents the brightness value of a pixel. The value ranges from 0 to 100, where 0 represents pure black, and 100 represents pure white. A represents the relative value between green and red. The value ranges from -128 to 127, where negative values represent the green range and positive values represent the red range. B represents the relative value between yellow and blue. The value ranges from -128 to 127, where negative values represent the blue range, and positive values represent the yellow range. Then, we can obtain a wider colour gamut from the input feature map. Finally, we can use a five-dimensional vector composed of L, a, b, x, y to represent each pixel of the feature map [45]. Among these, L, A, and B correspond to the three components in the colour space, and x and y correspond to the relative coordinates of the pixel in the feature map.

The superpixel centre is first determined. The N number of superpixel blocks is specified, and n reference points are generated in the map according to the proportion of the feature map. If the number of pixels in the entire image is M, the size of the pre-segmented superpixel blocks in the image is M/N pixels, and the side length of each superpixel block is $S = \sqrt{M/N}$. The gradient of all pixels in the surrounding 3×3 range is calculated with the superpixel reference point as the centre. The centre of the superpixel is then moved to the minimum gradient value in the range, which is the centre point of the superpixel. The formula of the gradient is defined as:

$$G(x,y) = (V(x+1,y) - V(x-1,y)^2 + (V(x,y+1) - V(x,y-1)^2 \quad (1)$$

$$V_i = [L_i, a_i, b_i, x_i, y_i]^{\mathrm{T}} \quad (2)$$

A class label is then assigned to each pixel within the 2S range of the superpixel centre point. For each pixel, the colour distance D_c and space distance D_s between the pixel and the centre point is calculated:

$$D_c = \sqrt{(L_i - L_j)^2 + (a_i - a_j)^2 + (b_i - b_j)^2} \tag{3}$$

$$D_s = \sqrt{(x_i - x_j)^2 + (y_i - y_j)^2} \tag{4}$$

$$D = \sqrt{(\frac{D_c}{N_c})^2 + (\frac{D_s}{N_s})^2} \tag{5}$$

where N_c is the maximum colour distance, representing the gap between two colours. In this study, N_c takes a fixed constant of 15. N_s represents the maximum intra-class space distance, and the value of N_s in this study is the side length S of the superpixel. Multiple centre points around the pixel will search each pixel. When distance D between the point and a certain centre point is minimum, the centre point will be the cluster centre of the superpixel block.

The integrity of superpixel blocks is subsequently optimized. The above steps are then iterated through until the cluster centre of each superpixel remains constant. If a superpixel size is too small, it will merge into other adjacent superpixel blocks in this area Figure 3 shows the results when a different number of superpixel blocks are set in an image

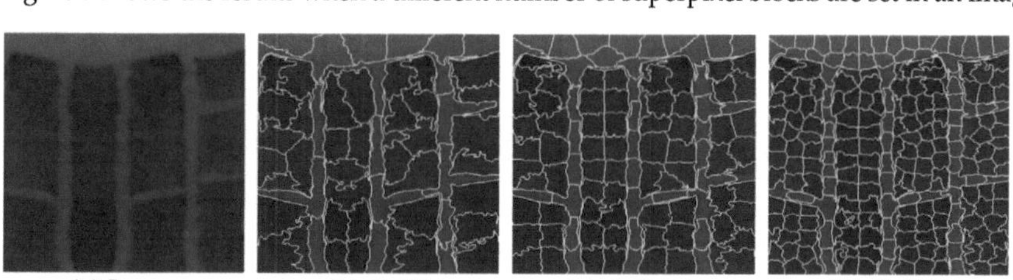

(a) Image (b) Cell64 (c) Cell128 (d) Cell256

Figure 3. The visualized results when a different number of superpixel blocks are set in an image. (**a**) Remote sensing image. (**b**) Superpixel blocks set as 64. (**c**) Superpixel blocks set as 128 (**d**) Superpixel blocks set as 256.

2.5.3. Network Architecture

The semantic segmentation results obtained based on the baseline network have completely classified semantic information, but the boundary accuracy is relatively poor The result of superpixel segmentation based on high-level features has accurate boundary information. However, due to the limitations of unsupervised segmentation, no semantic information is gathered. Therefore, each superpixel block cannot be automatically classified We use the semantic information of semantic segmentation to vote on superpixel blocks The dominant class in each superpixel block is used as the classification category of the whole superpixel block. After this, we obtain precise optimization of the classification boundary details. We call this network architecture a coastal aquaculture network (CANet) Figure 4 shows the network detail diagram of CANet.

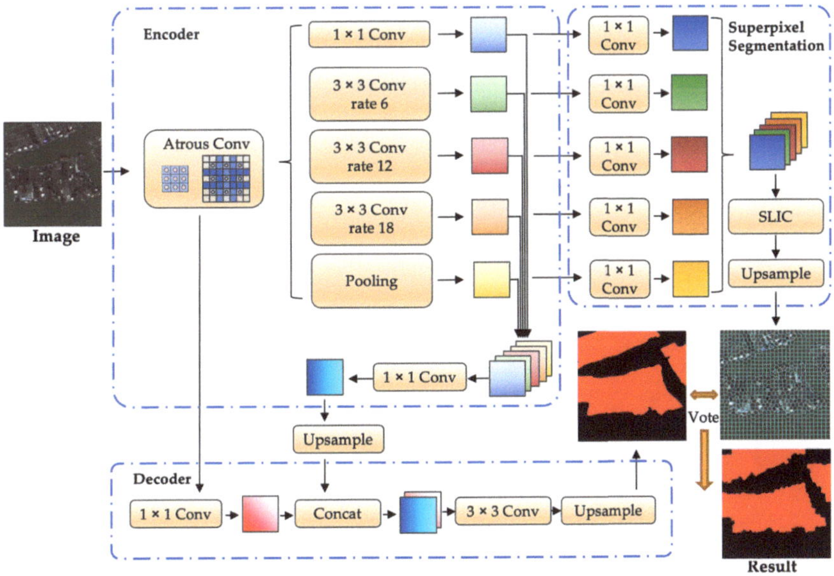

Figure 4. The network detail diagram of the coastal aquaculture network.

2.5.4. Loss Function

To calculate the loss value of the CANet, as the learning motivation of the network, it is necessary to normalize the model's output with the softmax function and use cross-entropy as the loss function. The cross-entropy formula for multi-classification problems is as follows:

$$L = -\sum_{i=1}^{n} y^{(i)} \cdot \log \hat{y}^{(i)} \tag{6}$$

$$\hat{y} = P(\hat{y} = i|x) = \frac{e^{X(i)}}{\sum_{j=0}^{n} e^{X(j)}} \tag{7}$$

For the above formula, $\hat{y}$ is the conditional probability when the pixel label is a certain category. Single-category information uses one-hot encoding as the label input, and the input vector is normalized and exponentially transformed using the softmax function. Then we obtain the predicted probability of each category. The probability values are all non-negative and the sum to 1.

2.6. Training Settings

We experimentally verified CANet based on the PyTorch deep learning framework [46]. We used the ResNet-101 [47] pre-trained model to initialize the network encoder, the initial learning rate was set to 0.001, and a warm-up strategy was used to optimize the learning rate. AdamW [48] was used as the optimizer for training, the weight decay was set to 0.0001, and the momentum was set to 0.9. We used four NVIDIA TITAN Xp GPUs and set the batch size to 32.

2.7. Evaluation Metrics

We used the F1 score to evaluate the extraction accuracy of cages, rafts, and the background, and used the mean F1 score and mean intersection over union (IoU) values to evaluate the overall accuracy. The F1 score can balance the missed and false detection of extracted results with the accuracy of the image, and the IoU is a more intuitive and universal metric.

The formula for the F1 score is as follows:

$$\mathrm{F1} = 2 \cdot \frac{Precision \cdot Recall}{Precision + Recall} \tag{8}$$

where

$$Precision = \frac{TP}{TP + FP} \tag{9}$$

$$Recall = \frac{TP}{TP + FN} \tag{10}$$

where TP is the number of pixels that are classified as breeding areas and are correct, FP is the number of pixels that are classified as breeding areas but are incorrect, and FN is the number of pixels that are not classified as breeding areas but are actually breeding areas. It is known that $TP + FP$ is the total number of pixels classified as breeding areas, and $TP + FN$ is the actual number of pixels that cover the breeding area.

The formula for the IoU is as follows:

$$\mathrm{IoU} = \frac{TP}{TP + FP + FN} \tag{11}$$

The mean F1 score and mean IoU represent the average F1 score and IoU for each category, respectively.

3. Experimental Results

3.1. Ablation Study

To evaluate the performance of the negative sample technology, superpixel optimization module, and CANet, we gradually added various modules and methods proposed in this paper to the baseline network DeepLabV3+ and conducted experiments on the coastline aquaculture dataset produced in this paper. First, we used the baseline network DeepLabV3+ for training using it as a benchmark for comparison. Then the negative sample technology was added based on the baseline to test the effect of the negative sample technology on aquaculture extraction. Next, a superpixel optimization module was added based on the baseline to check the module's ability to extract aquaculture details. Finally, based on the baseline, the negative sample technology and superpixel optimization module were added to verify the aquaculture extraction performance of the proposed CANet.

It can be seen from Table 2 that when using the baseline network, the overall accuracy mean F1 is 91.98%. When adding confusing negative samples as target resistance in the sample database, the overall accuracy rose to 93.22%. The single-category accuracy of aquaculture cages and rafts also increased by over 1%. After adding the superpixel optimization module to the baseline network, the overall accuracy reached 92.94%. The superpixel optimization module optimizes the extraction result details, so the accuracy improvement is smaller than the negative sample technology. After adding negative samples and superpixel optimization modules simultaneously, the overall accuracy reached 94.64% Additionally, the accuracy of breeding cages and raft identification further improved. The quantitative accuracy comparison shows that the proposed CANet can significantly improve the accuracy of aquaculture extraction with the assistance of thee negative sample technology and superpixel optimization module.

Table 2. Quantitative comparison of the accuracy of ablation learning in different modules of the CANet. "Baseline" represents the baseline DeepLabV3+; "+ns" represents the negative sample technology; "+sp" represents the superpixel optimization module; "+ns+sp" represents our proposed CANet with the negative sample technology and superpixel optimization module. Bold values indicate best precision.

Methods	Background	Cage Culture	Raft Culture	Mean F1	Mean IoU
baseline	92.71	90.72	92.51	91.98	88.66
+ns	93.91	91.91	93.85	93.22	89.86
+sp	93.39	91.76	93.66	92.94	89.78
+ns+sp	**95.49**	**92.55**	**95.87**	**94.64**	**90.91**

Figure 5 is the visualization result obtained using the above four methods to perform model inference on the sample dataset, used to intuitively evaluate the performance of the negative sample technology and superpixel optimization module. In the first row, due to the significant colour difference and small distance between one breeding cage and the other cages, the result of the baseline network connects the two cages together; thus, a false detection occurs. After adding the negative sample technology, using negative samples for adversarial learning effectively distinguished the background area. In the second row, false detections occurred in the culture cages on the left, and were effectively resolved by adding the negative sample technology. After introducing superpixel optimization, the edge details of the contour of the culture raft on the right were more accurate and richer. In the third row, small false detections are present for slender culture cages after introducing the negative sample technique. However, using the superpixel optimization module reduced false detection. In the fourth row, the boundary of the culture cage label is relatively smooth, but in the remote sensing image the boundary is relatively rough. In this complex situation, both the baseline network and negative sample technology have a large number of missing detections. Combining the superpixel optimization module and negative sample technology can effectively solve the problem of missed detection and ensure that the boundary is more in line with the original image. In summary, CANet, which integrates negative sample technology and the superpixel optimization module, performs excellently in coastal aquaculture extraction tasks.

3.2. Comparing Methods

We selected some mainstream semantic segmentation networks for comparison to verify whether our proposed CANet network can achieve state-of-the-art performance. The following introduces the mainstream models used for comparison: UNet [23], a classic encoder–decoder network using a symmetrical U-shape structure and skip-layer connections to fuse deep semantic information with shallow spatial information. DeepLabV3 [49], a classic backbone network using different dilation convolution rates to solve multi-scale information extraction problems. FPN [50] uses a feature pyramid structure to fuse the semantic information of features of different scales. PAN [51] uses the attention mechanism to accurately filter effective feature information. PSPNet [26] uses the spatial pooling pyramid structure to solve multi-scale feature aggregation problems. Our proposed CANet integrates the negative sample technique and superpixel optimization module.

It can be seen from Table 3 that PSPNet has the highest overall accuracy compared with the other mainstream networks, and the mean F1 reaches 93.28%, while the proposed CANet reaches 94.64%. The background category CANet exceeds PSPNet by about 1.2%, the culture cage category exceeds PAN by about 0.4%, and the culture raft exceeds PSPNet by about 1.4%. This shows that the proposed CANet surpasses the other mainstream semantic segmentation networks from the perspective of quantitative comparison and reaches state-of-the-art performance.

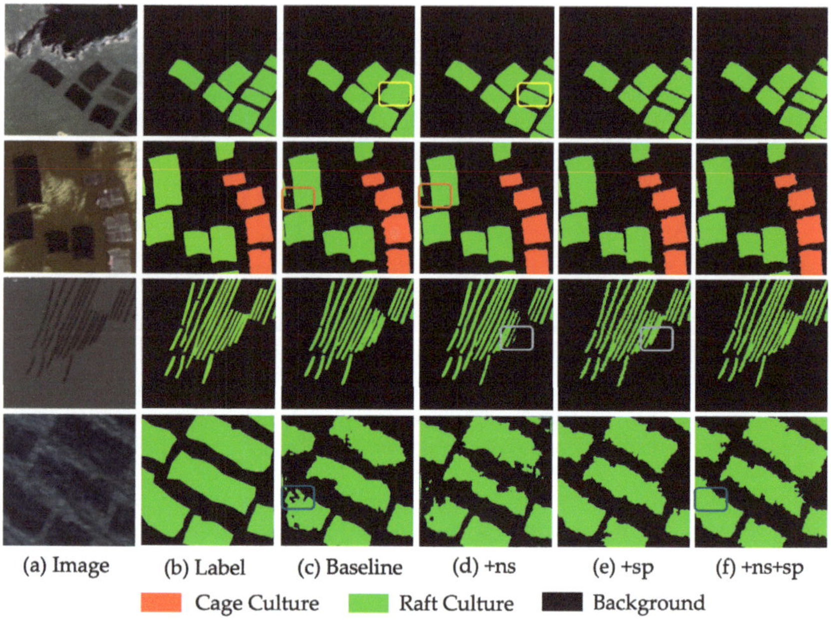

Figure 5. Visual comparison of the results of ablation learning of the different modules of the CANet (**a**) Remote sensing image. (**b**) Label for coastal aquaculture. Results of (**c**) the baseline DeepLabV3+. (**d**) baseline with the negative sample technology, (**e**) baseline with the superpixel optimization module, and (**f**) baseline with the negative sample technology and superpixel optimization module (proposed CANet).

Table 3. Quantitative comparison of the accuracy between our proposed CANet and other mainstream networks. Bold values indicate best precision.

Methods	Background	Cage Culture	Raft Culture	Mean F1	Mean IoU
UNet	92.38	89.66	92.71	91.58	86.77
DeepLabV3	92.77	89.55	92.97	91.76	86.82
FPN	93.09	86.19	94.08	91.12	85.88
PAN	93.66	92.17	92.66	92.83	87.92
PSPNet	94.21	91.15	94.47	93.28	88.43
CANet	**95.49**	**92.55**	**95.87**	**94.64**	**90.91**

Figure 6 compares the visualization results of CANet and other mainstream semantic segmentation networks for aquaculture extraction. In the first row, the FPN produces many false detections, and the other mainstream networks possess false concatenations and small false detections, while the CANet performs better. In the second row, almost all the other mainstream networks have certain false detections, while the extraction results of the proposed CANet are more accurate. In the third row, there are many missed detections in the FPN. PSPNet mistakenly connects breeding cages into one piece, while the result of proposed CANet is closest to the ground truth of the label. In the fourth row, the aquaculture cages and background in the original image are very complicated. Other mainstream networks struggle to effectively extract the aquaculture cages. PSPNet can extract a few, but there are still many missed detections. However, CANet can effectively extract all breeding cages, and the edge is more in line with the actual image. In summary, our proposed CANet achieved the best performance.

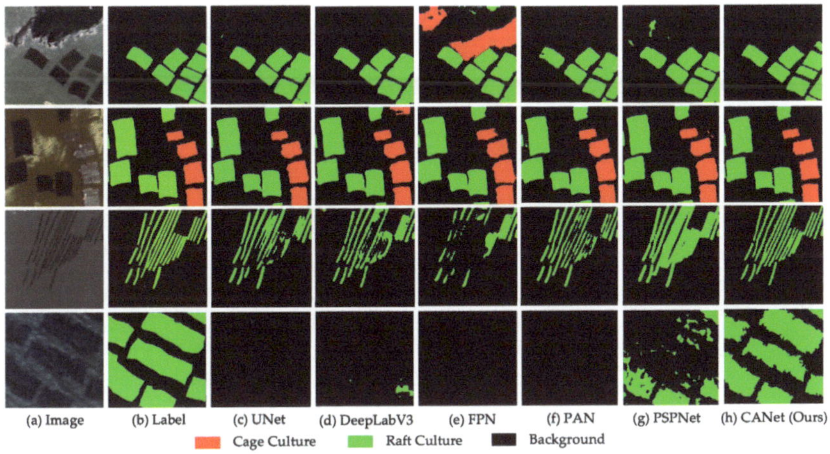

Figure 6. Visual comparison of the results between our proposed CANet and the other mainstream networks. (**a**) Remote sensing images. (**b**) Labels for coastal aquaculture. The results of (**c**) UNet, (**d**) DeepLabV3, (**e**) FPN, (**f**) PAN, (**g**) PSPNet, and (**h**) our proposed CANet.

3.3. Large-Scale Mapping and Statistics

We use the trained CANet model with the best performance to automatically extract aquaculture from Gaofen-2 satellite data covering 640 coastal scenes of mainland China. After this, we obtained a regional distribution map of the cage and raft cultures within 30 km of the coast, as shown in Figure 7. Based on the coastal aquaculture map, we calculated statistics to obtain the area of breeding areas and the number of breeding targets in each province, as shown in Table 4.

Table 4. The area of aquaculture areas and the number of aquaculture targets in each province.

Province	Aquaculture Area (km^2)			Number of Aquaculture Targets		
	Cage	Raft	Total	Cage	Raft	Total
Liaoning	3.99	671.20	675.19	947	51,285	52,232
Hebei & Tianjin	0.07	0.01	0.08	92	37	129
Shandong	6.79	564.16	570.95	2906	59,643	62,549
Jiangsu	2.29	653.93	656.21	2018	88,045	90,063
Zhejiang & Shanghai	1.89	54.78	56.67	1708	9552	11,260
Fujian	54.46	462.28	516.74	21,990	79,348	101,338
Guangdong	42.41	189.43	231.84	46,301	115,965	162,266
Guangxi	24.61	18.21	42.82	21,992	15,466	37,458
Hainan	3.13	0.73	3.86	2590	577	3167

3.4. Discussion

In this study, samples were created based on the panchromatic-fused Gaofen-2 images, and a rapid extraction method for two major marine aquaculture areas near the coast was explored based on high-resolution images. Compared with the Sentinel image, the Gaofen-2 images have a higher spatial resolution. The method improved the edge extraction and accuracy of the breeding area to varying degrees, providing a new benchmark method for the fast and high-precision extraction of small areas. Zhangzidao island, located in Liaoning Province, has been previously investigated for fraudulent aquaculture operations. The investigation used the Beidou navigation system to analyse the fishing operations of its fishing boats for several years. The actual production of aquaculture in Zhangzidao island can be evaluated with the help of single-phase high-resolution images using the method proposed in this work. Regarding data processing and interpretation efficiency, the advantages of using deep learning methods to extract aquaculture areas rapidly are demonstrated.

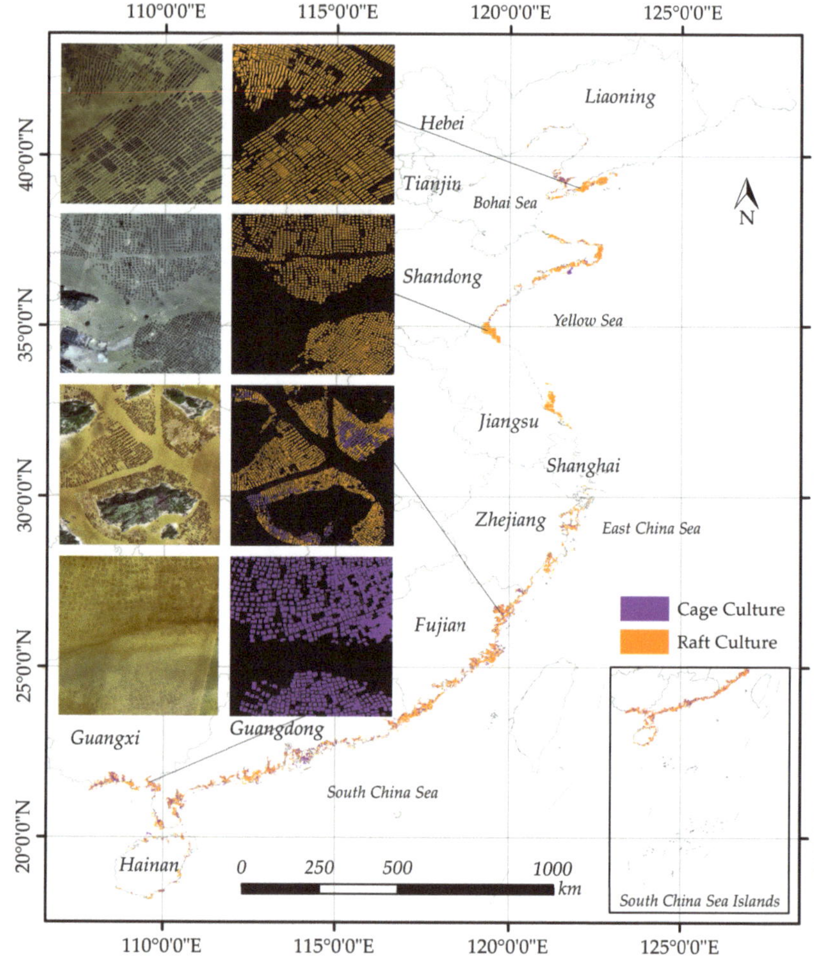

Figure 7. The coastal aquaculture mapping of the coastline in mainland China.

However, optical remote sensing images still have limitations [9]. Due to the characteristics of its sensors, the Gaofen-2 images have limited visibility to water bodies. The temporal resolution also results in a large amount of cloud coverage. For areas of turbulent sea, images may be jagged or contain disconnected phenomena in marine plant breeding areas [33], which will cause errors in the extraction results. Our superpixel optimization method can solve the impact of small-area disconnected problems. However, low-latitude regions with abundant clouds and rain often face problems of cloud coverage, making it difficult to obtain large-area images of the same time phase. Therefore, although the extraction of aquaculture areas by this method can maintain good accuracy, the actual extraction results will somewhat differ from human field survey data. Human field survey data depends on the date of data collection, whereas the automatic extraction algorithm depends on the imaging date. Although aquaculture in different periods does not change based on time or seasons, small changes cannot be avoided due to the influence of tides. Moreover, the model's accuracy could not reach 100%, so there are slight difference between the automatically extracted aquaculture area and human field surveys. The proposed CANet model avoids the influence of location and time on the extraction results as much as possible by labelling samples of different locations and dates, achieving strong and robust generalization capabilities. Compared with other deep learning-based aquaculture extraction research [30,31,33–37], our method adopts the negative sample technique and multi-scale superpixel optimization method, which has a stronger generalization ability and robustness across time and location. Therefore, using the proposed CANet achieves good extraction results in large-scale aquaculture extraction, providing technical support for the automation of fishery resource censuses and the sustainable development of marine resources.

4. Conclusions

In this work, we proposed a convolutional neural network for coastal aquatic extraction from high-resolution remote sensing images. We constructed a sample database balancing spatial distribution and solving the model generalization problem for large-scale aquaculture extraction. We expanded the sample database by using confused land features as negative samples, thus solving the interference of confusing land features on aquaculture extraction. We designed a multi-scale-fusion superpixel segmentation optimization module based on the baseline DeepLabV3+, and designed the CANet network architecture. CANet effectively optimizes boundary identification and improves the overall accuracy of aquaculture extraction. Based on CANet, we extracted cage and raft culture areas near the coastline of mainland China with an overall accuracy of 94.64%, reaching state-of-the-art performance. The results obtained in this work can provide scientific, technical, and data support for the spatial planning and regulation of China's coastal fisheries. In future research, we will aim to introduce more intensive time-series images to explore the relationship and laws between aquaculture and seawater flow, climate, economy, and other factors.

Author Contributions: Conceptualization, J.D. and Z.C.; methodology, J.D. and X.Y.; software, X.Y.; validation, J.D. and Y.B.; formal analysis, J.D. and X.Y.; investigation, J.D.; resources, C.L.; data curation, J.D. and Y.B.; writing—original draft preparation, J.D.; writing—review and editing, X.Y.; visualization, J.D. and Y.B.; supervision, T.S.; project administration, Z.C. and Y.B.; funding acquisition, Z.C. All authors have read and agreed to the published version of the manuscript.

Funding: This research was supported by the Strategic Priority Research Program of the Chinese Academy of Sciences under Grant No. XDA23100304.

Institutional Review Board Statement: Not applicable.

Informed Consent Statement: Not applicable.

Data Availability Statement: Not applicable.

Acknowledgments: The authors thank the editors and anonymous reviewers for their valuable comments, greatly improving the quality of the paper.

Conflicts of Interest: The authors declare no conflict of interest.

Abbreviations

The following abbreviations are used in this manuscript:

CANet	coastal aquaculture network
SLIC	simple linear iterative clustering
F1	F1 score
TP	true positive
FP	false positive
FN	false negative
IoU	intersection over union
ns	negative sample technology
sp	superpixel optimization technology

References

1. Food and Agriculture Organization of the United Nations. *The State of World Fisheries and Aquaculture 2020: Sustainability in Action*; Food and Agriculture Organization of the United Nations: Rome, Italy, 2020.
2. Hao, J.; Xu, G.; Luo, L.; Zhang, Z.; Yang, H.; Li, H. Quantifying the relative contribution of natural and human factors to vegetation coverage variation in coastal wetlands in China. *Catena* **2020**, *188*, 104429. [CrossRef]
3. Rico, A.; Van den Brink, P.J. Probabilistic risk assessment of veterinary medicines applied to four major aquaculture species produced in Asia. *Sci. Total. Environ.* **2014**, *468*, 630–641. [CrossRef] [PubMed]
4. Zheng, Y.; Jin, R.; Zhang, X.; Wang, Q.; Wu, J. The considerable environmental benefits of seaweed aquaculture in China. *Stoch. Environ. Res. Risk Assess.* **2019**, *33*, 1203–1221. [CrossRef]
5. Gao, Q.; Li, Y.; Qi, Z.; Yue, Y.; Min, M.; Peng, S.; Shi, Z.; Gao, Y. Diverse and abundant antibiotic resistance genes from mariculture sites of China's coastline. *Sci. Total. Environ.* **2018**, *630*, 117–125. [CrossRef] [PubMed]
6. Ottinger, M.; Clauss, K.; Kuenzer, C. Aquaculture: Relevance, distribution, impacts and spatial assessments—A review. *Ocean. Coast. Manag.* **2016**, *119*, 244–266. [CrossRef]
7. Zhang, J.; Wu, W.; Ren, J.S.; Lin, F. A model for the growth of mariculture kelp Saccharina japonica in Sanggou Bay, China. *Aquac. Environ. Interact.* **2016**, *8*, 273–283. [CrossRef]
8. Maiti, S.; Bhattacharya, A.K. Shoreline change analysis and its application to prediction: A remote sensing and statistics based approach. *Mar. Geol.* **2009**, *257*, 11–23. [CrossRef]
9. Campbell, J.B.; Wynne, R.H. *Introduction to Remote Sensing*; Guilford Press: New York, NY, USA, 2011.
10. Karthik, M.; Suri, J.; Saharan, N.; Biradar, R. Brackish water aquaculture site selection in Palghar Taluk, Thane district of Maharashtra, India, using the techniques of remote sensing and geographical information system. *Aquac. Eng.* **2005**, *32*, 285–302 [CrossRef]
11. Kapetsky, J.M.; Aguilar-Manjarrez, J. *Geographic Information Systems, Remote Sensing and Mapping for the Development and Management of Marine Aquaculture*; Number 458; Food & Agriculture Organization: Rome, Italy, 2007.
12. Seto, K.C.; Fragkias, M. Mangrove conversion and aquaculture development in Vietnam: A remote sensing-based approach for evaluating the Ramsar Convention on Wetlands. *Glob. Environ. Chang.* **2007**, *17*, 486–500. [CrossRef]
13. Saitoh, S.I.; Mugo, R.; Radiarta, I.N.; Asaga, S.; Takahashi, F.; Hirawake, T.; Ishikawa, Y.; Awaji, T.; In, T.; Shima, S. Some operational uses of satellite remote sensing and marine GIS for sustainable fisheries and aquaculture. *ICES J. Mar. Sci.* **2011**, *68*, 687–695. [CrossRef]
14. Zhang, T.; Yang, X.; Hu, S.; Su, F. Extraction of coastline in aquaculture coast from multispectral remote sensing images: Object-based region growing integrating edge detection. *Remote Sens.* **2013**, *5*, 4470–4487. [CrossRef]
15. Zhang, X.; Ma, S.; Su, C.; Shang, Y.; Wang, T.; Yin, J. Coastal oyster aquaculture area extraction and nutrient loading estimation using a GF-2 satellite image. *IEEE J. Sel. Top. Appl. Earth Obs. Remote Sens.* **2020**, *13*, 4934–4946. [CrossRef]
16. Sun, Z.; Luo, J.; Yang, J.; Yu, Q.; Zhang, L.; Xue, K.; Lu, L. Nation-scale mapping of coastal aquaculture ponds with sentinel-1 SAR data using google earth engine. *Remote Sens.* **2020**, *12*, 3086. [CrossRef]
17. Zhu, H.; Li, K.; Wang, L.; Chu, J.; Gao, N.; Chen, Y. spectral characteristic analysis and remote sensing classification of coastal aquaculture areas based on GF-1 data. *J. Coast. Res.* **2019**, *90*, 49–57. [CrossRef]
18. Kang, J.; Sui, L.; Yang, X.; Liu, Y.; Wang, Z.; Wang, J.; Yang, F.; Liu, B.; Ma, Y. Sea surface-visible aquaculture spatial-temporal distribution remote sensing: A case study in Liaoning province, China from 2000 to 2018. *Sustainability* **2019**, *11*, 7186. [CrossRef]
19. Du, Y.; Wu, D.; Liang, F.; Li, C. Integration of case-based reasoning and object-based image classification to classify SPOT images: A case study of aquaculture land use mapping in coastal areas of Guangdong province, China. *Gisci. Remote Sens.* **2013**, *50*, 574–589. [CrossRef]
20. Wei, Z. Analysis on the Relationship between Mangrove and Aquaculture in Maowei Sea Based on Object-Oriented Method. *E3S Web Conf.* **2020**, *165*, 03022. [CrossRef]
21. Zhang, B.; Chen, Z.; Peng, D.; Benediktsson, J.A.; Liu, B.; Zou, L.; Li, J.; Plaza, A. Remotely sensed big data: Evolution in model development for information extraction [point of view]. *Proc. IEEE* **2019**, *107*, 2294–2301. [CrossRef]
22. LeCun, Y.; Bengio, Y.; Hinton, G. Deep learning. *Nature* **2015**, *521*, 436–444. [CrossRef]
23. Ronneberger, O.; Fischer, P.; Brox, T. U-net: Convolutional networks for biomedical image segmentation. In Proceedings of the International Conference on Medical Image Computing and Computer-Assisted Intervention, Munich, Germany, 5–9 October 2015; Springer: Berlin/Heidelberg, Germany, 2015; pp. 234–241.
24. Badrinarayanan, V.; Kendall, A.; Cipolla, R. Segnet: A deep convolutional encoder-decoder architecture for image segmentation *IEEE Trans. Pattern Anal. Mach. Intell.* **2017**, *39*, 2481–2495. [CrossRef]
25. Chen, L.C.; Papandreou, G.; Kokkinos, I.; Murphy, K.; Yuille, A.L. Deeplab: Semantic image segmentation with deep convolutional nets, atrous convolution, and fully connected crfs. *IEEE Trans. Pattern Anal. Mach. Intell.* **2017**, *40*, 834–848. [CrossRef] [PubMed]
26. Zhao, H.; Shi, J.; Qi, X.; Wang, X.; Jia, J. Pyramid scene parsing network. In Proceedings of the IEEE Conference on Computer Vision and Pattern Recognition, San Juan, PR, USA, 17–19 June 1997; pp. 2881–2890.
27. Chen, L.C.; Zhu, Y.; Papandreou, G.; Schroff, F.; Adam, H. Encoder-decoder with atrous separable convolution for semantic image segmentation. In Proceedings of the European Conference on Computer Vision (ECCV), Munich, Germany, 8–14 September 2018, pp. 801–818.

28. Chen, J.; Lu, Y.; Yu, Q.; Luo, X.; Adeli, E.; Wang, Y.; Lu, L.; Yuille, A.L.; Zhou, Y. Transunet: Transformers make strong encoders for medical image segmentation. *arXiv* **2021**, arXiv:2102.04306.
29. Liu, Y.; Wang, Z.; Yang, X.; Zhang, Y.; Yang, F.; Liu, B.; Cai, P. Satellite-based monitoring and statistics for raft and cage aquaculture in China's offshore waters. *Int. J. Appl. Earth Obs. Geoinf.* **2020**, *91*, 102118. [CrossRef]
30. Cheng, B.; Liang, C.; Liu, X.; Liu, Y.; Ma, X.; Wang, G. Research on a novel extraction method using Deep Learning based on GF-2 images for aquaculture areas. *Int. J. Remote Sens.* **2020**, *41*, 3575–3591. [CrossRef]
31. Su, H.; Wei, S.; Qiu, J.; Wu, W. RaftNet: A New Deep Neural Network for Coastal Raft Aquaculture Extraction from Landsat 8 OLI Data. *Remote Sens.* **2022**, *14*, 4587. [CrossRef]
32. Ottinger, M.; Clauss, K.; Kuenzer, C. Large-scale assessment of coastal aquaculture ponds with Sentinel-1 time series data. *Remote Sens.* **2017**, *9*, 440. [CrossRef]
33. Cui, B.; Fei, D.; Shao, G.; Lu, Y.; Chu, J. Extracting raft aquaculture areas from remote sensing images via an improved U-net with a PSE structure. *Remote Sens.* **2019**, *11*, 2053. [CrossRef]
34. Fu, Y.; Ye, Z.; Deng, J.; Zheng, X.; Huang, Y.; Yang, W.; Wang, Y.; Wang, K. Finer resolution mapping of marine aquaculture areas using worldView-2 imagery and a hierarchical cascade convolutional neural network. *Remote Sens.* **2019**, *11*, 1678. [CrossRef]
35. Cui, B.G.; Zhong, Y.; Fei, D.; Zhang, Y.H.; Liu, R.J.; Chu, J.L.; Zhao, J.H. Floating raft aquaculture area automatic extraction based on fully convolutional network. *J. Coast. Res.* **2019**, *90*, 86–94. [CrossRef]
36. Lin, H.; Shi, Z.; Zou, Z. Maritime semantic labeling of optical remote sensing images with multi-scale fully convolutional network. *Remote Sens.* **2017**, *9*, 480. [CrossRef]
37. Feng, Q.; Yang, J.; Zhu, D.; Liu, J.; Guo, H.; Bayartungalag, B.; Li, B. Integrating multitemporal Sentinel-1/2 data for coastal land cover classification using a multibranch convolutional neural network: A case of the Yellow River Delta. *Remote Sens.* **2019**, *11*, 1006. [CrossRef]
38. Ferriby, H.; Nejadhashemi, A.P.; Hernandez-Suarez, J.S.; Moore, N.; Kpodo, J.; Kropp, I.; Eeswaran, R.; Belton, B.; Haque, M.M. Harnessing machine learning techniques for mapping aquaculture waterbodies in Bangladesh. *Remote Sens.* **2021**, *13*, 4890. [CrossRef]
39. Zhang, Y.; Wang, C.; Ji, Y.; Chen, J.; Deng, Y.; Chen, J.; Jie, Y. Combining segmentation network and nonsubsampled contourlet transform for automatic marine raft aquaculture area extraction from sentinel-1 images. *Remote Sens.* **2020**, *12*, 4182. [CrossRef]
40. Liu, S.; Gao, K.; Qin, J.; Gong, H.; Wang, H.; Zhang, L.; Gong, D. SE2Net: Semantic segmentation of remote sensing images based on self-attention and edge enhancement modules. *J. Appl. Remote Sens.* **2021**, *15*, 026512.
41. Wang, Y.; Aubrey, D.G. The characteristics of the China coastline. *Cont. Shelf Res.* **1987**, *7*, 329–349. [CrossRef]
42. Agency, C.I. *The World Factbook 2011*; Central Intelligence Agency: McLean, VI, USA, 2011.
43. Li, W.; Zhao, S.; Chen, Y.; Wang, L.; Hou, W.; Jiang, Y.; Zou, X.; Shi, S. State of China's climate in 2021. *Atmos. Ocean. Sci. Lett.* **2022**, *15*, 100211. [CrossRef]
44. Boulay, S.; Colin, C.; Trentesaux, A.; Pluquet, F.; Bertaux, J.; Blamart, D.; Buehring, C.; Wang, P. Mineralogy and sedimentology of Pleistocene sediment in the South China Sea (ODP Site 1144). *Proc. Ocean. Drill. Program Sci. Results* **2003**, *184*, 1–21.
45. Achanta, R.; Shaji, A.; Smith, K.; Lucchi, A.; Fua, P.; Süsstrunk, S. SLIC superpixels compared to state-of-the-art superpixel methods. *IEEE Trans. Pattern Anal. Mach. Intell.* **2012**, *34*, 2274–2282. [CrossRef]
46. Paszke, A.; Gross, S.; Massa, F.; Lerer, A.; Bradbury, J.; Chanan, G.; Killeen, T.; Lin, Z.; Gimelshein, N.; Antiga, L.; et al. PyTorch: An imperative style, high-performance deep learning library. In Proceedings of the Advances in Neural Information Processing Systems, Vancouver, BC, Canada, 8–14 December 2019; pp. 8024–8035.
47. He, K.; Zhang, X.; Ren, S.; Sun, J. Deep residual learning for image recognition. In Proceedings of the IEEE Conference on Computer Vision and Pattern Recognition, Las Vegas, NV, USA, 26 June–1 July 2016; pp. 770–778.
48. Loshchilov, I.; Hutter, F. Decoupled weight decay regularization. *arXiv* **2017**, arXiv:1711.05101.
49. Chen, L.C.; Papandreou, G.; Schroff, F.; Adam, H. Rethinking atrous convolution for semantic image segmentation. *arXiv* **2017**, arXiv:1706.05587.
50. Kirillov, A.; Girshick, R.; He, K.; Dollár, P. Panoptic feature pyramid networks. In Proceedings of the IEEE/CVF Conference on Computer Vision and Pattern Recognition, Long Beach, CA, USA, 16–17 June 2019; pp. 6399–6408.
51. Li, H.; Xiong, P.; An, J.; Wang, L. Pyramid attention network for semantic segmentation. *arXiv* **2018**, arXiv:1805.10180.

sustainability

MDPI

Article

Resource-Based Port Material Yard Detection with SPPA-Net

Xiaoyong Zhang [1], Rui Xu [1], Kaixuan Lu [2], Zhihang Hao [1], Zhengchao Chen [2] and Mingyong Cai [3,4,*]

1 Beijing Key Laboratory of High Dynamic Navigation, Beijing Information Science and Technology University, Beijing 100101, China
2 Airborne Remote Sensing Center, Aerospace Information Research Institute, Chinese Academy of Sciences, Beijing 100094, China
3 State Key Laboratory of Geohazard Prevention and Geoenvironment Protection, Chengdu University of Technology, Chengdu 610059, China
4 Satellite Application Center for Ecology and Environment, MEE, Beijing 100094, China
* Correspondence: caimingyong@e-mail.secmep.cn

Abstract: Since the material yard is a crucial place for storing coal, ore, and other raw materials, accurate access to its location is of great significance to the construction of resource-based ports, environmental supervision, and investment and operating costs. Its extraction is difficult owing to its small size, variable shape, and dense distribution. In this paper, the SPPA-Net target detection network was proposed to extract the material yard. Firstly, a Dual-Channel-Spatial-Mix Block (DCSM-Block) was designed based on the Faster R-CNN framework to enhance the feature extraction ability of the location and spatial information of the material yard. Secondly, the Feature Pyramid Network (FPN) was introduced to improve the detection of material yards with different scales. Thirdly, a spatial pyramid pooling self-attention module (SPP-SA) was established to increase the global semantic information between material yards and curtail false detection and missed detection. Finally, the domestic GF-2 satellite data was adopted to conduct extraction experiments on the material yard of the port. The results demonstrated that the detection accuracy of the material yard reached 88.7% when the recall rate was 90.1%. Therefore, this study provided a new method for the supervision and environmental supervision of resource-based port material yards.

Keywords: material yard detection; deep learning; attention mechanism

check for updates

Citation: Zhang, X.; Xu, R.; Lu, K.; Hao, Z.; Chen, Z.; Cai, M. Resource-Based Port Material Yard Detection with SPPA-Net. *Sustainability* **2022**, *14*, 16413. https://doi.org/10.3390/su142416413

Academic Editor: Ripon Kumar Chakrabortty

Received: 28 October 2022
Accepted: 3 December 2022
Published: 8 December 2022

1. Introduction

Ports, as a hub of maritime logistics, undertake the function of the temporary storage of coal, ore, grain, and other raw materials. After entering ports, these raw materials will be transferred to the material yard for temporary stacking and then transported to other destinations by train or ship [1]. In 2018, the Ministry of Transport announced The Action Plan for Further Promoting Green Port Construction (2018–2022) [2], which was proposed to integrate the concept of green development into the construction of ports. However, the ports themselves and the surrounding environment are seriously influenced due to the considerable amount of dust generated in the daily construction process of open-air material yards of ports. Given the above reasons, the detection of port material yards can not only assist in the planning of their development but also facilitate the timely and effective supervision of open material yards in ports by environmental protection departments in various regions.

Compared with the two port targets (ships and oil storage tanks), material yards are characterized by small targets, dense distribution, and variable scale. The shape of ships and storage tanks is fixed, while the shape and size of material yards usually change with the stacking process of the workers. Some have an excessively large scale, while some have an excessively small scale. Some are rectangular, some are elliptical, and some are irregular. These factors lead to the difficulty in detecting material yard targets.

In recent years, deep learning, as a sample-driven data analysis method, has been extensively used in the field of remote sensing. The deep convolutional neural network (CNN) considerably improves the recognition accuracy of the image. It does not require tedious manual design while being able to autonomously perceive the feature information in the image and present better universality and expansibility. There are two main types of object detection methods based on deep learning: (1) single-stage target detection algorithms, including SSD [3], YOLO series [4–7], and RetinaNet [8]; (2) two-stage target detection algorithms, such as R-CNN [9], Fast R-CNN [10], and Faster R-CNN [11]. The composition of the two-stage target detection algorithms mainly consists of the feature extraction network and the region recommendation network in the first stage, as well as the classifier and regressor in the second stage. Feature extraction networks are employed to extract the feature information of the target. The commonly used feature extraction networks comprise VGGNet [12], GoogleNet [13], and ResNet [14]. ResNet tackles the phenomenon of the exploding gradient or vanishing gradient of the network with the increase in the number of layers to a certain extent [14]. The region proposal network is the core of two-stage target detection algorithms, and this is the fundamental difference from one-stage algorithms. Its role is to generate a series of candidate boxes that may contain targets and roughly screen the original image. The effect of dense detection can be achieved by laying a large number of candidate boxes on the original image. There is no region proposal network in the one-stage algorithm, but feature extraction is directly performed on the image to predict the location and classification of the target object. Therefore, the accuracy of the two-stage algorithm is higher than that of the one-stage algorithm. An attention mechanism [15] is needed to effectively extract image features. It is the embodiment of selective attention in the field of computer vision. It assigns different weights to the feature maps, and the positions with more weights represent more attention. Among the current attention mechanisms, channel attention [16], spatial attention [17–19], channel-attention attention [20,21] and self-attention [22] have been widely used.

Several target detection algorithms based on deep learning have been proposed to handle small, densely distributed, and multi-scale remote sensing image targets. For example, Lu et al. [23] utilized the hybrid attention mechanism of spatial attention and channel attention mechanism in parallel to effectively suppress the background noise of the image and strengthen the feature extraction ability. The mAP of small ground targets such as vehicles and ships reached 52.6%. Hua et al. [24], Ying et al. [25], and Zhu et al. [26] added self-attention mechanisms to different network structures to improve the extraction accuracy of small densely distributed targets. Huang Zhipeng et al. [27] enhanced the Faster R-CNN by sending the feature maps generated in different stages of the feature extraction network into the RPN to obtain the feature information of targets with different sizes. Its accuracy was boosted by 5% compared with the original Faster R-CNN. Lin Zhijie et al. [28] further improved the Faster R-CNN based on Huang Zhipeng et al. and employed the feature maps generated by the last three stages of the feature extraction network to construct a feature pyramid, contributing to reinforcing the detection ability of multi-scale targets. Its mAP on the PASCAL VOC 2007 and 2008 datasets reached 74.8%. Li et al. [29] improved the feature pyramid and proposed a saliency-based pyramid combining the feature pyramid and the saliency algorithm, which augmented the ability to reduce background noise. The mAP of the aerial image data set reached 72.96%. Zhong et al. [30] added the structure of the feature pyramid and self-attention mechanism to YOLOv3 to strengthen the detection ability of multi-scale and densely distributed targets, and the mAP reached 87.41% on the UCAS-AOD dataset.

Although the improved methods mentioned above have improved the detection accuracy of different targets, material fields with different shapes and complex background information will cause the problem of missed detection and false detection under the conditions of multi-scale and densely distributed remote sensing images. Therefore, an SPPA-Net target detection algorithm was proposed in this paper for the detection of port material yards based on the domestic GF-2 satellite remote sensing image. This method adopts the Faster R-CNN and the ResNet-50 as the basic framework and the feature extraction network, respectively. Firstly, the dual hybrid attention module was proposed in this paper to increase the effectiveness of the channel and spatial information extraction in the feature extraction stage. Secondly, the feature pyramid structure was introduced, and then the spatial pyramid pooling self-attention module proposed in this paper was integrated to enrich the semantic information of the feature map for subsequent network detection. Finally, the target detection data set of self-built port material yards was adopted to train the algorithm. The experimental results suggested that the proposed method effectively extracted the material field target, with a recall rate and accuracy rate of 90.1% and 88.7%, respectively.

The main contributions of this paper are as follows:

(1) A deep learning target detection algorithm was constructed for port stockyard targets, and the algorithm was used to verify with self-built port material yard datasets;
(2) The Dual-Channel-Spatial-Mix Block was proposed, to improve the feature extraction ability of densely arranged and multi-scale stockyard targets;
(3) The spatial pyramid pooling attention module was designed to globally model the features of each position in the feature map for obtaining more abstract global features

2. Principles and Methods

2.1. *Sppa-Net*

The frame foundation of the SPPA-Net is the Faster R-CNN, as shown in Figure 1, which consists of four parts: feature extraction network, Region Proposal Networks (RPN), ROI Pooling layer, and parallel classifiers and regressors. The algorithm flow is detailed, as follows. First, the image is input into the feature extraction network to obtain the feature map; second, the feature map is input into the RPN to acquire a candidate box that may be the target; third, the matrix of the feature map of the image area where the candidate box is located is scaled to a 7 × 7 feature map through the ROI Pooling layer; finally, the scaled feature map matrix is input into the classifier and regressor to generate the predicted results

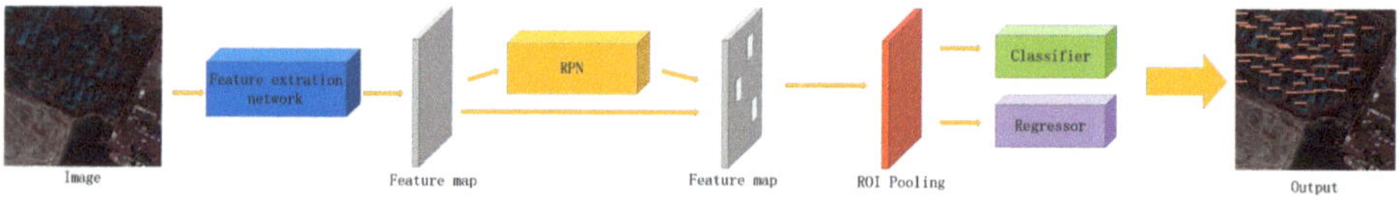

Figure 1. The overall framework of the SPPA-Net.

In this paper, the ResNet-50 was selected as the feature extraction network, which consists of five stages. Specifically, stage 1 is composed of convolution and maximum pooling, and the remaining stages are stacked by residual structures. The region proposal network consists of a fully convolutional network. The classifier and regressor comprise a fully connected layer.

In this paper, the feature extraction network was improved from the following aspects to enhance the extraction ability of densely distributed, multi-scale, and small targets (1) The dual hybrid attention mechanism module proposed in this paper was added after each residual structure of the ResNet-50; (2) the feature pyramid was constructed using the feature map generated by the ResNet-50 in each stage, and a spatial pooling self-attention mechanism was added after the feature map output by the feature pyramid. The detailed structure is illustrated in Figure 2.

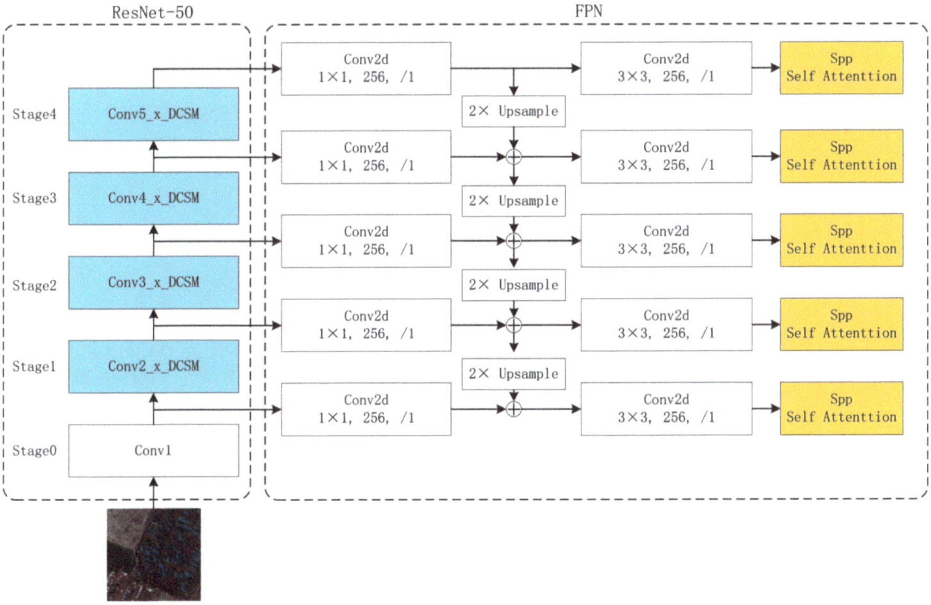

Figure 2. Improved feature extraction network.

2.2. Improved Feature Extraction Network

2.2.1. Double-Mixed Attention Mechanism

The shape of various material yards is noticeably different. Moreover, the number of material yards, the types of materials stacked, and the arrangement of materials significantly vary in different scenarios. Hence, a dual-mixed attention mechanism was proposed in this paper to improve the network's ability to extract features of material yards.

Mixed attention is composed of channel attention and spatial attention. Generally, there are two ways of combination: series and parallel. On this basis, a dual-mixed attention module was constructed. The structure of the DCSM-Block is exhibited in Figure 3. It consists of two horizontal attention modules and two spatial attention modules, which are combined in series and parallel simultaneously.

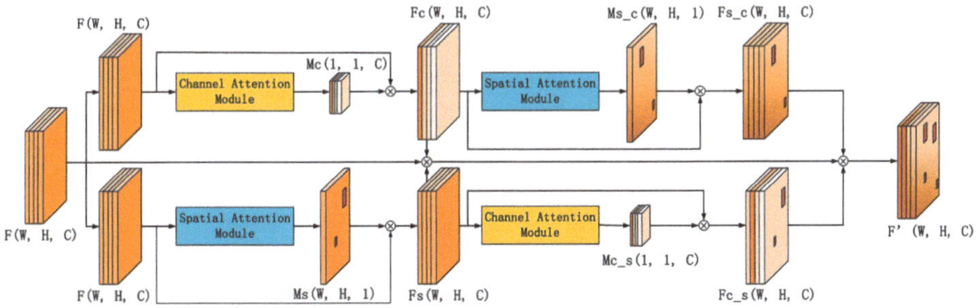

Figure 3. Structure of Dual-Channel-Spatial-Mix Block.

After entering the DCSM-Block, the feature map F with a size of W×H×C entered the upper and lower lines, as presented in Figure 3. On the upper line, the feature map *F* will first enter the channel attention module to perform feature extraction on the channel dimension information, and finally obtain the feature map *Fc* with different channel weight information. Then, *Fc* will be sent to the spatial attention module. The spatial attention module will perform further feature extraction on the feature map Fc with channel information, so as to obtain the position feature information of the target to be detected, and finally generate a feature map Fs_c with both the target position information and channel information. The specific process is as follows:

$$M_c = \sigma(MLP(MaxPool(F)) + MLP(AvgPool(F))), \quad (1)$$

$$F_c = F \otimes M_c, \quad (2)$$

$$M_{s\_c} = \sigma(f^{7\times7}(AvgPool(F_c); MaxPool(F_c)) = \sigma(f^{7\times7}(F_{c\_pool})), \quad (3)$$

$$F_{s\_c} = F_c \otimes M_{s\_c}, \quad (4)$$

where M_c represents the channel attention map; $\otimes$ represents the dot multiply operator, σ represents the Sigmiod function; MLP represents the multi-layer perceptron; MaxPool represents the global maximum pooling; AvgPool represents the global average pooling, M_{s_c} represents the spatial attention map with channel information; $f^{7\times7}$ represents 7×7 convolution layer.

On the lower line, the spatial attention module and the channel attention module performed the same operation on the feature map F as on the upper line. Firstly, the feature map F generated the feature map Fs with spatial information through the spatial attention module. Secondly, the feature map Fs entered the channel attention module to generate a feature map F_{c_s} with channel and spatial information. Finally, the feature maps F, F_c, F_s, F_{s_c}, and F_{c_s} were multiplied to obtain the feature map F′, so as to further strengthen the feature information of the space and channel.

$$F' = F \otimes F_c \otimes F_s \otimes F_{s\_c} \otimes F_{c\_s}, \quad (5)$$

In this paper, DCSM-Block was added after each residual structure of ResNet-50 to increase the content of critical parts of the feature map and curtail the features of useless information.

2.2.2. Spatial Pyramid Pooling Self-Attention Module

In ResNet-50, a feature map with a gradually decreasing size and a gradually increasing number of channels was generated. The feature map of the early stage contained the feature information of the small target. Nevertheless, the feature information of the small target in the feature map generated in the later stage was ignored, while the feature information of the large target was retained. Therefore, a feature pyramid structure based on the feature maps generated by ResNet-50 at each stage was constructed for the semantic information of small targets to strengthen the extraction ability of small-size material yards Although the feature pyramid structure can significantly improve the detection effect of small targets in material yards, it does not wrestle from the non-target false detection of the network. Therefore, in this paper, spatial pyramid pooling was introduced into the self-attention mechanism, and this improved self-attention mechanism was added after the feature map output by the feature pyramid to make up for the shortcomings of the local perception of convolutional neural networks.

The core content of the self-attention module, as displayed in Figure 4a, is to calculate the relationship between the pixels in the feature map and to achieve global context modeling. Different from the spatial attention mechanism, the self-attention mechanism not only assigns weights to each pixel in the feature map using a single-layer convolution structure but also maps the original feature image into three vector branches (Query, Key, and Value). The self-attention mechanism is different from the one-dimensional vector in natural language processing, though it can effectively model the relationship between each pixel in the feature map. Images are the basic input in computer vision, and the generated one-dimensional vector is too long in the process of vectorization, resulting in a serious waste of computing resources. With the feature map size of 200 × 200 input as an example, the red box in Figure 4a will generate dimensions of (40,000, 40,000), which will take up a lot of computing resources. This not only puts forward higher requirements for hardware but also makes it impossible to train in batches ascribed to the large occupation of memory, reducing the speed of model convergence.

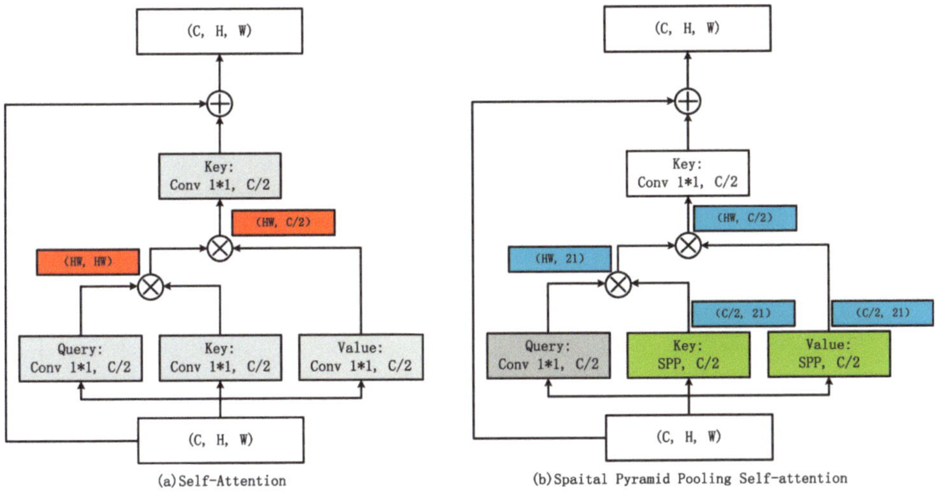

Figure 4. Comparison between self-attention and spatial pyramid pooling self-attention module: (**a**) Self-attention module. (**b**) Spatial pyramid pooling self-attention module. (SPP represents Spatial pyramid pooling; the red rectangular box represents the change in the dimension of the feature matrix in the self-attention module; the blue rectangle represents the dimension change of the feature matrix of the spatial pyramid pooling self-attention module).

Aiming at the above problems of self-attention, the dimension was lowered in this paper without losing feature information by improving the two branches of Key and Value to reduce the amount of computation and memory usage, so as to grapple with the above complications of the self-attention module. As suggested in Figure 4b, the 1×1 convolutional layers of the Key and Value branches were replaced by the spatial pyramid pooling layers, and then the feature maps were extracted at different resolutions through pooling windows of different sizes to form a one-dimensional feature vector. Compared with the 1×1 convolution in the original self-attention module, it is easier for the spatial pyramid pooling network to extract the global semantic information of the feature map, and the feature dimension obtained was much smaller than the result after 1×1 convolution processing. Compared with the self-attention module, the spatial pyramid pooling self-attention module has significantly reduced the amount of computation. From the experimental results, the precision rate and recall rate have been significantly improved.

3. Results and Discussion

3.1. Dataset

In this paper, GF-2 satellite remote sensing images of Tianjin Port and Tangshan Port were used to prepare the data set. The specific workflow is displayed in Figure 5. Firstly, ArcGIS software was employed to remove the non-port area and thus obtain the overall remote sensing images of the two ports. Secondly, Python was adopted to write a script to cut the original remote sensing image into slices of the same size, and the size of the slices was 1024×1024. Finally, the number of slices in the Tangshan port area and the Tianjin port area was 1872 and 9828, respectively.

Figure 5. Material yard sample preparation.

The slices containing material field targets were selected from the slices of the Tianjin Port and Tangshan Port with manual interpretation as the data set. Finally, 1362 images were obtained, and the number of material fields was 10,191. Deep learning should use labeled training data to train the algorithm. In this paper, LabelMe software was utilized to mark the sample pictures, and polygons were adopted to mark the material yard to obtain accurate edge information. Each sample image corresponded to a tag file. Finally, the tag files were summarized using Python to generate training files and verification files.

Through the above steps, 1362 material yard data sets with a size of 1024 × 1024 were obtained, among which 1225 and 137 were used as the training set and the test set, respectively, at a ratio of 9:1.

3.2. Environment Configuration and Training Methods

The environment hardware device used in this training is an NVIDIA graphics card, with a model of Titan XP and a memory size of 12196MiB. The software involves CUDA (version 11.2), Python (version 3.7), and PyTorch (deep learning framework 1.10). Random data enhancement was firstly performed on the read training samples to expand the training samples. The main data enhancement methods include random cropping, flipping, brightness transformation, and contrast transformation. Finally, the data enhanced was made into variables in PyTorch for gradient calculation of backpropagation. The hyperparameters of the network are comprised of the optimizer, Lr config, and epoch. The details are shown in Table 1.

Table 1. Hyper-parameters setting.

The Hyperparameters	Parameters Setting	
optimizer	Type	SGD
	Learning base	0.02
	Momentum	0.9
	Weight decay	0.0001
Lr config	Policy	Step
	Warmup	Linear
	Warmup iters	500
	Warmup ratio	0.001
	Step	8
epoch	120,000	

The metrics including *Precision*(*P*), *Recall* (*R*), and *mAP* were adopted to evaluate the network performance constructed in this paper, expressed as:

$$Precision = \frac{TP}{TP + FP} \tag{6}$$

$$Recall = \frac{TP}{TP + FN} \tag{7}$$

$$mAP = \int_0^1 P(R)dR \tag{8}$$

where TP (True Positives) represents the number of actual material yard targets correctly identified as material yards, FP (False Positives) indicates the number of the actual backgrounds but are mistakenly identified as material yards, and FN (False Negatives) denotes the number of actual material yard targets but are mistakenly classified as the background. Precision reflects the correct proportion of all material yard targets predicted by the model. Recall implies what proportion of all material yard targets is predicted by the model.

In addition, mAP (mean Average Precision) signifies the average accuracy rate, which comprehensively evaluates the accuracy and recall rates of the model. The calculation method is to calculate the accuracy and recall rates under different IOU thresholds, and then draw a curve with abscissa and ordinate. Finally, the area enclosed by the curve and abscissa and ordinate is calculated, where the IOU threshold is from 0.5 to 0.95, increasing every 0.05.

3.3. Ablation Experimental Results and Analysis

In this paper, ablation experiments were conducted to verify the effectiveness of the mixed attention mechanism and the FPN structure with the improved self-attention mechanism. Firstly, the Faster R-CNN network was used as the baseline network, with the mixed attention mechanism DCSM-Block and the FPN with SPP-SA module (SPP-SA FPN) being added, respectively. As shown in Table 2, the results revealed that the mAP score and recall rate were significantly increased by 3.2% and 5.2%, respectively, with the superposition and improvement of the two modules.

Table 2. Precision, Recall and mAP after superposition of DCSM and SPP-SA FPN modules.

DCSM	SPP-SA FPN	Precision	Recall	mAP
-	-	0.860	0.824	0.881
✓	-	**0.888**	0.845	0.897
✓	✓	0.887	**0.901**	**0.913**

Note: - represents that the module was not added to the experiment; ✓represents the module was added to the experiment.

In Figure 6(1), the original Faster R-CNN falsely detected targets similar in color to the material yard; in Figure 6(2)–(5), the Faster R-CNN had a significant missed detection when the material yard was arranged too densely and the shape was changed. After the Faster R-CNN of the DCSM-Block was added, the false detection and missed detection rates of the material yard were significantly reduced, and the accuracy and recall rates were improved by 3.2% and 2.1%, respectively. This was in that the DCSM-Block module enhanced the ability of the backbone network to extract the characteristics of the material field and focused the attention of the algorithm on faceted objects of the material yard.

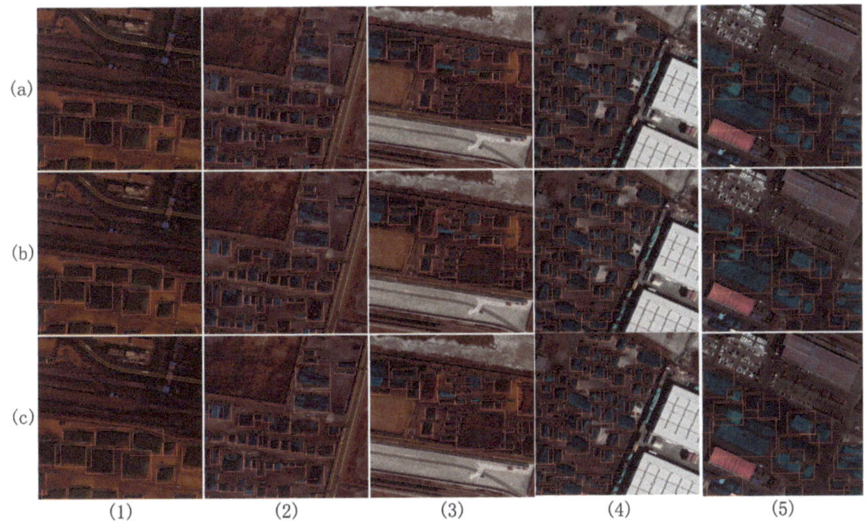

Figure 6. Results after superposition of DCSM and SPP-SA FPN modules: (**a**) Faster R-CNN. (**b**) Faster R-CNN + DCSM Block. (**c**) Faster R-CNN + DCSM module + SPP-SA FPN module.

After the addition of SPP-SA FPN, the accuracy slightly decreased compared with the DCSM-Block, the recall rate was significantly improved by 5.6%, and the detection ability of small-area material yards was further improved, as illustrated in Figure 6(2)–(5). The reason for this phenomenon is that the self-attention mechanism boosted the network's perception of global information and reinforced the classification ability of the algorithm while weakening the positioning ability of the algorithm, leading to a decrease in the accuracy of the model and an increase in the recall rate.

3.4. Comparative Experimental Results and Analysis

SSD300, SSD512, YOLOv7, RetinaNet, Faster R-CNN, and the proposed method were compared to verify the effectiveness of the algorithm. The experimental results are listed in Table 3. The proposed algorithm achieved the highest recall rate and mAP (0.901 and 0.913), respectively. SSD512 reached the highest precision and parameter (0.907 and 46.04 M).

Table 3. Comparison of Precision, Recall, mAP and Parameters of different algorithms.

Algorithms	Precision	Recall	mAP	Parameters
SSD300	0.906	0.701	0.857	34.31 M
SSD512	**0.907**	0.785	0.906	**46.04 M**
YOLOv7	0.890	0.816	0.905	37.20 M
RetinaNet	0.782	0.707	0.740	37.74 M
Faster R-CNN	0.860	0.824	0.881	41.12 M
Ours	0.887	**0.901**	**0.913**	41.53 M

The specific test results are demonstrated in Figure 7, from which the following findings can be obtained.

(1) These six methods can extract large and rectangular material fields. However, there are some material fields with irregular shapes in Figure 7(3). Except for the method in this paper, a certain degree of missed detection occurred, among which RetinaNet was the most significant. This suggested that the DCSM-Block in this paper effectively extracted the feature information of the material field, reduced the useless feature information, and thus enhanced the network's ability to extract the characteristics of the material field.

(2) Concerning small and densely-distributed material yards, as illustrated in Figure 7(2) and (4), the most severe missed detection of RetinaNet and SSD300 also missed a large area of the material yard. In SSD512 and YOLOv7, the phenomenon of missed detection of large-area stockyards occurred less, while small-area stockyards when the distance between stockyards was close was not detected. Additionally, YOLOv7 caused the false detection of targets, whose color was similar to that of the material field. The method in this paper demonstrated superiority in this respect. As the FPN structure was constructed in this paper and the SPP-SA module was added after the feature map, it improved the global perception ability, effectively enhanced the extraction ability of multi-scale and small-area material fields, and compensated for the local perception of the convolutional neural network.

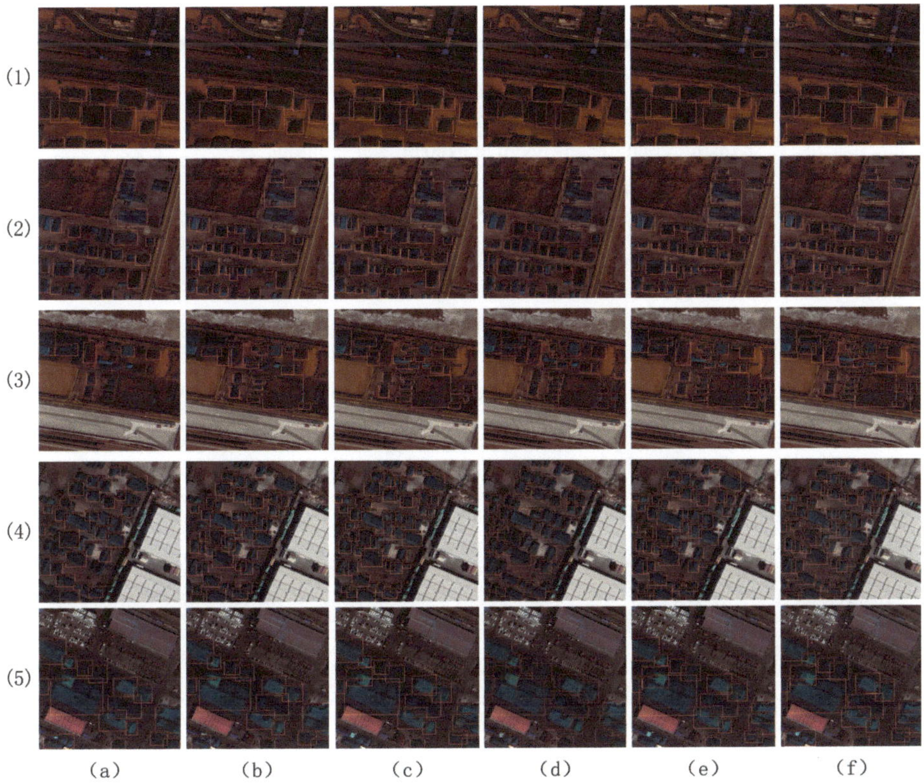

Figure 7. Comparison of different algorithms: (**a**) RetinaNet. (**b**) SSD300. (**c**) SSD512. (**d**) YOLOv7. (**e**) Faster R-CNN. (**f**) Ours.

4. Conclusions

In this paper, the SPPA-Net algorithm was proposed to detect material yard targets. Based on the original Faster R-CNN, ResNet-50 was selected as the feature extraction network, and a dual mixed attention module was embedded to enhance the extraction of the material field features. Subsequently, the feature pyramid was constructed using the feature maps generated by the ResNet-50 at each stage. The spatial pyramid pooling self-attention module was embedded to globally model the features of each position in the feature map, so as to compensate for the limitations of the local perception of the convolutional neural networks and expand the universality of the network. Compared with the original attention mechanism, the computational complexity was reduced by 19 times. Finally, a material field data set was established with the GF-2 satellite. The experiment revealed that the proposed method enabled fast and efficient extraction of the material fields within ports with high accuracy. Compared with other methods, this paper effectively improved the extraction of densely distributed and variable-scale stockyard targets, while curtailing the probability of false detection and missed detection. The recall rate reached 90.1%, and the accuracy rate reached 88.7%. In this paper, the single objective of the material yard was only employed to verify the effectiveness of the proposed algorithm. In future research, target detection will be performed on important targets in other ports, such as ships, containers, oil storage tanks and wharves, so as to further demonstrate the performance of the method proposed in this paper.

Author Contributions: Conceptualization, R.X.; Methodology, R.X. and Z.H.; Data curation, K.L.; Investigation, R.X.; Resources, X.Z.; Visualization, R.X.; Writing—original draft, R.X. and X.Z.;

Review and editing, X.Z., Z.C. and M.C. All authors have read and agreed to the published version of the manuscript.

Funding: The work was supported by the National Key Research and Development Program of China (Grant No. 2021YFB3901202).

Institutional Review Board Statement: Not applicable.

Informed Consent Statement: Not applicable.

Data Availability Statement: Not applicable.

Conflicts of Interest: The authors declare no conflict of interest.

References

1. Zhang, Q.; Wang, S.; Zhen, L. Yard truck retrofitting and deployment for hazardous material transportation in green ports. *Ann. Oper. Res.* **2022**, *in press*. [CrossRef]
2. Xi, Y. The Action Plan for Further Promoting Green Port Construction. *China Logist. Purch.* **2018**, *2*, 33–34. 10.16079/j.cnki.issn1671-6663.2018.08.005. [CrossRef]
3. Liu, W.; Anguelov, D.; Erhan, D.; Szegedy, C.; Reed, S.; Fu, C.Y.; Berg, A.C. Ssd: Single shot multibox detector. In Proceedings of the European Conference on Computer Vision, Amsterdam, The Netherlands, 11–14 October 2016; pp. 21–37.
4. Redmon, J.; Divvala, S.; Girshick, R.; Farhadi, A. You only look once: Unified, real-time object detection. In Proceedings of the IEEE Conference on Computer Vision and Pattern Recognition, Las Vegas, NV, USA, 27–30 June 2016; pp. 779–788.
5. Redmon, J.; Farhadi, A. Yolov3: An incremental improvement. *arXiv* **2018**, arXiv:1804.02767.
6. Bochkovskiy, A.; Wang, C.Y.; Liao, H.Y.M. Yolov4: Optimal speed and accuracy of object detection. *arXiv* **2020**, arXiv:2004.10934
7. Redmon, J.; Farhadi, A. YOLO9000: Better, faster, stronger. In Proceedings of the IEEE Conference on Computer Vision and Pattern Recognition, Honolulu, HI, USA, 21–26 July 2017; pp. 7263–7271.
8. Lin, T.Y.; Goyal, P.; Girshick, R.; He, K.; Dollár, P. Focal loss for dense object detection. In Proceedings of the the IEEE International Conference on Computer Vision, Venice, Italy, 22–29 October 2017; pp. 2980–2988.
9. Girshick, R.; Donahue, J.; Darrell, T.; Malik, J. Rich feature hierarchies for accurate object detection and semantic segmentation In Proceedings of the IEEE Conference on Computer Vision and Pattern Recognition, Columbus, OH, USA, 23–28 June 2014 pp. 580–587.
10. Girshick, R. Fast r-cnn. In Proceedings of the IEEE International Conference on Computer Vision, Santiago, Chile, 7–13 December 2015; pp. 1440–1448.
11. Ren, S.; He, K.; Girshick, R.; Sun, J. Faster r-cnn: Towards real-time object detection with region proposal networks. *arXiv* **2015**, arXiv:1506.01497.
12. Simonyan, K.; Zisserman, A. Very deep convolutional networks for large-scale image recognition. *arXiv* **2014**, arXiv:1409.1556.
13. Szegedy, C.; Liu, W.; Jia, Y.; Sermanet, P.; Reed, S.; Anguelov, D.; Erhan, D.; Vanhoucke, V.; Rabinovich, A. Going deeper with convolutions. In Proceedings of the IEEE Conference on Computer Vision and Pattern Recognition, Boston, MA, USA, 7–12 June 2015; pp. 1–9.
14. He, K.; Zhang, X.; Ren, S.; Sun, J. Deep residual learning for image recognition. In Proceedings of the IEEE Conference on Computer Vision and Pattern Recognition, Las Vegas, NV, USA, 27–30 June 2016; pp. 770–778.
15. Vaswani, A.; Shazeer, N.; Parmar, N.; Uszkoreit, J.; Jones, L.; Gomez, A.N.; Kaiser, Ł.; Polosukhin, I. Attention is all you need *arXiv* **2017**, arXiv:1706.03762.
16. Wang, Q.; Wu, B.; Zhu, P.; Li, P.; Zuo, W.; Hu, Q. ECA-Net: Efficient Channel Attention for Deep Convolutional Neural Networks In Proceedings of the 2020 IEEE/CVF Conference on Computer Vision and Pattern Recognition (CVPR), Seattle, WA, USA, 13–19 June 2020; pp. 11531–11539. [CrossRef]
17. Xu, Y.; Zhang, Y.; Yu, C.; Ji, C.; Yue, T.; Li, H. Residual Spatial Attention Kernel Generation Network for Hyperspectral Image Classification with Small Sample Size. *IEEE Trans. Geosci. Remote Sens.* **2022**, *60*, 3175494. [CrossRef]
18. Praveen, B.; Menon, V. Dual-Branch-AttentionNet: A Novel Deep-Learning-Based Spatial-Spectral Attention Methodology for Hyperspectral Data Analysis. *Remote Sens.* **2022**, *14*, 3644. [CrossRef]
19. Peñaloza, B.; Ogmen, H. Effects of spatial attention on spatial and temporal acuity: A computational account. *Attention, Percept. Psychophys.* **2022**, *84*, 1886–1900. [CrossRef] [PubMed]
20. Song, C.H.; Han, H.J.; Avrithis, Y. All the attention you need: Global-local, spatial-channel attention for image retrieval. In Proceedings of the IEEE/CVF Winter Conference on Applications of Computer Vision, Waikoloa, HI, USA, 4–8 January 2022 pp. 2754–2763.
21. Liu, T.; Luo, R.; Xu, L.; Feng, D.; Cao, L.; Liu, S.; Guo, J. Spatial Channel Attention for Deep Convolutional Neural Networks *Mathematics* **2022**, *10*, 1750. [CrossRef]
22. Cao, F.; Lu, X. Self-attention technology in image segmentation. In Proceedings of the International Conference on Intelligent Traffic Systems and Smart City (ITSSC 2021), Nanjing, China, 28–30 October 2022; Volume 12165, pp. 271–276.

23. Lu, X.; Ji, J.; Xing, Z.; Miao, Q. Attention and feature fusion SSD for remote sensing object detection. *IEEE Trans. Instrum. Meas.* **2021**, *70*, 1–9. [CrossRef]
24. Hua, X.; Wang, X.; Rui, T.; Zhang, H.; Wang, D. A fast self-attention cascaded network for object detection in large scene remote sensing images. *Appl. Soft Comput.* **2020**, *94*, 106495. [CrossRef]
25. Ying, X.; Wang, Q.; Li, X.; Yu, M.; Jiang, H.; Gao, J.; Liu, Z.; Yu, R. Multi-attention object detection model in remote sensing images based on multi-scale. *IEEE Access* **2019**, *7*, 94508–94519. [CrossRef]
26. Zhu, X.; Lyu, S.; Wang, X.; Zhao, Q. TPH-YOLOv5: Improved YOLOv5 based on transformer prediction head for object detection on drone-captured scenarios. In Proceedings of the IEEE/CVF International Conference on Computer Vision, Montreal, QC, Canada, 10–17 October 2021; pp. 2778–2788.
27. Huang, J.; Shi, Y.; Gao, Y. Multi-Scale Faster-RCNN Algorithm for Small Object Detection. *Comput. Res. Dev.* **2019**, *56*, 319–327.
28. Lin, Z.; Luo, Z.; Zhao, L.; Lu, D. Multi-scale convolution target detection algorithm with feature pyramid. *J. Zhejiang Univ.* **2019**, *53*, 533–540.
29. Li, C.; Luo, B.; Hong, H.; Su, X.; Wang, Y.; Liu, J.; Wang, C.; Zhang, J.; Wei, L. Object detection based on global-local saliency constraint in aerial images. *Remote Sens.* **2020**, *12*, 1435. [CrossRef]
30. Li, Z.; Wang, H.; Zhong, H.; Dai, Y. Self-attention module and FPN-based remote sensing image target detection. *Arab. J. Geosci.* **2021**, *14*, 1–18. [CrossRef]

sustainability

MDPI

Article

A Prior Semantic Network for Large-Scale Landcover Change of Landsat Imagery

Xuan Yang [1], Yongqing Bai [2], Pan Chen [1,3], Cong Li [2], Kaixuan Lu [2] and Zhengchao Chen [2,*]

[1] Key Laboratory of Digital Earth Science, Aerospace Information Research Institute, Chinese Academy of Sciences, Beijing 100094, China
[2] Airborne Remote Sensing Center, Aerospace Information Research Institute, Chinese Academy of Sciences, Beijing 100094, China
[3] University of Chinese Academy of Sciences, Beijing 100049, China
* Correspondence: chenzc@aircas.ac.cn; Tel.: +86-10-82178775

Abstract: Landcover change can reflect changes in the natural environment and the impact of human activities. Remotely sensed big data with large-scale and multi-temporal key characteristics provide the data support for landcover change information extraction. The development of deep learning provides technical method support for information extraction from remotely sensed big data. However, the current mainstream deep learning change detection methods only establish the changing relationship between two phases of images. They cannot directly extract the ground object categories before and after the change. It is easily affected by pseudo-changes caused by the color difference of multi-temporal images, resulting in many false detections. In this paper, we propose a prior semantic network and a difference enhancement block module to establish prior guidance and constraints on changing features to solve the pseudo-change problem. We propose a semantic-change integrated single-task network, which can simultaneously extract multi-temporal landcover classification and landcover change. On the self-made, large-scale multi-temporal Landsat dataset, we have performed multi-temporal landcover change information extraction, reaching an overall accuracy of 83.1% and achieving state-of-the-art performance. Finally, we thoroughly analyzed the landcover change results in the study area from 2005 to 2020.

Keywords: landcover change; deep learning; prior constraint; difference enhancement; single-task network; Landsat; multi-temporal

check for updates

Citation: Yang, X.; Bai, Y.; Chen, P.; Li, C.; Lu, K.; Chen, Z. A Prior Semantic Network for Large-Scale Landcover Change of Landsat Imagery. *Sustainability* **2022**, *14*, 13167. https://doi.org/10.3390/su142013167

Academic Editor: Mohammad Valipour

Received: 12 September 2022
Accepted: 10 October 2022
Published: 13 October 2022

1. Introduction

Landcover classification from remote sensing images is a significant application in remote sensing [1]. The classification of and changes in landcover can be obtained by using multi-temporal images. Landcover change can intuitively show the dynamic changes of the land surface, which has significant application value in land resource monitoring, ecological protection, urban expansion, returning farmland to forest, etc. [2–6]. However, the multi-temporal landcover classification also has more requirements on the amount of remote sensing data. With the development of remote sensing technology, remote sensing data is developing towards remotely sensed big data [7,8]. Remotely sensed big data with 4V characteristics volume, variety, velocity, and veracity provide massive data for remote sensing information extraction [9,10]. Traditional information extraction methods need to artificially design feature extractors according to data characteristics, such as index-based methods (NDVI, NDWI, NDBI, etc.) [11–13], texture-based methods (edge detection, keypoint extraction, etc.) [14–16], or statistical-based methods (support vector machines, random forests, etc.) [17–19]. When these methods face massive data, due to the limitations of artificial design, it is not easy to design a model that perfectly fits all the data. Therefore, the generalization ability of the model is limited, and the model may not be able to fit the new data. The emergence of deep learning technology has solved the data explosion problem of remotely sensed big data. Deep learning allows the network to independently exploit and learn effective features through deep convolutional neural networks (DCNNs) [20]. The DCNN has a stronger fitting ability to massive data and better generalization ability. At the same time, the deep learning method can quickly and accurately produce large-scale products, significantly saving time, human resources, and financial resources. Therefore, the deep learning method has gradually become the mainstream remote sensing information extraction method [21].

The traditional methods of landcover change detection are mainly based on two ideas [22–26]. The first is to set a change threshold based on the difference between the two images to detect the changing area. This method is largely limited by the quality of image processing. Differences in the colors of the two images are caused by differences in the imaging season and atmospheric conditions. As a result, pseudo-changes may appear in areas that have not changed between the two images. The second is to first classify the landcover of the two phases of images, then make the difference between the two phases of classification results to obtain the changed areas. This method is subject to the landcover classification accuracy for each phase of images. It may also be affected by the color difference between the images of the two phases. There may be errors in the classification results of the two phases, which will also cause pseudo-changes. These pseudo-changes and errors will cause changes in the ground objects to be inconsistent with the real natural scenes. The change detection methods based on deep learning establish a complex correlation between the two images by building a complex DCNN. The network learns the changing area's features through supervised learning, eliminates the pseudo-changing area's interference, and finally obtains a more accurate landcover change result.

Change detection methods based on deep learning have made significant development [27]. FC-EF [28] stacks two-phase images into multi-channel images, inputs images into a fully convolutional neural network, and outputs the changed regions. FC-Siam-conc [28] uses two encoders to extract the features of the two-phase images, respectively. It stacks each stage's features, transfers them to the decoder for feature fusion, and outputs the changed regions. FC-Siam-diff [28] is similar to FC-Siam-conc. The only difference is that the two encoder feature fusion methods are replaced from stacking to difference operation. Similar to FC-EF, CDNet [29] stacks two-phase images into multi-channel images as input. It is a classic encoder–decoder architecture. Based on the dual encoder network, DSIFN [30] adds channel attention and spatial attention mechanisms to improve the detection accuracy of changing regions. UNet++MSOF [31] and DDCNN [32] stack two-phase images as input. The backbone of them adopts the densely connected UNet++ [33] network. These change detection methods can only find the changing area, which can be regarded as a changed/unchanged binary classification problem. For multi-class landcover change detection, the above methods cannot exploit semantic information in the changing area and often need to use additional semantic segmentation networks to supplement the semantic categories. Semantic segmentation in deep learning is pixel-level classification in remote sensing. Since the change detection of landcover requires pixel-by-pixel labels of multi-temporal registered images, there are few practical applications of large-scale landcover change detection based on deep learning methods. However, many studies still indirectly implement the change trend analysis after landcover classification of the multi-temporal images [34]. This paper is completely based on the deep learning method to detect the landcover changes of the multi-temporal images. The category information before and after the change in the ground objects can also be accurately classified.

We call change detection with semantic information before and after the change as semantic change detection. Landcover change is a semantic change detection task. In this paper, we propose a prior semantic network that integrates the difference enhancement block module and compresses the multi-task network into the single-task network, which implements high-precision Landcover change mapping and change details analysis.

In summary, the main contributions of this paper are as follows:

- We propose a prior semantic network architecture. Based on the two-phase data the third-phase data and labels are introduced as prior constraint knowledge. It can solve the problem of pseudo changes caused by differences in color distribution and greatly improve the stability and robustness of change detection and semantic classification.
- We propose a difference enhancement block module, which weights the differences between the two branches, enhances true changes with large differences, and suppresses pseudo-changes with small differences.
- We compress the multi-task network, which is relatively independent of change detection and semantic segmentation, into a single-task network, which can simultaneously obtain the area of landcover change and the category of ground objects before and after the change in the network output.
- Extensive experiments on our self-made, large-scale, multi-temporal Landsat dataset achieve state-of-the-art performance. Through our proposed network, multi-temporal landcover change detection and specific change trend analysis were carried out for the large-scale study area from 2005 to 2020.

2. Methodology

This chapter mainly introduces the prior semantic network, the difference enhancement block module, and the single-task semantic change integration. We take the two-encoder Siamese UNet as the benchmark network. First, based on Siamese UNet, we add an additional encoder branch as a prior semantic knowledge constraint to build a prior semantic network. Then, the change feature fusion module in the prior semantic network is replaced with a difference enhancement block module to build PSNet-DBB. Finally, we combine the semantic segmentation and change detection multi-task decoders in PSNet into a single-task decoder to build PSNet-ST.

As shown in Figure 1, although the deep learning method can overcome the pseudo-change problem caused by color difference to a certain extent, the semantic information extraction is still unstable. The changing amplitude jitters seriously, and the change in the ground object category does not conform to the actual situation. The pseudo-changes may make it impossible to correctly obtain the change trend of the landcover and lose the practical application value. The current deep learning change detection network only extracts relevant information from the two-phase images, which is greatly affected by the color distribution of the images themselves. Based on the two-phase images, we can introduce another phase of images. The classification labels corresponding to the images of the new phase are also input into the network as auxiliary reference data for semantic information. It can improve the stability of the original two-phase image semantic information extraction, thereby improving the accuracy of landcover changes and ensuring the accuracy of changing trends. Therefore, we propose a prior semantic network architecture to achieve change detection under the constraints of additional reference branches.

Currently, in the mainstream change detection network using dual encoders, the feature fusion of the two branches is mainly performed through concatenation and difference operations. The concatenation operation simply stacks the features without enhancing the changing features. The difference operation expresses the feature difference of the two branches, but it causes the decoder to only have the change features and lose the semantic features. Therefore, we propose a difference enhancement block module, which enhances the features after the feature map difference is weighted to the original feature map as a weight. The module can amplify the obvious change features and suppress the features with very small differences. The reason for this is that such small changes are likely pseudo-changes caused by color differences. The module can also keep the original semantic classification information.

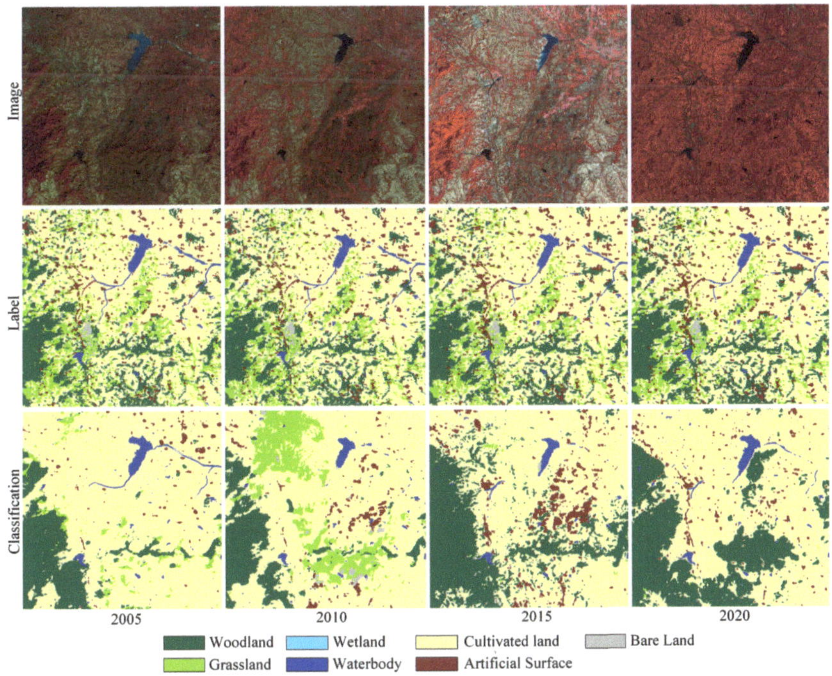

Figure 1. Color differences makes the landcover change results unstable and inconsistent.

2.1. *Prior Semantic Network*

UNet is currently the most widely used fully convolutional neural network [35] and is often used as a baseline network in semantic segmentation and change detection. UNet is an encoder–decoder network with simple architecture, fast running speed, and low GPU memory overhead. We also choose UNet as the quasi-baseline network and ResNet-50 as the encoder. However, UNet has only one encoder branch. Two images must be stacked at the input end if the network is used for change detection. The network principle is similar to FC-EF. Therefore, we first add an encoder branch based on UNet, which is also ResNet-50 [36]. The weights of the two encoders are shared to build a Siamese UNet, named SiamUNet. Encoder weight sharing ensures that the feature positions in the two encoder branches are the same so that they are comparable to compute feature differences. The two branches of SiamUNet are fused by a concatenation operation to learn differential features. The network principle is similar to FC-Siam-conc. SiamUNet is the baseline network in this work, and the modules and structures proposed in this paper are gradually added based on the baseline network.

The two encoders of the dual-branch SiamUNet input image data of the T1 and T2 phases, respectively, and then detect the changing area between the T1 and T2 phases. We introduce an additional time-phase T0 of image and label data as prior semantic information, which is input into the network. Since the T0 phase requires label data as the additional input, the number of input channels is one more than that of the T1 and T2 phases. We add a new encoder branch, also ResNet-50, to extract T0 phase features. This branch does not share weights with the T1 and T2 phase branches. We named this branch the prior branch and named the T1 and T2 phase branches as the pre-change branch and post-change branch, respectively.

Unlike the direct fusion of the pre-change branch and post-change branch in SiamUNet, we first directly fuse the prior branch with the pre-change branch. The network builds a complex function map between T0 and T1 images by stacking many convolution operations and eliminates the effects of color differences by itself. The network will pay attention to the change feature information between the T0 and T1 images. With the help of the prior semantic information in the T0 label, the network will automatically establish more accurate semantic information for the T1 phase. The network will learn the differential features before and after the change in ground objects. This differential feature can be regarded as a change mapping feature. The network deduces the ground object category at T1 through T0 prior knowledge and change mapping features in changing areas. It will directly bring the prior semantics of T0 into T1 if there is no change between before and after. We obtain T1 features with T0 prior knowledge, named prior pre-change features. In the same way, we also fuse the prior branch with the post-change branch and establish the change feature association between T0 and T2 and the semantic information of the T2 phase. We obtain the T2 feature with T0 prior knowledge, named prior post-change feature.

Unlike SiamUNet, which can only learn T1 and T2 change features, prior pre-change and prior post-change features can learn the change features and use semantic information to assist in optimizing the change features. It is because the two change features contain the category semantic information of the ground objects, which can eliminate the interference caused by the pseudo-change phenomenon caused by the color difference in the image. In the decoder stage, the prior pre-change and prior post-change features are fused to calculate the changing area. This part of the decoder is called the change task decoder. In addition, the prior pre-change and prior post-change features independently calculate the semantic segmentation results. These two decoders are called segmentation task decoders. The entire network architecture is named the prior semantic network (PSNet), which can implement change detection and semantic segmentation at the same time. It is a multi-task network. The schematic diagram of the network architecture of PSNet is shown in Figure 2.

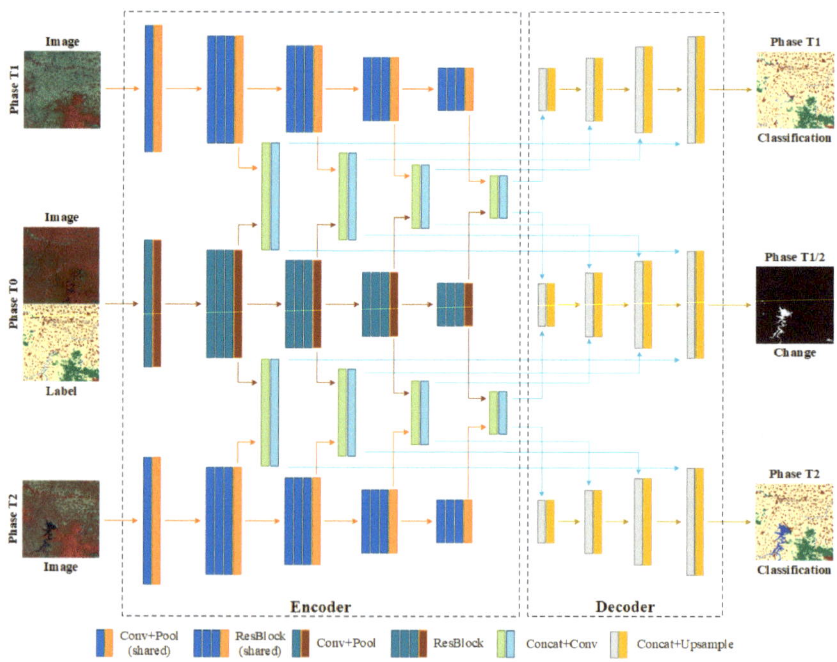

Figure 2. Schematic diagram of prior semantic network (PSNet).

In the training stage, the images and labels of the T0 phase are stacked into the $N + 1$ band data and input to the prior branch. T1-phase images are input to the pre-change branch as N-band data. T2-phase images are input to the post-change branch as N-band data. The pre-change branch shares weights with the post-change branch. The binary change label is used for the loss calculation at the end of the decoder of the change task, and the landcover classification labels of the T1 and T2 phases are respectively used for the loss calculation at the end of the decoder of the two shared weight segmentation tasks. All three loss values guide the backpropagation and gradient update of the network. In the inference stage, we only need to input the images and labels of the T0 phase and the images of the T1 and T2 phases. The changing area of T1 and T2 can be calculated, as well as the respective landcover classification results of T1 and T2.

2.2. Difference Enhancement Block Module

When calculating the changing area for the feature fusion of prior pre-change and prior post-change, if the common difference absolute value method is used to calculate the feature difference, a slight difference in the feature will be regarded as a change. This results in errors and pseudo-changes in the results. Because the features only contain differences, the decoder can only implement the change detection task. At this time, only a multi-decoder multi-task network architecture can be used for the semantic change detection task.

Therefore, we use the concatenation method to fuse the two features containing semantic segmentation information and keep all the feature information completely. At the same time, the square of the difference between the two features is calculated as the weight feature. Then, the fused features containing semantic segmentation information are weighted by the weight feature, which amplifies the changed features and suppresses the pseudo-changed features with minor changes. After the feature difference is squared, when the value is greater than 1, the feature difference weight will be amplified. When the value is less than 1, the feature difference weight will be reduced.

As shown in Figure 3, we first concatenate the prior pre-change and prior post-change features to obtain the fused features. Then we used 1×1 convolution to reduce the number of channels of the fused features by half, the same as the number of channels before fusion. Batch normalization [37] and sigmoid are used to normalize and activate features. Then we calculate the difference square of the prior pre-change and prior post-change features to obtain the difference weight feature. Next, the difference weight feature is weighted to the fusion feature to obtain the difference-enhanced fused feature. Finally, 1×1 convolution, batch normalization, and rectified linear unit (ReLU) are used to reintegrate, normalize, and activate the fused features to obtain the final difference-enhanced feature. We name it the difference enhancement block (DEB) module.

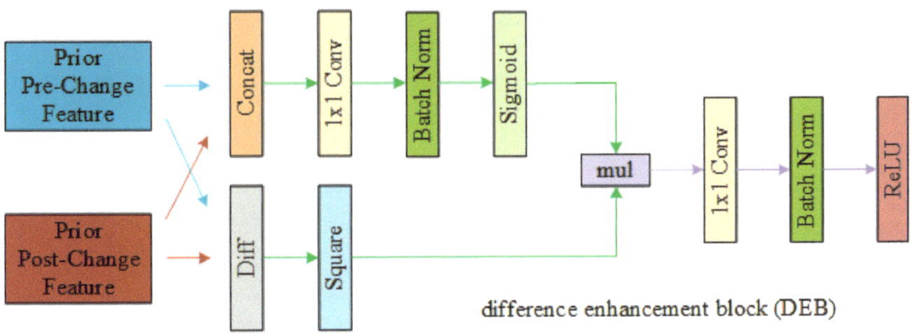

Figure 3. Schematic diagram of difference enhancement block (DEB) module.

We denote the convolution operation as:

$$\mathrm{W}^n(x) = \mathbf{W}^{n \times n} \odot x + \mathbf{b}, \quad (1)$$

where $\odot$ represents the convolution operator, $\mathbf{W}^{n \times n}$ represents the $n \times n$ convolutional kernel, $\mathbf{b}$ represents the vector of bias, and x represents the input data.

This section will perform the batch normalization operation after each convolution operation. To simplify the expression, $W^n(x)$ not only represents the convolution layer but also includes the batch normalization layer. Therefore, the DEB module can be expressed as

$$f_{DEB}(x_{pre}, x_{post}) = f_{ReLU}(W_2^1((f_{sigmoid}(W_1^1(x_{pre} \oplus x_{post}))) \otimes (x_{post} - x_{pre})^2)), \quad (2)$$

where $\oplus$ represents the concatenation operator; $\otimes$ represents the dot multiply operator, $f_{sigmoid}$ represents the sigmoid function; f_{ReLU} represents the ReLU function; W_1^1 and W_1^2 represents the first and second 1×1 convolution layer, respectively; x_{pre} represents the prior pre-change feature; and x_{post} represents the prior post-change feature.

We use the DEB module to replace the prior pre-change and prior post-change feature fusion modules in PSNet to build the PSNet-DEB network.

2.3. Single-Task Architecture for Semantic Change

There is certain independence between multiple decoders, which will prevent the features between multi-branches in the decoding stage from directly assisting and optimizing each other in the learning process. As a result, there will be minor contradictions between the semantic segmentation results and the change detection results. For example, the semantic segmentation results of the two phases have not changed, but the change detection results are considered to have changed. The single integrated decoder simultaneously implements semantic segmentation and change detection at the end of the network, which can optimize learning from each other and avoid conflicting problems.

The DEB module in Section 2.2 can highlight the change feature information while keeping the complete semantic segmentation information, which provides a theoretical basis for building a single decoder to directly implement the semantic change task. We remove the two segmentation-task decoders in PSNet-DEB and keep only one change-task decoder. However, we change the output of the change-task decoder from binary-value change to the form of multi-value classification. We choose a number to describe the changing state between every two categories. For example, we label the first class change to the third class as 13, the fourth class change to the second class as 42, and the fifth class remains unchanged as 55. We rename the upgraded decoder as the semantic change decoder and build the PSNet-ST network. The schematic diagram of the network architecture of PSNet-ST is shown in Figure 4.

In the training stage, the input data form of PSNet-ST is the same as the multi-task PSNet. We use the joint label, which can describe the semantic change information, to compute the loss value and guide the network's backpropagation and gradient updates. In the inference stage, we only need to input the images and labels of the T0 phase and the images of the T1 and T2 phases. Then, we can calculate the categories of the ground object before and after the change from T1 to T2 and deduce the changing area.

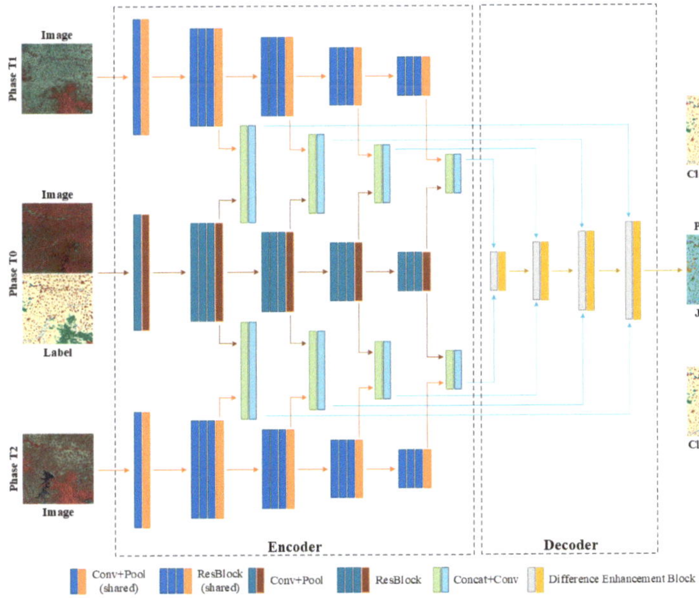

Figure 4. Schematic diagram of single-task prior semantic network (PSNet-ST).

3. Experimental Results

3.1. Datasets

We can easily download multi-temporal Landsat imagery, which can be used to study landcover classification and change. However, no public Landsat dataset currently contains both semantic segmentation and change detection labels. Therefore, to test our proposed method's performance on semantic change detection through experiments, we made a multi-temporal semantic change detection Landsat dataset.

We selected part of central and southern China as the study area, covering an area of 360,000 km^2, located between 110.933° E∼116.622° E and 28.444° N∼34.133° N. We downloaded images of five time phases in 2000, 2005, 2010, 2015, and 2020. Each phase needs 26 images to cover the whole research area. The data path is between 121∼126, and the data row is between 36∼40. Among them, the images of 2000, 2005, and 2010 use Landsat-5 data, equipped with a thematic mapper (TM) sensor, including seven bands. Except for the thermal infrared band with a spatial resolution of 120 m, the other bands have a spatial resolution of 30 m. The images of 2015 and 2020 use Landsat-8 data, equipped with the operational land imager (OLI) sensor, including nine bands. Except for the panchromatic band, which has a spatial resolution of 15 m, the other bands have a spatial resolution of 30 m. All downloaded images are at the L1TP level. We only used six bands of data, including blue, green, red, near-infrared, shortwave infrared 1, and shortwave infrared 2. We then mosaicked the downloaded images by year and cropped them according to the latitude and longitude of the study area, removing redundant images outside the study area. Finally, we obtained 20,480 × 20,480 pixels of Landsat image in five phases. The spatial resolution is 30 m.

As shown in Figure 5, to train the semantic change model, we annotated the images from 2000, 2005, and 2010 at the pixel level, including seven categories: woodland, grassland, wetland, waterbody, cultivated land, artificial surface, and bare land. The 2015 and 2020 images shown in Figure 6 are not labeled, and the landcover classification and change results will be inferred through the deep learning method. All labels were visually interpreted in ArcGIS software by a team of 10. Controversial ground objects that cannot be identified on the image are labeled by high-resolution remote sensing images or field surveys. All samples were randomly cross-checked three times, and disputed samples were uniformly determined. Although there is a certain possibility of error in manual labeling, we try our best to minimize it and make the label's accuracy as close to 100% as possible. High-resolution imagery is the primary reference for the edge of ground objects prone to mislabeling. Using the labels on the high-resolution images to downsample to the medium resolution can eliminate the label errors at the edge of the ground objects. To verify the model's accuracy more accurately, we randomly selected 1000 points for each category of the ground object in the study area. Then, we obtained the ground truth corresponding to the 2020 images through a field survey and high-resolution image reference.

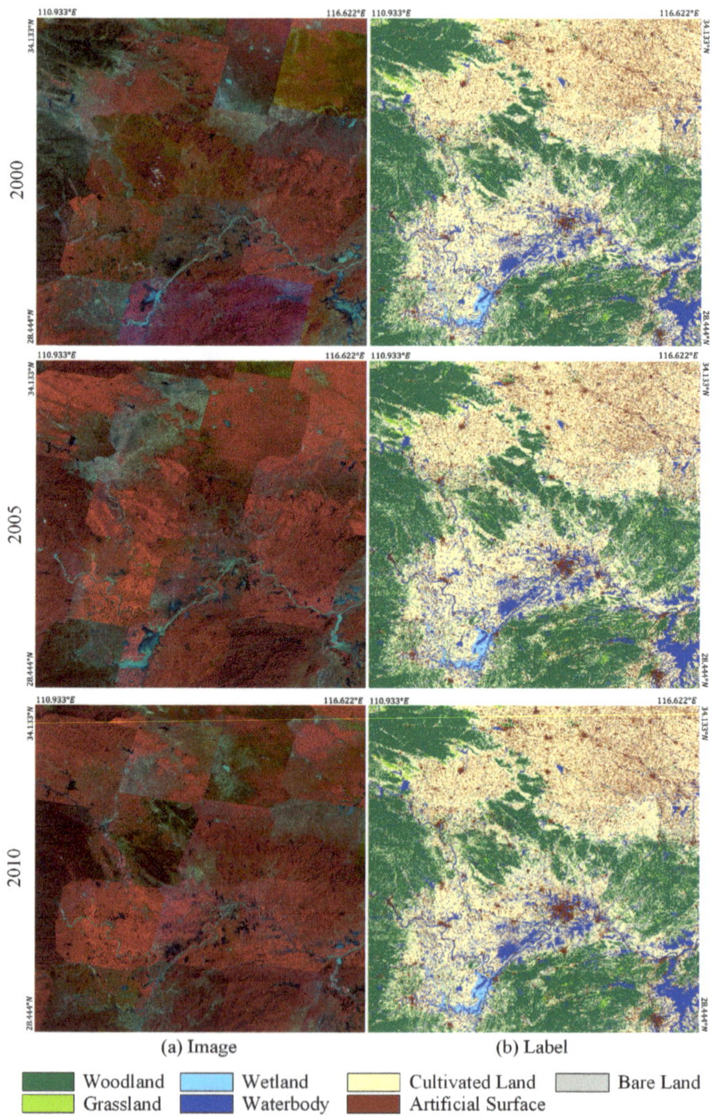

Figure 5. Raw false color composite images and the corresponding classification labels in 2000, 2005, and 2010: (**a**) images; (**b**) labels.

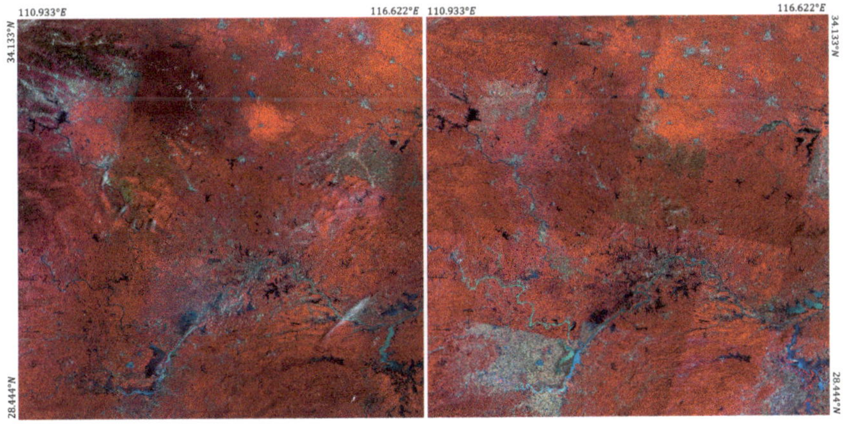

(a) Image in 2015 (b) Image in 2020

Figure 6. Raw false color composite images in (**a**) 2015 and (**b**) 2020.

3.2. Implementation Details

3.2.1. Data Preprocessing

The PSNet network proposed in this paper requires two samples for change detection and one for prior knowledge input. Therefore, we take the sample in 2000 as prior knowledge and the samples in 2005 and 2010 as the data for actual semantic change detection. Since only two samples are needed for the published mainstream change detection network, we use samples in 2005 and 2010 for training to make the evaluation data the same.

We use the sliding window to crop each image with 20,480 $\times$ 20,480 pixels into 1600 small tiles with 512 $\times$ 512 pixels without overlapping. To evaluate the model's generalization ability more objectively, we keep a large proportion of the data out of training. Therefore, we divide the dataset into a training set, validation set, and test set according to the ratio of 4:1:5. That is, 640 tiles are used for training the model, 160 tiles are used to validate intermediate model accuracy during the training stage to pick the best model, and 800 tiles do not participate in the training stage and are only used for prediction and accuracy evaluation.

As shown in Figure 7, in addition to keeping semantic segmentation labels, the label data obtain binary change detection samples by comparing two-phase samples with different values. We adopt the form of joint labels to make semantic change labels. We use a two-digit number to denote each pixel's categories before and after changes. The first digit is the category number before the change, and the second is the category number after the change. When the two digits are the same, there is no change. The category numbers in semantic change map are shown in Table 1. The columns indicate the categories before the change, and the rows indicate the categories after the change.

Table 1. The category number of semantic changes in joint label. The columns indicate the categories before the change, and the rows indicate the categories after the change.

	Woodland #1	**Grassland #2**	**Wetland #3**	**Waterbody #4**	**Cultivated Land #5**	**Artificial Surface #6**	**Bare Land #7**
Woodland #1	11	12	13	14	15	16	17
Grassland #2	21	22	23	24	25	26	27
Wetland #3	31	32	33	34	35	36	37
Waterbody #4	41	42	43	44	45	46	47
Cultivated land #5	51	52	53	54	55	56	57
Artificial surface #6	61	62	63	64	65	66	67
Bare land #7	71	72	73	74	75	76	77

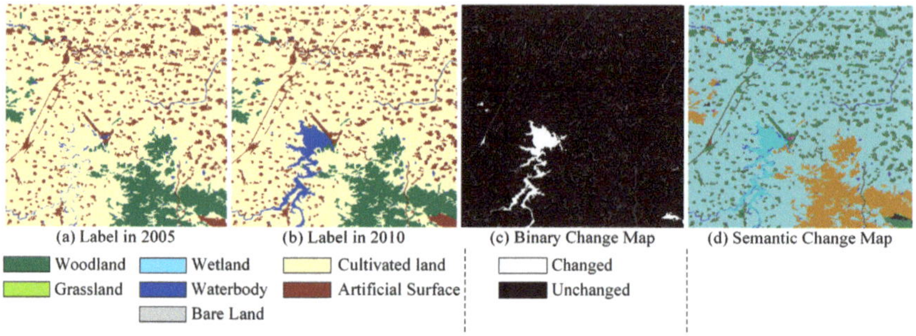

Figure 7. Examples of binary change detection samples and semantic change detection samples (**a**) Label in 2005. (**b**) Label in 2010. (**c**) Binary change map, calculated from the difference between the two-phase labels. (**d**) Semantic change map, where the joint label was obtained by merging the two-phase label category numbers. (Colored randomly for visual display).

To ensure a more reasonable distribution of data input to the model, we normalize the input data first. The data normalization is defined as:

$$I' = \frac{I - mean}{stddev}, \quad (3)$$

where I' represents the normalized image data, I represents the original input image data *mean* represents the mean value of each band in the image data, and *stddev* represents the standard deviation value of each band in the image data.

3.2.2. Training Settings

We use the PyTorch deep learning framework [38] to implement the PSNet proposed in this paper and other mainstream change detection networks published. We used four NVIDIA RTX 3090 GPUs with 24 GB memory to train the model. Data augmentation operations include random horizontal flips, random vertical flips, and random rotation The optimizer is AdamW [39], and the batch size is set to 32. The initial learning rate is 1×10^{-5} and gradually increases to 1×10^{-3} during the first 10 epochs. Then, the learning rate is automatically adjusted using the model validation accuracy. When the accuracy has not improved for 20 consecutive epochs, the learning rate is multiplied by the drop coefficient of 0.3. When the learning rate drops to 1×10^{-7}, the training process ends.

The formula of the early learning rate increasing stage is:

$$lr = lr_0 \cdot (\frac{lr^*}{lr_0})^{\frac{t}{k \cdot e}}, \quad (4)$$

where lr represents the real-time learning rate, lr_0 represents the initial learning rate, lr^* represents the maximum learning rate, t represents the real-time training iterations, k represents the number of iterations per epoch, and e represents the number of training epochs when the learning rate reaches the maximum value.

The formula of the later learning rate automatic decreasing stage is:

$$lr' = \alpha \cdot lr, \quad (5)$$

where lr' represents the decreased learning rate, lr represents the learning rate before decreasing, and α represents the drop coefficient.

The cross entropy loss can optimize the model by pixel level. Lovász-softmax [40] loss can optimize the model by region level and from intra-class and inter-class differences Therefore, we choose cross-entropy loss and Lovász-softmax loss as loss functions to train the network. The binarized version corresponding to the two losses is selected in the binary change detection task. The task loss is calculated as follows:

$$L = L_{ce} + L_{Lovász} \tag{6}$$

For the single-task semantic change network PSNet-ST, Equation (6) is the final loss value. For multi-task PSNet, multiple branch losses need to be added according to Equation (7):

$$L_{MT} = L_{change} + L_{T1} + L_{T2} \tag{7}$$

3.2.3. Evaluation Metrics

We mainly adopt three evaluation metrics: overall accuracy (OA), intersection over union (IoU), and F1 score. OA represents the proportion of correctly classified pixels among all pixels. IoU is used to evaluate the accuracy of a certain class, where intersection refers to the number of correctly classified pixels, and union refers to the sum of the number of correctly classified and misclassified pixels. The F1 score is also used to evaluate the accuracy of a certain class, taking into account both precision and recall.

We denote all pixels according to the following rules: TP means the label is true and the prediction is true. FP means the label is false and the prediction is true. FN means the label is true and the prediction is false. TN means the label is false and the prediction is false.

The formula for OA is as follows:

$$OA = \frac{TP + TN}{TP + FP + FN + TN} \tag{8}$$

The formula for IoU is as follows:

$$IoU = \frac{TP}{TP + FP + FN} \tag{9}$$

F1 score is calculated by *precision* and *recall*:

$$F1 = 2 \cdot \frac{precision \cdot recall}{precision + recall}, \tag{10}$$

where *precision* and *recall* are calculated by TP, FP, and FN:

$$precision = \frac{TP}{TP + FP}, \tag{11}$$

$$recall = \frac{TP}{TP + FN}. \tag{12}$$

For multi-task PSNet, the binary change detection accuracy is evaluated with IoU and F1 scores. The semantic segmentation task evaluates single-class accuracy using the F1 score and overall accuracy using mean F1 score (mF1) and OA. Semantic change detection accuracy for single-task PSNet-ST using F1 score was used to evaluate the accuracy of each change.

3.3. Comparing Methods for Binary Change Detection

Since other published mainstream change detection networks can only implement binary change detection, to verify the effect of the prior semantic branch and DEB module proposed in this paper, we use the PSNet series for binary change detection tasks for a fair comparison. Table 2 shows the quantitative accuracy comparison of mainstream change detection networks and the PSNet series used in this paper. It can be seen that the F1 score of the mainstream change detection network can only reach 44.81%. The accuracy of most networks can exceed 30%, and the accuracy of FC-Siam-diff and DDCNN is lower. The baseline SiamUNet built based on the idea of UNet and Siamese encoder in this paper can reach 49.06%. A prior branch is added to the baseline, and the images and labels of the prior reference phase are used for additional constraints. The accuracy of the PSNet can reach 62.79%. It can be seen that prior knowledge is very effective in improving accuracy. Based on PSNet, we replace the multi-branch encoder feature fusion module with the DEB module proposed in this paper. The DEB module performs filtering, weighting, and optimizing the changing features. The PSNet-DEB network achieves 68.62% accuracy. Therefore, using feature differences to weight, optimize and fuse the features has a certain effect on improving accuracy. Finally, based on PSNet-DEB, the semantic segmentation and change detection decoder are combined to build PSNet-ST. The single-task decoder is used to directly learn the categories of ground objects before and after the change. Since the two tasks can be optimized for each other after the decoder is integrated, the accuracy of change detection is greatly improved, reaching 80.91%.

Table 2. The quantitative binary change detection accuracy comparison between the PSNet series network and other mainstream change detection networks.

Methods	IoU	F1
FC-EF	21.81	32.25
FC-Siam-diff	13.36	16.61
FC-Siam-conc	22.01	32.60
DSIFN	29.16	44.81
DDCNN	16.99	23.49
CDNet	25.08	37.95
UNet++MSOF	23.44	35.11
SiamUNet	31.77	49.06
PSNet	44.01	62.79
PSNet-DEB	51.87	68.62
PSNet-ST	65.75	80.91

We visualize the change detection results of the mainstream change detection network and the PSNet series network proposed in this paper. Figure 8 shows comparison charts of the change detection results. In the first group, it can be seen that the errors of the three methods, FC-Siam-diff, DDCNN, and UNet++MSOF, are very obvious, and the change detection fails. The error of the CDNet result is also more conspicuous. FC-EF, FC-Siam-conc, and DSIFN missed changes in waterbodies. SiamUNet can detect changes in waterbodies, but there are many false detections. After adding prior knowledge constraints, PSNet can reduce some false detections. After using the DEB module, the waterbody changes detected by PSNet-DEB are more accurate. However, many small changes are missed since DEB modules inhibit small changes, and multi-task decoders cannot directly assist each other in the learning stage. For the single-task network PSNet-ST after multi-decoder integration, the change results of waterbody are very accurate, and other small changing objects can also be detected smoothly.

In the second group, the errors of FC-Siam-diff, UNet++MSOF, and DDCNN are very obvious. It can be seen from the images that there is no major change in the two phases of the ground objects. However, due to the impact of imaging conditions and seasonal factors, the cultivated land shows completely different colors in the images of the two phases. It also brings more significant challenges to change detection. Without the constraints of prior knowledge, baseline SiamUNet, like other mainstream networks, has many false detections. While the prior knowledge assists PSNet in reducing the false detection rate, the DEB module further eliminates small patches with false detections. After the PSNet-ST integrated decoders, the changes in the waterbody can be correctly detected, and the very small changes that have been eliminated in PSNet-DEB can also be successfully detected. More comparisons can be found in Appendix A.1.

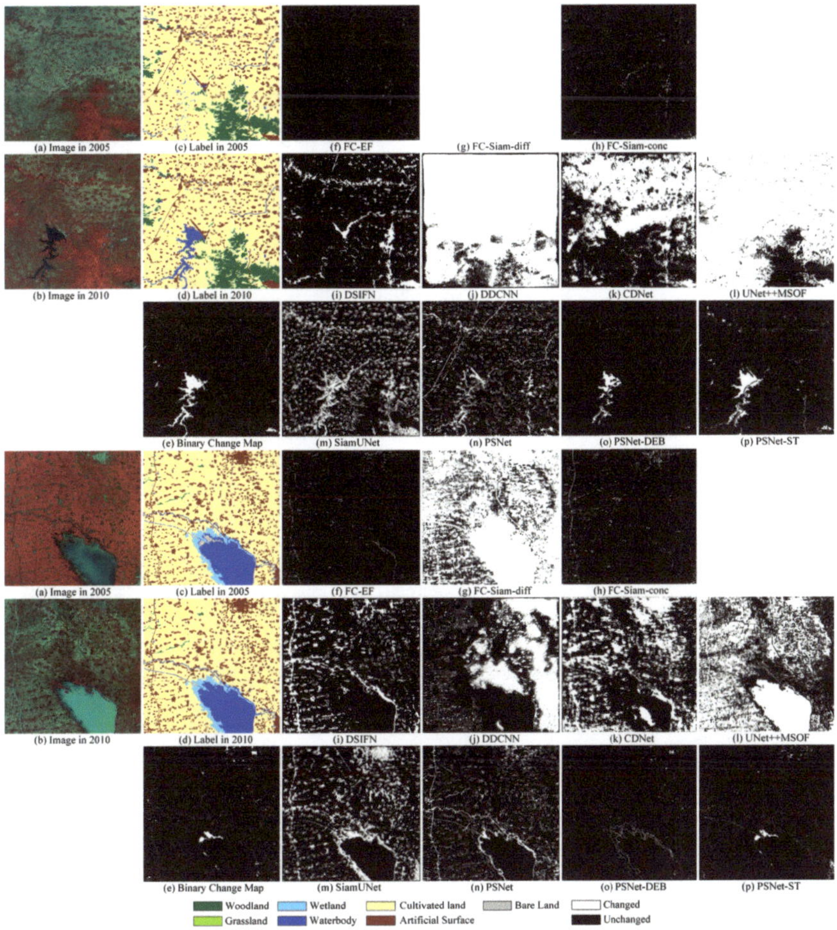

Figure 8. Comparison of the binary change detection results between PSNet and other methods: (**a**) Images in 2005. (**b**) Images in 2010. (**c**) Labels in 2005. (**d**) Labels in 2010. (**e**) Binary change labels. Inference result of (**f**) the FC-EF, (**g**) the FC-Siam-diff, (**h**) the FC-Siam-conc, (**i**) the DSIFN, (**j**) the DDCNN, (**k**) the CDNet, (**l**) UNet++MSOF, (**m**) the SiamUNet, (**n**) our proposed PSNet, (**o**) our proposed PSNet-DEB, and (**p**) our proposed PSNet-ST.

To sum up, FC-Siam-diff and DDCNN almost completely fail for change detection on 30 m resolution Landsat images. FC-EF and FC-Siam-conc mainly show more missed detections, while DSIFN, CDNet, and UNet++MSOF show more false detections. In our proposed method, the effect of adding one phase image as prior knowledge is pronounced. Although some false detections exist, the detected change contours are gradually approaching the labels. The DEB module can suppress a large number of false detections. However, semantic segmentation and change detection tasks are independent of each other. The semantic segmentation results cannot be used to optimize the change detection results. Therefore, there will be over-suppression, and small change areas will be missed. After the semantic segmentation and change detection tasks are combined into a single task, the features are optimized for each other. The advantages of the prior semantic information and the DEB module are integrated, and the shortcomings are overcome. The best change detection performance is achieved.

3.4. Comparing Methods for Semantic Change Detection

Since other mainstream change detection networks cannot achieve change detection tasks with semantic information, we only compare the baseline SiamUNet, PSNet, PSNet-DEB, and PSNet-ST networks. SiamUNet has no prior knowledge constraints, and the two decoders for semantic segmentation and change detection tasks are independent and belong to a multi-task network. PSNet has prior knowledge constraints, the two decoders of semantic segmentation and change detection tasks are separated, and it is also a multi-task network. PSNet-DEB is basically the same as PSNet, except that the DEB module replaces the multi-branch feature fusion module, and the feature difference is used for feature weighting enhancement. PSNet-ST has prior knowledge constraints and the DEB module for feature optimization. The single decoder implements the simultaneous extraction of the ground object categories of the two phases. Therefore, semantic segmentation and change detection tasks can be performed simultaneously, which belongs to a single-task network.

Table 3 shows the quantitative accuracy comparison of landcover classification between SiamUNet and PSNet series networks. The OA of baseline SiamUNet is 83.45% and the accuracy of grassland, artificial surface, and bare land is relatively low. After adding prior knowledge constraints, the OA of PSNet reaches 92.77%, and the accuracy of each category is significantly improved. After the DEB module is integrated with the multi-decoder, the accuracy is further improved. The OA of the final single-task network PSNet-ST can reach 94.26%. Table 4 shows the accuracy comparison of SiamUNet and PSNet series network with ground truth in 2020. All methods showed a downward trend in accuracy. Without prior knowledge constraints, SiamUNet has the most severe drop in accuracy. The PSNet series network has only a slight decrease in accuracy, which shows that with the prior knowledge constraints, the network's generalization ability has been significantly improved. The OA of the best-performance network PSNet-ST can reach 92.99%, which means that out of 7000 ground truth points, 6509 points are correctly classified.

Table 3. The quantitative landcover classification accuracy comparison between the SiamUNet and the PSNet series network.

Method	Woodland	Grassland	Wetland	Waterbody	Cultivated Land	Artificial Surface	Bare Land	Mean F1	OA
SiamUNet	87.46	38.18	70.30	78.22	86.47	56.62	33.20	64.35	83.45
PSNet	90.32	77.62	75.27	88.88	94.07	83.10	70.66	82.84	92.77
PSNet-DEB	94.64	77.11	82.94	89.12	94.99	83.71	74.55	85.29	93.21
PSNet-ST	96.23	82.49	90.88	91.51	95.33	86.04	77.49	88.56	94.26

Table 4. The quantitative accuracy comparison with ground truth in 2020 between the SiamUNet and the PSNet series network.

Method	Woodland	Grassland	Wetland	Waterbody	Cultivated Land	Artificial Surface	Bare Land	Mean F1	OA
SiamUNet	63.24	13.89	54.37	61.41	70.15	32.83	16.92	44.69	63.57
PSNet	84.34	72.43	70.79	83.57	88.29	78.49	65.77	77.67	88.81
PSNet-DEB	89.66	72.57	78.19	84.55	89.54	78.26	69.63	80.34	91.20
PSNet-ST	91.82	78.39	85.90	86.10	90.94	81.15	72.29	83.80	92.99

For the quantitative accuracy evaluation of semantic change detection, we construct a matrix indicating the mutual conversion to describe the change accuracy between any two categories. As shown in Table 5, the column represents the object category before the change, and the Mean Out column represents the mean F1 score calculated based on the category before the change. The row represents the object category after the change and the Mean In row represents the mean F1 score calculated based on the category after the change. Diagonal elements indicate that the ground object has not changed. Table 5 shows the semantic change detection accuracy between any two categories in the results of SiamUNet, and the overall mean F1 score is 63.1%. Table 6 shows the semantic change detection accuracy between any two categories in the results of PSNet, and the overall mean F1 score is 73.25%. Table 7 shows the semantic change detection accuracy between any two categories in the results of PSNet-DEB, and the overall mean F1 score is 79.07%. Table 8 shows the semantic change detection accuracy between any two categories in the results of PSNet-ST, and the overall mean F1 score is 83.10%. It can be seen that prior constraint knowledge, DEB module, and single-task integration can significantly improve the semantic change detection task.

Table 5. The semantic change detection accuracy of SiamUNet. (Metric: F1 score).

	Woodland	Grassland	Wetland	Waterbody	Cultivated Land	Artificial Surface	Bare Land	Mean Out
Woodland	87.21	62.42	59.17	62.85	65.29	65.90	61.66	66.36
Grassland	62.35	38.25	89.92	59.67	61.64	59.46	59.84	61.59
Wetland	59.05	59.00	70.55	67.81	63.75	60.31	61.00	63.07
Waterbody	62.62	60.07	61.82	78.51	65.63	65.18	61.13	65.00
Cultivated Land	64.82	61.73	63.83	67.23	85.68	74.80	63.13	68.75
Artificial Surface	59.66	59.00	59.41	61.98	64.39	55.08	59.84	59.91
Bare Land	60.76	59.09	59.79	62.52	60.34	66.54	34.33	57.62
Mean In	65.21	57.08	66.36	65.79	66.67	63.89	57.28	63.18

Table 6. The semantic change detection accuracy of PSNet. (Metric: F1 score).

	Woodland	Grassland	Wetland	Waterbody	Cultivated Land	Artificial Surface	Bare Land	Mean Out
Woodland	97.00	71.22	69.05	71.50	72.11	71.52	72.56	74.99
Grassland	73.85	83.29	69.10	70.65	72.60	70.71	71.15	73.05
Wetland	69.20	69.03	65.05	72.64	70.66	69.84	70.07	69.50
Waterbody	71.69	69.77	70.34	92.21	72.88	72.49	72.10	74.50
Cultivated Land	72.12	70.95	69.24	74.14	96.01	77.34	71.60	75.91
Artificial Surface	70.55	69.83	69.19	70.97	71.83	85.73	70.18	72.61
Bare Land	71.17	70.80	69.28	71.53	70.23	71.06	81.07	72.16
Mean In	75.08	72.13	68.75	74.81	75.19	74.10	72.68	73.25

Table 7. The semantic change detection accuracy of PSNet-DEB. (Metric: F1 score).

	Woodland	Grassland	Wetland	Waterbody	Cultivated Land	Artificial Surface	Bare Land	Mean Out
Woodland	96.35	90.12	60.00	86.54	75.69	82.13	86.39	82.46
Grassland	86.68	83.32	60.00	75.97	86.42	73.72	60.00	75.16
Wetland	81.52	60.00	92.88	84.28	84.47	75.01	60.00	76.88
Waterbody	87.45	75.83	86.74	92.07	80.02	85.03	80.83	84.00
Cultivated Land	76.76	87.13	84.76	86.35	95.15	86.40	81.49	85.43
Artificial Surface	85.03	77.41	83.61	85.99	75.95	83.75	60.00	78.82
Bare Land	82.68	60.00	60.00	82.05	77.72	60.00	72.52	70.71
Mean In	85.21	76.26	75.43	84.75	82.20	78.01	71.61	79.07

Table 8. The semantic change detection accuracy of PSNet-ST. (Metric: F1 score).

	Woodland	Grassland	Wetland	Waterbody	Cultivated Land	Artificial Surface	Bare Land	Mean Out
Woodland	97.28	82.52	80.70	81.68	82.46	81.43	82.28	84.05
Grassland	82.66	87.58	79.15	81.92	82.31	80.20	82.14	82.28
Wetland	84.26	81.17	89.37	81.50	82.50	83.58	79.70	83.15
Waterbody	82.25	80.86	80.58	94.14	82.68	83.74	81.20	83.64
Cultivated Land	82.53	81.83	80.85	84.38	96.46	87.50	81.96	85.07
Artificial Surface	81.29	79.85	79.99	80.88	81.81	87.67	82.33	81.97
Bare Land	81.47	81.63	79.17	81.64	80.94	82.42	83.59	81.55
Mean In	84.53	82.20	81.40	83.73	84.16	83.79	81.89	83.10

We visualize the semantic change detection results of baseline SiamUNet and the PSNet series networks proposed in this paper. Figure 9 is a comparison chart of the semantic change detection results. In each set of examples, the first line is the landcover classification results in the pre-change phase. The second line is the classification results before the change in the changing area. The third line is the classification results after the change in the changing area. The fourth line is the landcover classification results in the post-change phase.

In the first group, the main change is that the cultivated land becomes the waterbody, with some other minor changes. SiamUNet's landcover classification results are not detailed enough, and small objects are missed. Due to false detections in the change detection, there are errors in the changing area that the before-and-after phases do not in fact change. PSNet has fewer false detections with the help of prior knowledge constraints. However, due to the impact of multi-task independent decoders, the change detection result contradicts the result of semantic segmentation. That is, the change detection branch believes there has been a change, and the semantic segmentation branch believes the ground objects in the before-and-after phases are the same. Smaller fragmented changes are missed from the PSNet-DEB results. PSNet-ST has excellent landcover classification results and change results. After integrating the multi-task decoder into a single-task decoder, the inconsistency between the two results has been eliminated.

In the second group, similar to the first group, the classification results of SiamUNet's landcover are not detailed enough, the change detection error is obvious, and a large number of unchanged ground objects are placed in the changing area. With the help of prior knowledge constraints, PSNet has fewer false detections, and the changing area is still too large. The result of PSNet-DEB is close to the label, but the bare land's change in the middle of the image is fragmented and missed. PSNet-ST completely extracts the undetected bare land in PSNet-DEB. It can all be extracted, whether it is a large or a small change. The landcover classification results of the two phases are also very accurate. More comparisons can be found in Appendix A.2.

In summary, the prior semantic knowledge constraints, DEB module, and single-task integrated decoder strategy proposed in this paper have achieved state-of-the-art performance in Landsat's semantic change detection task.

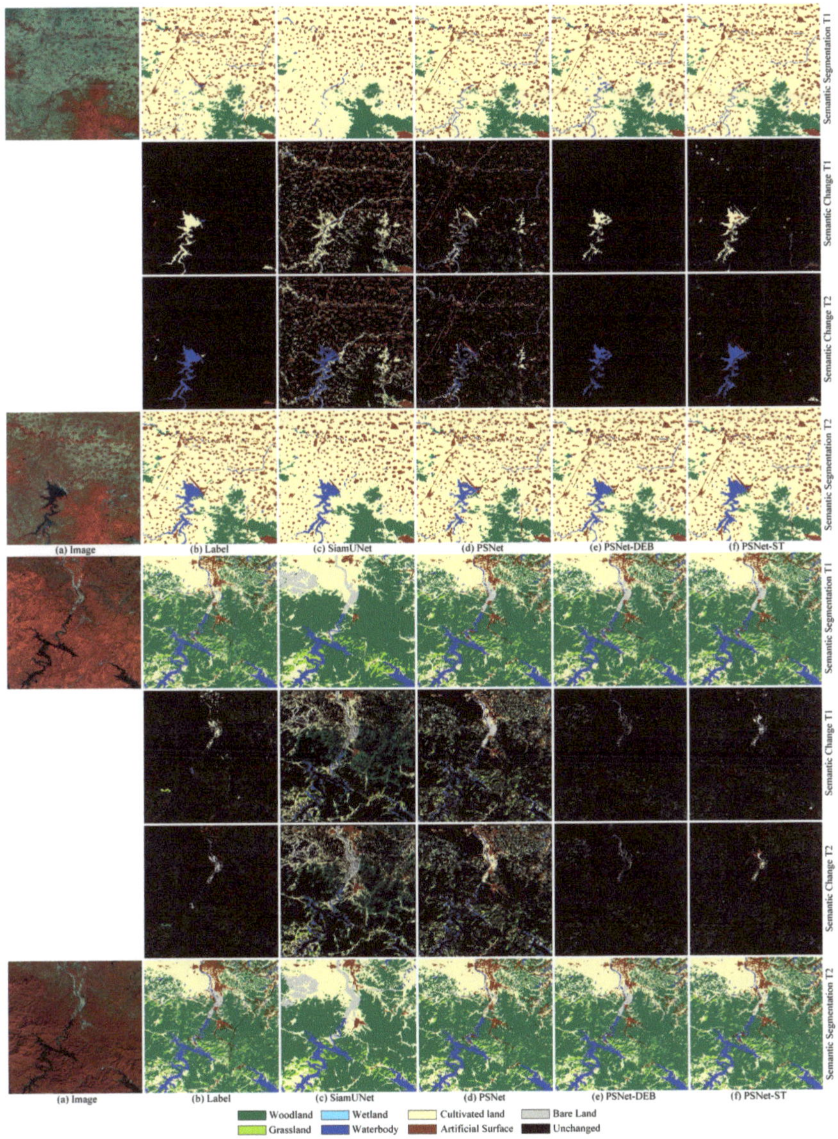

Figure 9. Comparison of the semantic change detection results between the SiamUNet and PSNet series networks: (**a**) Images. (**b**) Labels. Inference result of (**c**) the SiamUNet, (**d**) our proposed PSNet, (**e**) our proposed PSNet-DEB, and (**f**) our proposed PSNet-ST.

4. Discussion

In this section, based on the best-performing PSNet-ST model in Section 3, we perform semantic change detection on the four-phase images of the study area in 2005, 2010, 2015, and 2020 and obtain large-scale landcover change results. In addition, the samples in 2000 are used as prior knowledge to assist and constrain the other two phases of data for training and prediction. In addition to this section, more details on these results can be found in Appendix B.

To count the process state of the mutual changes between the ground objects, we adopted a category transition matrix to represent the mutual change areas. As shown in Table 9, the column represents the area of a certain category becoming other categories. The Total Out column represents the total area of a certain category turning into other categories, which can be regarded as the transfer-out. The row represents the area of each other category turned into a certain category. The Total In row represents the total area of other categories turned into a certain category, which can be regarded as the transfer-in. The Total Change row represents the overall area change of each category, which is calculated by combining the transferred-out and transferred-in areas of the category.

4.1. Analysis of Landcover Change from 2005 to 2010

The landcover change in the whole study area from 2005 to 2010 is shown in Figure 10. The figure shows the landcover classification results in 2005, the landcover classification results in 2010, and the two-phase corresponding ground object classes in the changing area. Equivalent to the ground objects in Figure 10e becomes the ground objects in Figure 10f. Table 9 is the category transition matrix of the two-phase landcover changes. It can be seen that from 2005 to 2010 in the study area, the woodland, grassland, and bare land changed very little. The wetland area becomes smaller, and the water body area increases. The cultivated land area decreased more, while the artificial surface area increased more. A more specific analysis shows a mutual exchange between woodland and cultivated land. It is caused by the interaction between returning farmland to forest and cutting down trees for reclamation. As the main feature of greening, the woodland area remains unchanged. There is less exchange between grassland and cultivated land, as deserted arable land grows weeds, which can also be reclaimed for cultivation. There is a mutual exchange between wetlands and waterbodies. This is due to the similarities between wetlands and waterbodies. Shallow tidal flats submerged by water will be classified as wetlands. Affected by the imaging season, there will be a mutual conversion between wetlands and waterbodies. There is also a small exchange between waterbodies and cultivated land. This is because when there is more water storage in paddy fields, it looks similar to waterbodies. Since urban development is on a trend of continuous expansion, artificial surfaces occupy more cultivated land. The cultivated land around the city is changed to artificial surfaces. With urbanization, the rural population and the area of cultivated land decrease, and the migration of the rural population to cities will make urban expansion a usual trend. Some villages were demolished to build new reservoirs. However, some of the demolished village lands were planted with trees and converted into woodland. At the same time, some urban artificial surfaces have been re-planned as forest parks. Therefore, in five years, many artificial surfaces have been converted into waterbodies and woodland. More changes details can be found in Appendix B.1.

Table 9. Landcover change statistics from 2005 to 2010. (Unit: square kilometers).

	Woodland	**Grassland**	**Wetland**	**Waterbody**	**Cultivated Land**	**Artificial Surface**	**Bare Land**	**Total Out**
Woodland	118,172.30	425.12	7.38	266.38	3978.30	402.38	51.21	5130.78
Grassland	423.47	3633.26	0.49	14.13	210.93	11.00	3.84	663.87
Wetland	14.87	1.96	2281.03	403.33	74.09	22.80	1.55	518.59
Waterbody	286.05	11.38	129.09	25,023.67	1560.86	349.48	20.33	2357.17
Cultivated Land	4105.40	206.11	56.00	1637.30	181,559.61	4757.88	57.84	10,820.53
Artificial Surface	294.72	7.63	14.76	248.93	3081.54	22,843.81	19.37	3666.94
Bare Land	55.41	2.34	0.10	34.79	63.18	24.71	635.28	180.52
Total In	5179.92	654.53	207.83	2604.85	8968.90	5568.25	154.14	23,338.41
Total Change	49.13	−9.34	−310.76	247.68	−1851.63	1901.31	−26.38	-

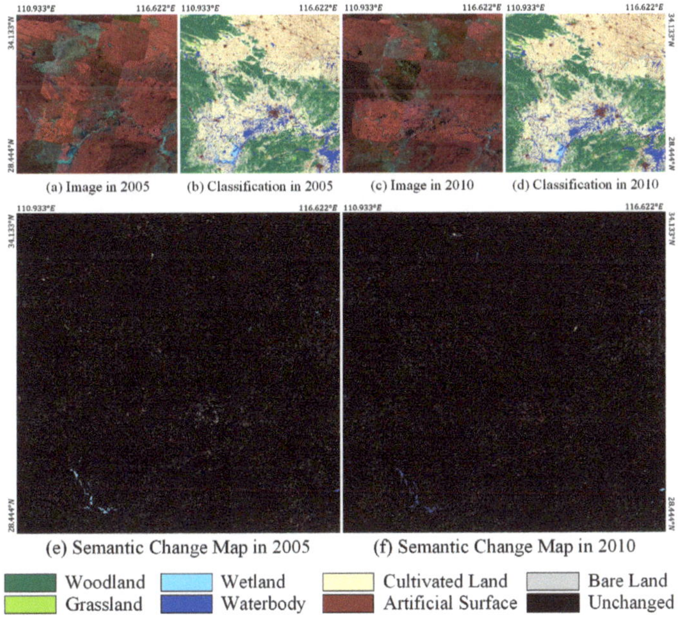

Figure 10. The landcover change map from 2005 to 2010: (**a**) Image in 2005. (**b**) Classification result in 2005. (**c**) Image in 2010. (**d**) Classification result in 2010. (**e**) Semantic change map in 2005. (**f**) Semantic change map in 2010.

4.2. Analysis of Landcover Change from 2010 to 2015

The landcover change in the whole study area from 2010 to 2015 is shown in Figure 11. The figure shows the landcover classification results in 2010, the landcover classification results in 2015, and the two-phase corresponding ground object classes in the changing area. The ground objects in Figure 11e become equivalent to the ground objects in Figure 11f. Table 10 is the category transition matrix of the two-phase landcover changes. It can be seen that from 2010 to 2015, the changes in grassland, wetland, and bare land were very small in the study area. The changes in woodlands and waterbodies are also not obvious enough. The area of cultivated land decreased more, while the artificial surface area increased more. A more specific analysis shows a mutual exchange between woodland and cultivated land. The mutual exchange between the waterbody and the cultivated land is also similar to the last five years because when there is more water in a paddy field, it looks similar to a waterbody. Cultivated land continues to be transformed into artificial surface, indicating that urbanization further devours the surrounding cultivated land. Compared with the last five-year changes, the area of cultivated land has decreased more, and the artificial surface area has also increased. This shows that the economy has developed faster in the past five years. The speed of urbanization has also accelerated. More details on these changes can be found in Appendix B.2.

Table 10. Landcover change statistics from 2010 to 2015. (Unit: square kilometers).

	Woodland	Grassland	Wetland	Waterbody	Cultivated Land	Artificial Surface	Bare Land	Total Out
Woodland	122,209.50	89.21	0.31	57.79	847.25	122.75	25.41	1142.71
Grassland	157.05	4061.13	0.20	5.83	59.67	3.28	0.62	226.66
Wetland	9.91	0.53	2419.57	39.37	10.83	8.63	0.03	69.29
Waterbody	81.57	3.66	27.43	26,873.41	486.73	147.75	7.97	755.11
Cultivated Land	965.14	55.31	15.53	462.25	186,001.79	3004.00	24.49	4526.72
Artificial Surface	42.62	1.60	1.87	28.02	374.65	27,957.45	5.84	454.60
Bare Land	2.51	0.34	0.03	1.41	2.29	7.56	775.28	14.14
Total In	1258.79	150.64	45.37	594.68	1781.42	3293.97	64.36	7189.22
Total Change	116.08	−76.02	−23.92	−160.43	−2745.30	2839.37	50.22	-

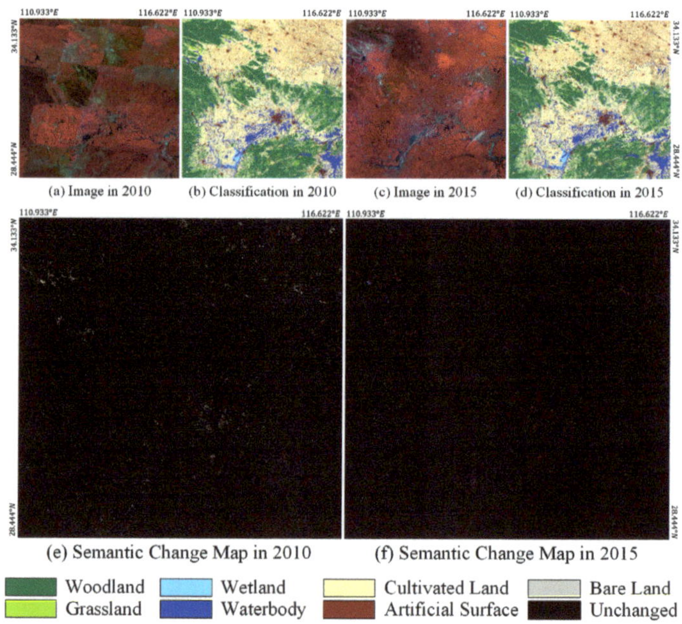

Figure 11. The landcover change map from 2010 to 2015: (**a**) Image in 2010. (**b**) Classification result in 2010. (**c**) Image in 2015. (**d**) Classification result in 2015. (**e**) Semantic change map in 2010. (**f**) Semantic change map in 2015.

4.3. Analysis of Landcover Change from 2015 to 2020

The landcover change in the whole study area from 2015 to 2020 is shown in Figure 12. The figure shows the landcover classification results in 2015, the landcover classification results in 2020, and the two-phase corresponding ground object classes in the changing area. The ground objects in Figure 12e become equivalent to the ground objects in Figure 12f. Table 11 is the category transition matrix of the two-phase landcover changes. It can be seen that from 2015 to 2020 in the study area, except for cultivated land and artificial surface, the area changes of other categories are very small. The area of cultivated land was significantly reduced, and the artificial surface area was significantly increased. In a more specific analysis, there is a small interchange between woodland and cultivated land. This is because there will be a small dynamic balance change at the boundaries. In addition, a small part of the woodland has been turned into artificial surface. The reason for this is the encroachment of some forest land by urban development. Wetlands are relatively stable, indicating that wetland protection policies have achieved practical results. The exchange of cultivated land and wetlands is also due to the similarity between paddy fields and waterbodies. A large area of cultivated land has become artificial surface, and the change is larger than in the previous ten years. It shows that the city expanded very rapidly from 2015 to 2020, occupying a large amount of cultivated land around the city, reflecting the acceleration of urbanization and the rapid development of the economic level. More details on these changes can be found in Appendix B.3.

Table 11. Landcover change statistics from 2015 to 2020. (Unit: square kilometers).

	Woodland	Grassland	Wetland	Waterbody	Cultivated Land	Artificial Surface	Bare Land	Total Out
Woodland	122,767.65	39.60	0.20	47.50	456.40	139.46	17.48	700.65
Grassland	119.15	4060.60	0.04	3.76	25.92	1.97	0.31	151.16
Wetland	5.78	0.14	2412.62	27.13	8.30	10.89	0.07	52.32
Waterbody	52.21	2.34	18.14	26,887.04	353.20	150.39	4.76	581.04
Cultivated Land	572.89	46.89	7.24	424.72	183,451.63	3264.78	15.06	4331.58
Artificial Surface	16.90	1.46	0.78	18.02	189.90	31,021.68	2.67	229.74
Bare Land	1.12	0.24	0.02	1.05	1.00	5.39	830.81	8.83
Total In	768.05	90.68	26.42	522.20	1034.73	3572.89	40.36	6055.33
Total Change	67.40	−60.49	−25.90	−58.85	−3296.85	3343.15	31.53	-

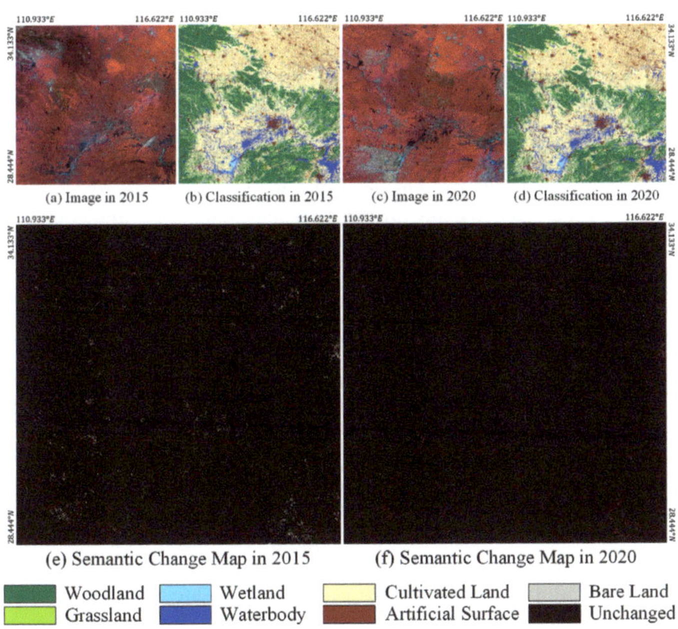

Figure 12. The landcover change map from 2015 to 2020: (**a**) Image in 2015. (**b**) Classification result in 2015. (**c**) Image in 2020. (**d**) Classification result in 2020. (**e**) Semantic change map in 2015. (**f**) Semantic change map in 2020.

4.4. Implications and Limitations

The PSNet proposed in this paper solves three critical problems encountered by the multi-temporal Landsat landcover changes. The first problem is the pseudo-changes caused by differences in color distribution. The second problem is the enhancement and suppression of true and false changes. The third problem is that multi-task networks cannot jointly optimize and constrain each other when performing change detection and semantic segmentation.

For the study of landcover changes, independent semantic segmentation of multi-temporal images is a mainstream method. This method is very sensitive to color differences between multi-temporal images, resulting in many pseudo-change errors in the results. However, the current method of remote sensing multi-temporal image change detection can only focus on the binary information of change and unchanged. It cannot obtain information on the mutual change process between categories. Therefore, a new idea is proposed to design a two-in-one single-task network for semantic segmentation and change detection to solve these three critical problems.

We add an additional encoder branch to the mainstream Siamese network for remote sensing change detection. The original two Siamese branches extract the image features of the two phases, respectively. Then, the data and labels of the third phase are introduced into the newly added additional encoder, which is used as prior knowledge to guide and constrain the feature learning of the original two Siamese branches. Under the constraints of prior knowledge, the two Siamese encoders are simplified from learning complete texture features to only learning their change information relative to the prior image. In the region that has not changed, the label of the a priori phase is directly brought in. Since the network learns two-phase change thresholds based on samples, this dramatically reduces the problem of pseudo-changes caused by color differences. At this time, PSNet adds two additional decoders to implement the multi-task semantic change.

In the Landsat image with a resolution of 30 m, most of the changed features are very small, maybe only one pixel wide. As a result, the distinction between true changes and false changes is not high enough. We redesigned the commonly used difference or concatenation operation and used the difference square to amplify the true change and reduce the false change. The optimized features are then weighted onto feature maps with complete semantic information. This way, pseudo-change errors that are difficult to eliminate can be suppressed. The kept complete semantic features can lay the foundation for the subsequent semantic change two-in-one single-task network.

When using multiple decoder branches to implement the tasks of change detection and semantic segmentation, only the features of the shallow encoder will be fused. However, the decoders are still relatively independent. Therefore, the multi-branch features at the end of the network cannot constrain each other for optimization, information cannot be shared and even contradictory errors may occur. We combine multiple decoders into a single decoder that converts two-phase semantic segmentation samples into a single semantic change sample with joint labels. With the help of feature learning enhancement and change feature optimization of the prior knowledge branch and DEB module, the single-task network directly implements semantic change detection in one step and can learn complex change states. While improving the efficiency of multi-task learning, the single-task network also solves the risk of conflicting multiple decoders, significantly improving the accuracy and reliability of semantic change detection.

The change detection method with complete semantic information provides powerful support for multi-temporal landcover changes. The mutual change information extracted by PSNet provides data support for analyzing landcover classification and change. Based on the landcover change data, it is possible to further discover and explore phenomena such as returning farmland to forests, wetland ecological protection, and urban expansion and mine information, such as the status quo of social and economic development and policy decision making and planning. The PSNet proposed in this paper can quickly extract and analyze the landcover change information in large-scale and multi-temporal dimensions and has significant application value. Based on PSNet, we also conduct a detailed analysis of the large-scale multi-temporal Landsat landcover changes from 2005 to 2020. Urban expansion and the devouring of cultivated land are the most critical keywords obtained in the land cover change analysis.

Compared with mainstream semantic segmentation and change detection networks, the PSNet proposed in this paper provides a new problem-solving idea. The idea is the introduction of prior knowledge constraints, which significantly improves the accuracy and has more practical application value. However, this also brings certain limitations. When performing semantic change detection on two-phase images, the third-phase images and labels must be sacrificed. That is to say, at least three registered multi-temporal samples are required for model training. However, the more time phases there are, the more difficult it is to label samples. Therefore, while PSNet achieves higher accuracy, the workload of manual interpretation in the early stage is also higher than other mainstream algorithms. In future research, we will explore unpaired labels as prior knowledge to improve network accuracy, which may avoid manual labeling work in additional phases and reduce the workload and difficulty of preparatory work. We will also apply our proposed PSNet to single-band images and high-resolution images.

5. Conclusions

In this paper, we propose a prior semantic network for the Landsat semantic change task. Based on the dual-branch Siamese network, we add a prior knowledge encoder branch to solve the problem of pseudo-changes caused by color distribution differences. We design a difference enhancement block module to replace the common difference or concatenation operation and solve the problem of the enhancement and suppression of true and false changes. We propose a single-task PSNet, which combines multiple decoders into one decoder, solving the problem that features cannot be jointly optimized and mutually constrained in multi-task networks. Our proposed method achieves state-of-the-art performance on large-scale multi-temporal Landsat landcover change datasets, far exceeding other change detection and semantic segmentation networks. Based on PSNet, we conducted a specific analysis and discussion on the landcover changes in an area of central and southern China. The acceleration of urbanization construction and the acceleration of economic development are the two keywords we have found from the results of landcover changes over 15 years. This work has a particular value for sustainable development goals. In future research, we will study constraint learning on unpaired samples and generalize PSNet to more multi-temporal remote sensing data of different resolutions.

Author Contributions: X.Y. wrote the manuscript, designed the methodology, and conducted experiments; Y.B. and P.C. validated and analyzed the results; C.L. and K.L. preprocessed the data of the study area and made the datasets; Z.C. supervised the study and reviewed the manuscript. All authors have read and agreed to the published version of the manuscript.

Funding: This research was supported by the Strategic Priority Research Program of the Chinese Academy of Sciences under Grant No. XDA23100304.

Institutional Review Board Statement: Not applicable.

Informed Consent Statement: Not applicable.

Data Availability Statement: Not applicable.

Acknowledgments: The authors thank the editors and anonymous reviewers for their valuable comments, which greatly improved the quality of the paper.

Conflicts of Interest: The authors declare no conflict of interest.

Abbreviations

The following abbreviations are used in this manuscript:

DCNN	deep convolutional neural network
DEB	difference enhancement block
F1	F1 score
mF1	mean F1 score
FN	false negative
FP	false positive
IoU	intersection over union
MT	multi-task
NDBI	normalized difference built-up index
NDVI	normalized difference vegetation index
NDWI	normalized difference water index
OA	overall accuracy
OLI	operational land imager
PSNet	prior semantic network
ReLU	rectified linear unit
SiamUNet	Siamese UNet
ST	single-task
TM	thematic mapper
TN	true negative
TP	true positive

Appendix A. More Comparisons for Experiments

Appendix A.1. More Comparisons for Binary Change Detection

Figure A1 is a comparison chart of the change detection results. In the first group, FC-Siam-diff and DDCNN cannot detect the changing regions normally. FC-EF and FC-Siam-conc have missed many obvious detections, while DSIFN, CDNet, and UNet++MSOF have many false detections. Baseline SiamUNet also has false detections, but they are relatively less. PSNet, assisted by prior knowledge, further reduces the false detection rate. The DEB module is over-filtered. The change detection results of the integrated decoder of PSNet-ST have the highest agreement with the label and the best performance. In the second group, none of FC-Siam-diff, DDCNN, DSIFN, CDNet, or UNet++MSOF can detect the change correctly. Although FC-EF and FC-Siam-conc have no obvious false detections, the phenomenon of missed detection is serious. Due to the absence of prior knowledge constraints, baseline SiamUNet performs similarly to mainstream networks. Most areas of the city are falsely detected as changing areas. PSNet dramatically reduces the false detection rate. With the help of the DEB module, the change detection results are very close to the ground truth labels, but there are some cases of missed detections. PSNet-ST with the integrated decoder can combine the advantages of PSNet and PSNet-DEB, and its change detection performance is the best.

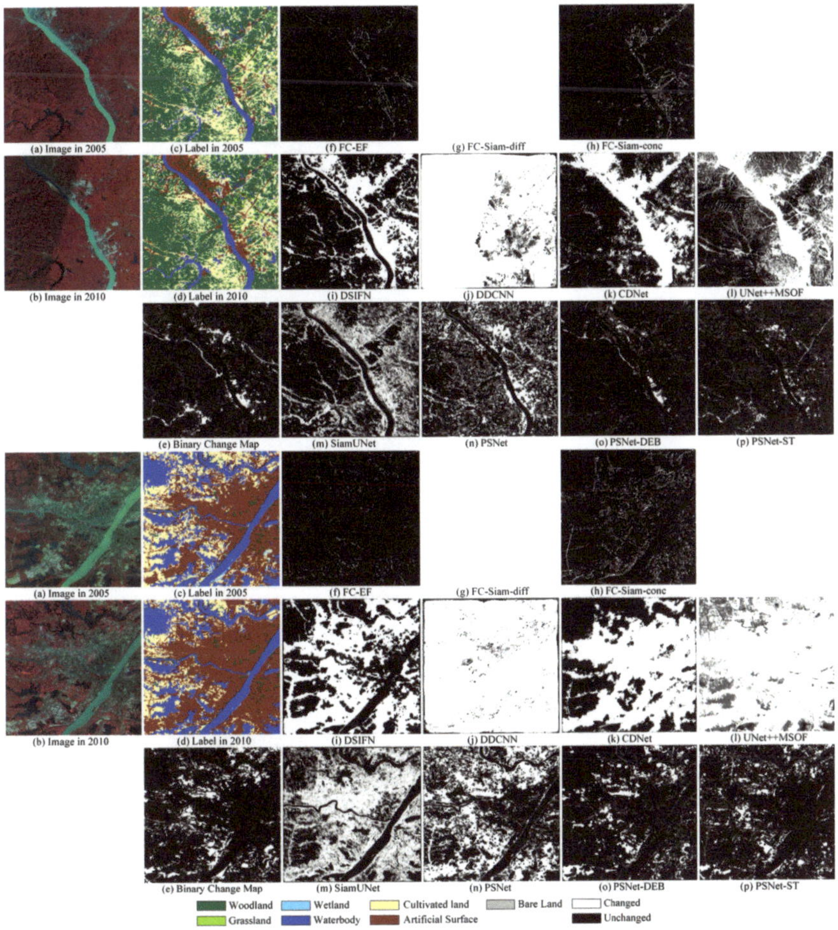

Figure A1. Comparison of the binary change detection results between PSNet and other methods: (**a**) Images in 2005. (**b**) Images in 2010. (**c**) Labels in 2005. (**d**) Labels in 2010. (**e**) Binary change labels. Inference result of (**f**) the FC-EF, (**g**) the FC-Siam-diff, (**h**) the FC-Siam-conc, (**i**) the DSIFN, (**j**) the DDCNN, (**k**) the CDNet, (**l**) UNet++MSOF, (**m**) the SiamUNet, (**n**) our proposed PSNet, (**o**) our proposed PSNet-DEB, and (**p**) our proposed PSNet-ST.

Appendix A.2. More Comparisons for Semantic Change Detection

Figure A2 is a comparison chart of the semantic change detection results. The main change is that cultivated land becomes the artificial surface as the city expands. The results of SiamUNet extracted many unchanged artificial surfaces as changed regions. The results of PSNet take a small part of the cultivated land as the changing area, but due to the effect of prior knowledge, the false detection rate has been greatly reduced. The results of PSNet-DEB and PSNet-ST are both close to the labels, and the semantic change results of PSNet-ST are more refined and detailed. It also reflects mutual assistance and guidance advantages after combining semantic segmentation and change detection tasks.

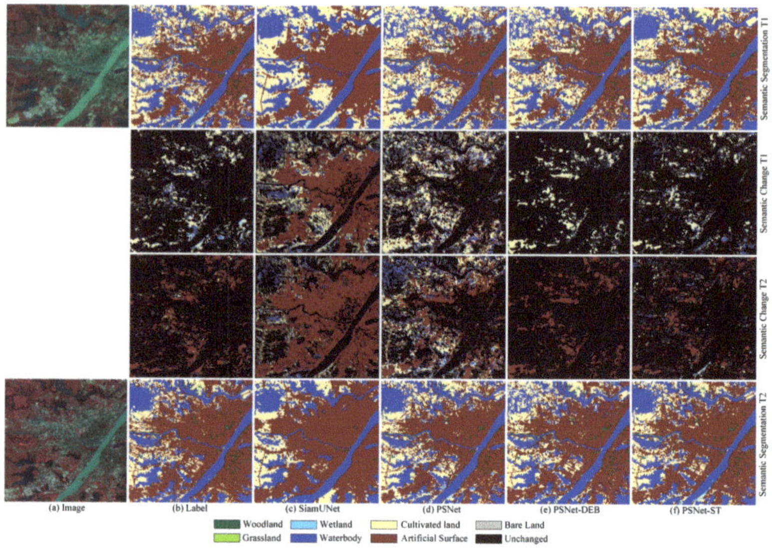

Figure A2. Comparison of the semantic change detection results between the SiamUNet and PSNet series networks. (**a**) Images. (**b**) Labels. Inference result of (**c**) the SiamUNet, (**d**) our proposed PSNet (**e**) our proposed PSNet-DEB, and (**f**) our proposed PSNet-ST.

Appendix B. Detailed Discussion for Landcover Change

Since the changing area of the adjacent two-phase images only accounts for about 6% of the total study area, most change patches are very small. The whole study area is divided into four small blocks of 10,240 × 10,240 pixels displayed separately. As shown in Figure A3, each small block is numbered 1-1, 1-2, 2-1, and 2-2. The actual area covered by each small image is 90,000 km^2.

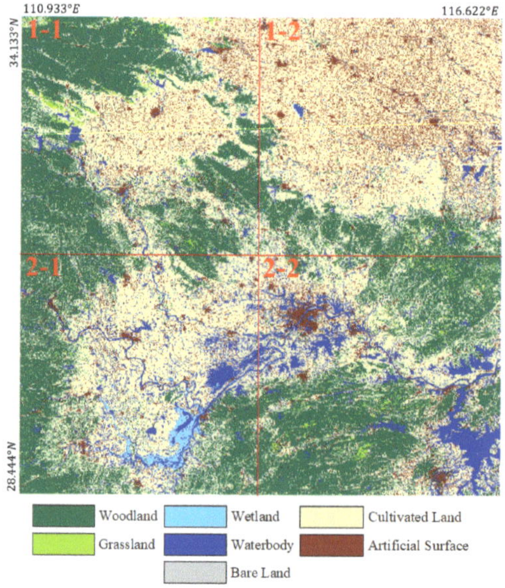

Figure A3. Geographical distribution diagram of four small blocks in study area. IDs 1-1, 1-2, 2-1 and 2-2.

Appendix B.1. Detailed Analysis of Landcover Change from 2005 to 2010

To show the geographical distribution of the changes' details, Figure A4 shows the large-scale details in block 1-1 of the study area. It can be seen that two large cultivated land became waterbodies. From the image, these two areas show the increase in the reservoir area or the newly built reservoir. A piece of the artificial surface has been turned into cultivated land and wetlands because this area is planned as a wetland ecological protection area. The other obvious changes are the transformation of cultivated land into artificial surface. This change is basically distributed around towns and cities. The change range is not large, and the distribution is scattered, which aligns with the urban development trend. Figure A5 shows the large-scale details in block 1-2 of the study area. It can be seen that there is a piece of woodland that has been turned into cultivated land, where excessive deforestation has occurred. There is an exchange phenomenon between a waterbody and a wetland. This is because the location is a reservoir and a wetland protection area. Part of the water area is shallow, and the difference is not obvious enough. Figure A6 shows the large-scale details in block 2-1 of the study area. It can be seen that a large piece of wetland has become a waterbody. This area is the Dongting Lake area. In the season when the water volume is large, the depth of the wetland becomes deeper, and it looks like a waterbody. Most of Dongting Lake is shallow, so wetlands dominate the waterbody. Exchanges between wetlands and waterbodies occur in different seasons. In addition, the most obvious change is the expansion of the city and the annexation of the surrounding cultivated land. Figure A7 shows the large-scale details in block 2-2 of the study area. It can be seen that there is an area in which the artificial surface has occupied a large amount of cultivated land. This area is Wuhan City, which shows that the development speed of the city is very fast. Another small piece of waterbody became a wetland. This area is located around Poyang Lake. It is normal for a waterbody to turn into a wetland during the dry season.

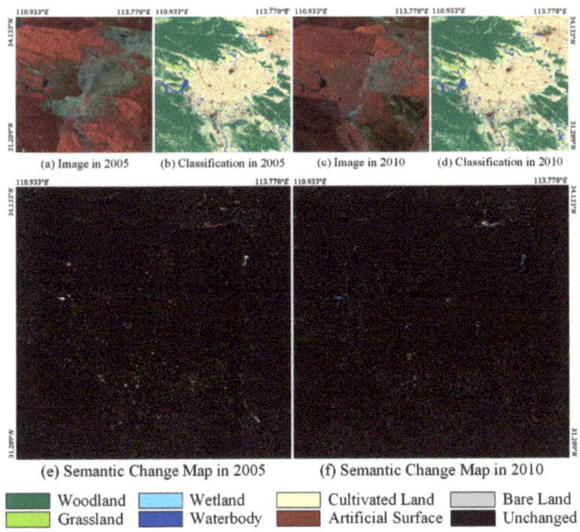

Figure A4. The landcover change map in block 1-1 from 2005 to 2010. (**a**) Image in 2005. (**b**) Classification result in 2005. (**c**) Image in 2010. (**d**) Classification result in 2010. (**e**) Semantic change map in 2005. (**f**) Semantic change map in 2010.

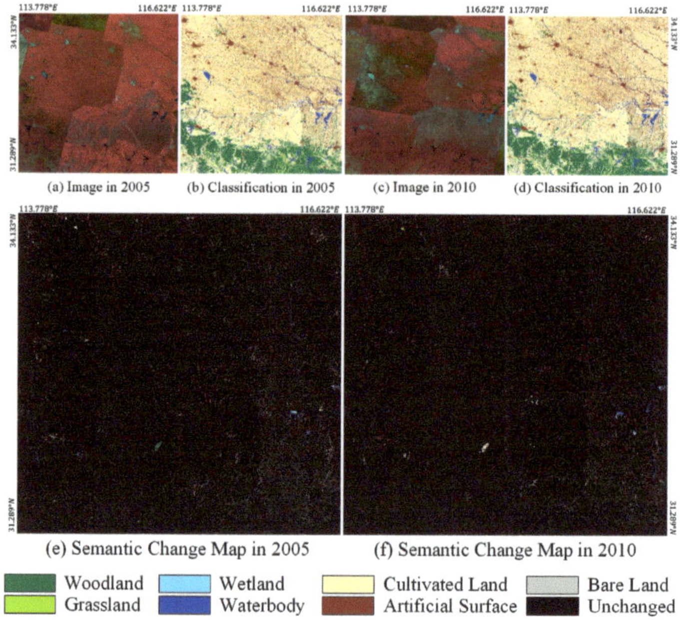

Figure A5. The landcover change map in block 1-2 from 2005 to 2010. (**a**) Image in 2005. (**b**) Classification result in 2005. (**c**) Image in 2010. (**d**) Classification result in 2010. (**e**) Semantic change map in 2005. (**f**) Semantic change map in 2010.

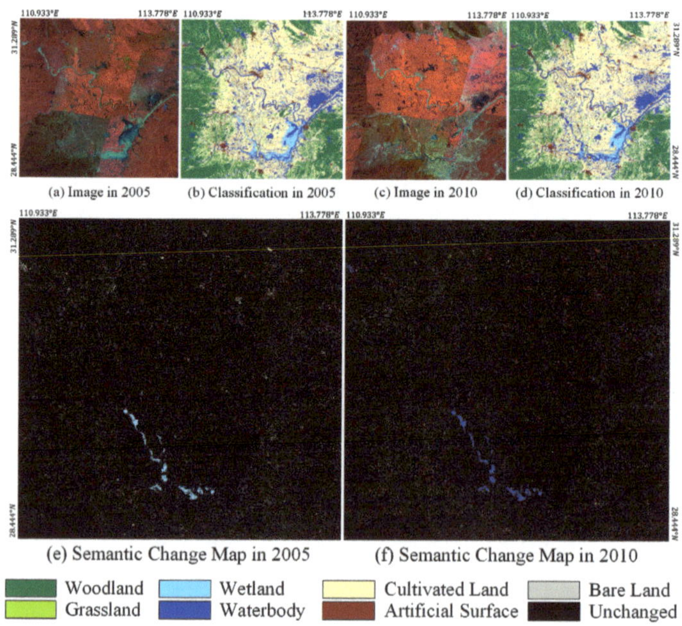

Figure A6. The landcover change map in block 2-1 from 2005 to 2010. (**a**) Image in 2005. (**b**) Classification result in 2005. (**c**) Image in 2010. (**d**) Classification result in 2010. (**e**) Semantic change map in 2005. (**f**) Semantic change map in 2010.

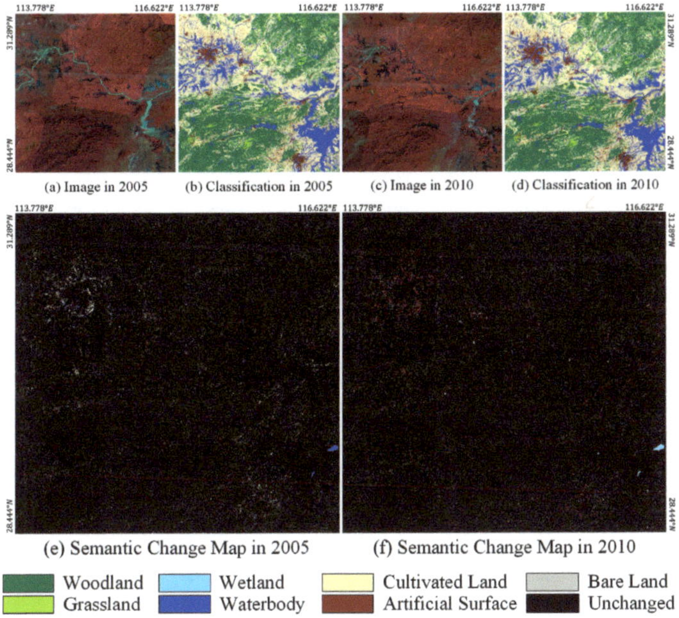

Figure A7. The landcover change map in block 2-2 from 2005 to 2010. (**a**) Image in 2005. (**b**) Classification result in 2005. (**c**) Image in 2010. (**d**) Classification result in 2010. (**e**) Semantic change map in 2005. (**f**) Semantic change map in 2010.

Appendix B.2. Detailed Analysis of Landcover Change from 2010 to 2015

To show the geographical distribution of the changes' details, Figure A8 shows the large-scale details in block 1-1 of the study area. It can be seen that a large amount of cultivated land has become artificial surface, and the speed of urbanization has been greatly accelerated. A few small pieces of cultivated land have become waterbodies, mainly because the water storage capacity of the reservoir has increased, and the original cultivated land has been adjusted for the reservoir. Figure A9 shows the large-scale details in block 1-2 of the study area. It can be seen that the most noticeable change is that the artificial surface has encroached on a large amount of surrounding cultivated land, which also reflects the speed of urbanization. Figure A10 shows the large-scale details in block 2-1 of the study area. Changes in this area are small. Since there are no megacities, there is an increase in artificial surfaces, but not very significant. In addition, there is a small increase in the waterbody, indicating that the water volume has increased in this changing region. Figure A11 shows the large-scale details in block 2-2 of the study area. It can be seen that the urban scale of Wuhan, a large city, is still expanding rapidly. On the whole, the changes in natural features from 2010 to 2015 were small, and the artificial surface expanded significantly. It shows that the urban development in the past five years is the main factor in landcover change.

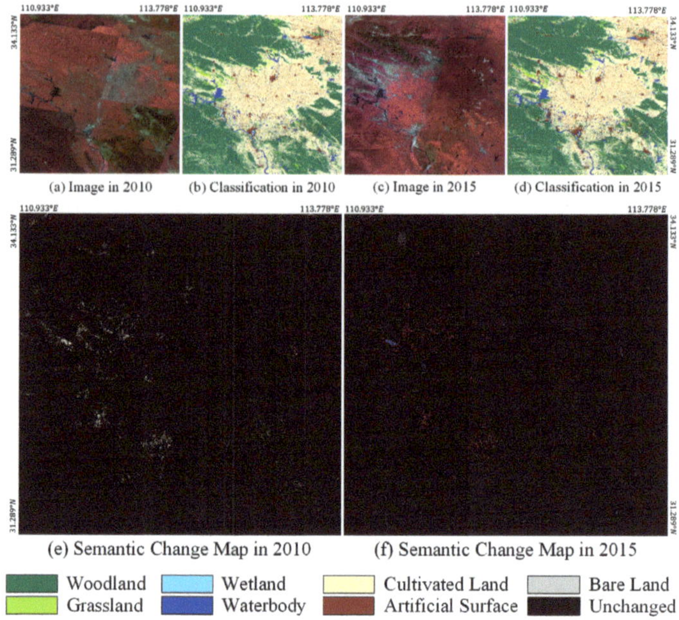

Figure A8. The landcover change map in block 1-1 from 2010 to 2015. (**a**) Image in 2010. (**b**) Classification result in 2010. (**c**) Image in 2015. (**d**) Classification result in 2015. (**e**) Semantic change map in 2010. (**f**) Semantic change map in 2015.

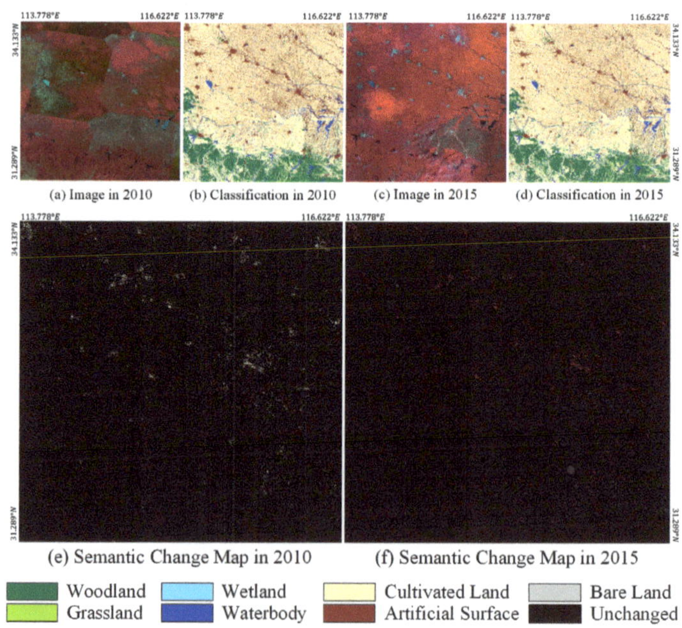

Figure A9. The landcover change map in block 1-2 from 2010 to 2015. (**a**) Image in 2010. (**b**) Classification result in 2010. (**c**) Image in 2015. (**d**) Classification result in 2015. (**e**) Semantic change map in 2010. (**f**) Semantic change map in 2015.

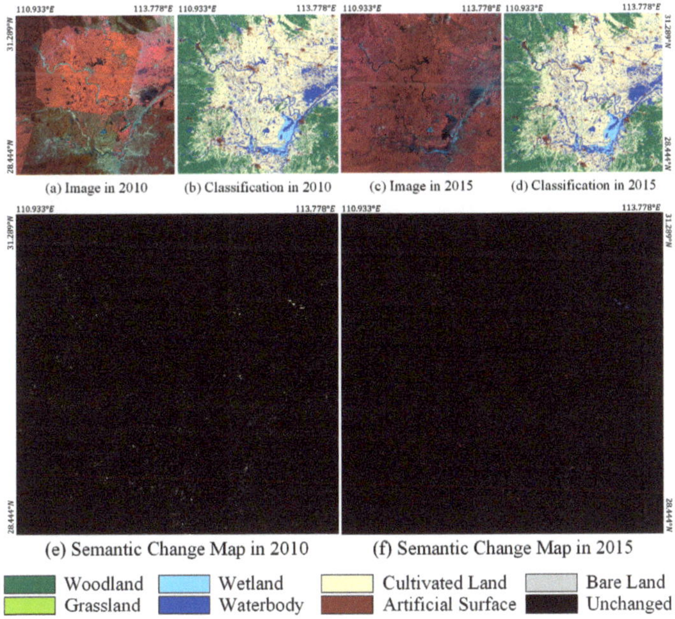

Figure A10. The landcover change map in block 2-1 from 2010 to 2015. (**a**) Image in 2010. (**b**) Classification result in 2010. (**c**) Image in 2015. (**d**) Classification result in 2015. (**e**) Semantic change map in 2010. (**f**) Semantic change map in 2015.

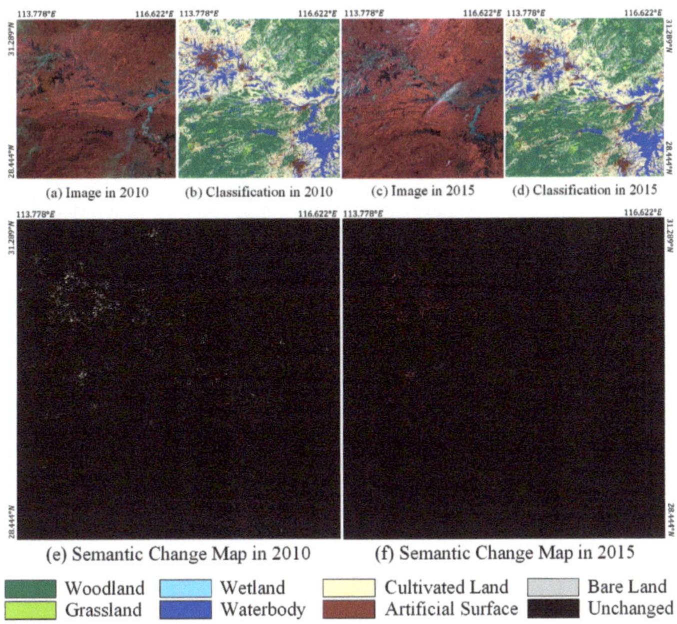

Figure A11. The landcover change map in block 2-2 from 2010 to 2015. (**a**) Image in 2010. (**b**) Classification result in 2010. (**c**) Image in 2015. (**d**) Classification result in 2015. (**e**) Semantic change map in 2010. (**f**) Semantic change map in 2015.

Appendix B.3. Detailed Analysis of Landcover Change from 2015 to 2020

To show the geographical distribution of the changes' details, Figure A12 shows the large-scale details in block 1-1 of the study area. It can be seen that there are mainly two changes. One is that the cultivated land around the town has become artificial surface, representing the expansion of the city. The other is that part of the cultivated land around the waterbody is swallowed up. This is because the water storage capacity of the reservoir increased, so the area of the waterbody expanded. Figures A13–A15 show the large-scale details in blocks 1-2, 2-1, and 2-2 of the study area, respectively. The changes are basically similar to block 1-1, and the urban expansion is pronounced. It shows that the leading development tone from 2015 to 2020 is urbanization construction, and the soil occupied during urban growth is basically cultivated land. It is also because, in most cases, towns are surrounded by cultivated land in the countryside. If the city is surrounded by woodland, the corresponding woodland will be converted into artificial surface. This trend exists in the study area, but it is much smaller than the area of cultivated land converted into artificial surface.

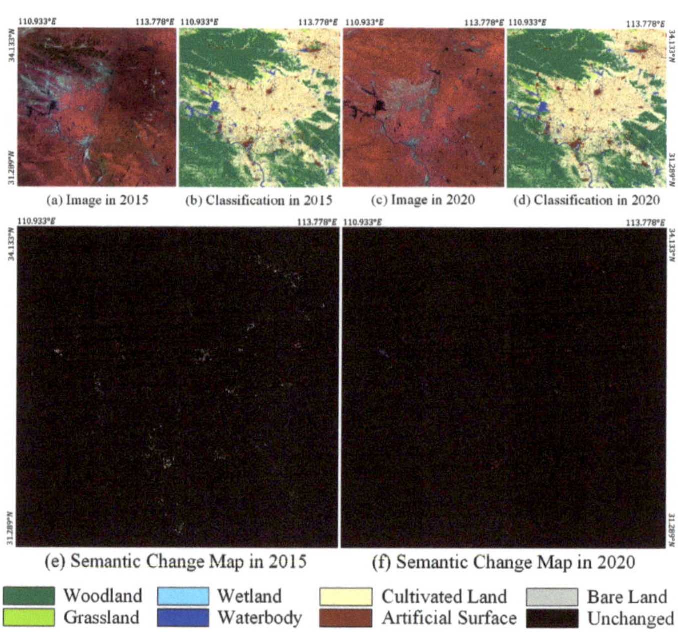

Figure A12. The landcover change map in block 1-1 from 2015 to 2020. (**a**) Image in 2015. (**b**) Classification result in 2015. (**c**) Image in 2020. (**d**) Classification result in 2020. (**e**) Semantic change map in 2015. (**f**) Semantic change map in 2020.

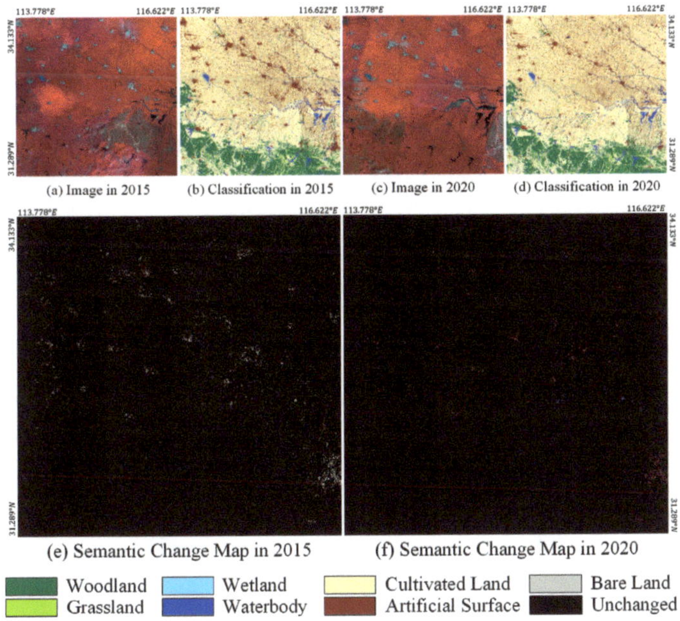

Figure A13. The landcover change map in block 1-2 from 2015 to 2020. (**a**) Image in 2015. (**b**) Classification result in 2015. (**c**) Image in 2020. (**d**) Classification result in 2020. (**e**) Semantic change map in 2015. (**f**) Semantic change map in 2020.

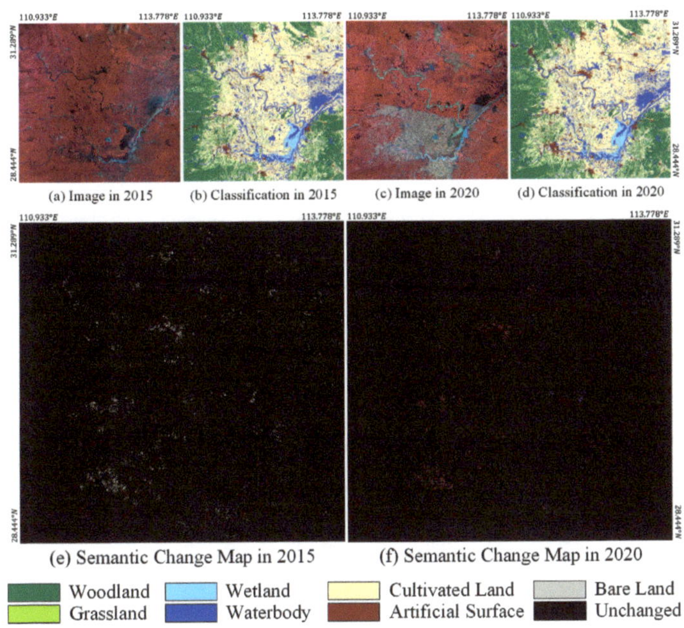

Figure A14. The landcover change map in block 2-1 from 2015 to 2020. (**a**) Image in 2015. (**b**) Classification result in 2015. (**c**) Image in 2020. (**d**) Classification result in 2020. (**e**) Semantic change map in 2015. (**f**) Semantic change map in 2020.

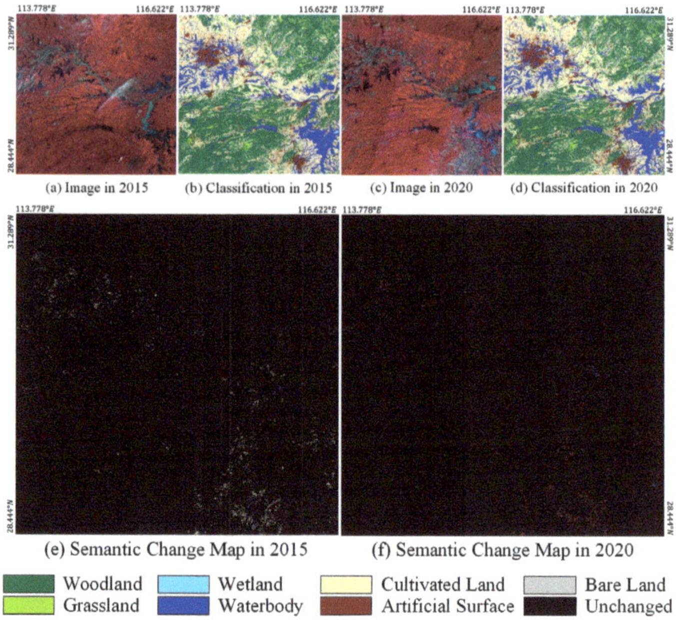

Figure A15. The landcover change map in block 2-2 from 2015 to 2020. (**a**) Image in 2015. (**b**) Classification result in 2015. (**c**) Image in 2020. (**d**) Classification result in 2020. (**e**) Semantic change map in 2015. (**f**) Semantic change map in 2020.

Appendix B.4. Analysis of Landcover Change in Fifteen Years

From Sections 4.1–4.3, it can be found that in the study area, from 2005 to 2020, the most obvious change is that a large amount of cultivated land has become artificial surface. The reduced area of cultivated land is approximately equal to the increased area of artificial surface, which is in line with the fact that the surrounding cultivated land is occupied during urban expansion. In the early days, the waterbody area increased more, which was related to the construction of water conservancy projects such as reservoirs in the area. The changes in other categories are not obvious.

We have accumulated the changes in the past 15 years, and the specific changes in landcover are shown in Table A1. The changing area within the study area amounted to 28,985.68 km^2. The woodland increased by 232.62 km^2 and has a small dynamic balance changed with grassland, waterbody, cultivated land, and artificial surface. Among them, the mutual change with cultivated land is the most because there are cases of returning farmland to forest and cutting down trees for reclamation in this area. Grassland has decreased by 145.85 km^2 and has a small dynamic balance changed with woodland and cultivated land. Affected by the season, the grass on the ground will appear in the image after the trees have fallen, and the grass will be blocked when the leaves are dense. Therefore, there will be changes in the dynamic balance between woodland and grassland. Similarly, weeds may grow when cultivated land is idle, and cultivated land can be restored after weeding. Therefore, there will be exchanges between grassland and cultivated land. Wetlands decreased by 360.58 km^2, mainly turning into waterbodies. When the water storage capacity of wetlands increases, wetlands may become lakes or reservoirs. In addition, there is a small amount of mutual change between wetlands and woodland, cultivated land, and artificial surfaces. This is because wetlands are relatively fragile and may become other categories if they are not well protected. The waterbody area increased by 28.4 km^2. It is very small. However, in the past 15 years, the waterbody has been in a state of dynamic change. The mutual change between waterbody and woodland areas indicates that the newly added reservoir in this area will submerge some woodland. At the same time, trees will become the main features in some dry water areas. The exchange between waterbodies and cultivated land can also reflect the cultivated land being submerged and newly reclaimed cultivated land brought about by the construction of new reservoirs and the drying up of waters. There are also minor changes in the waterbody and artificial surface because the small villages and towns will be relocated when the reservoir is newly built. The cultivated land decreased by 7893.78 km^2, mainly because the artificial surface gradually swallowed it up in urbanization. There is basically a state of balanced exchange between cultivated land and other categories. The artificial surface has increased by 8083.83 km^2. The speed of urbanization also reflects the outstanding economic level in the past 15 years. The newly added artificial surface area has reached one-third of the unchanged artificial surface area. In other words, the urban area has increased by about 34.1% in 15 years. The bare land area increased by 55.37 km^2, with relatively little change between categories. Bare land accounts for a small proportion, and due to conditions such as imaging seasons and human factors such as urban planning and construction, bare land will appear or disappear briefly.

Table A1. Landcover change statistics in the past 15 years. (Unit: square kilometers).

	Woodland	**Grassland**	**Wetland**	**Waterbody**	**Cultivated Land**	**Artificial Surface**	**Bare Land**	**Total Out**
Woodland	117,813.68	381.17	3.10	270.82	4112.36	643.07	78.87	5489.40
Grassland	508.33	3532.18	0.25	16.03	219.32	16.59	4.44	764.95
Wetland	26.53	1.02	2232.25	424.07	71.03	43.02	1.70	567.38
Waterbody	330.94	9.09	135.38	24,587.92	1692.60	598.83	26.08	2792.92
Cultivated Land	4549.51	218.68	55.46	1870.65	176,007.48	9600.65	77.70	16,372.67
Artificial Surface	262.49	6.79	12.50	208.82	2334.08	23,665.93	20.13	2844.81
Bare Land	44.21	2.34	0.11	30.93	49.48	26.48	662.25	153.55
Total In	5722.02	619.10	206.79	2821.32	8478.88	10,928.64	208.92	28,985.68
Total Change	232.62	−145.85	−360.58	28.40	−7893.78	8083.83	55.37	-

Figure A16 shows the change details in the study area from 2005 to 2020. It can be seen that the transformation of cultivated land into artificial surface is the most critical change, and it is mainly distributed around towns and cities. It is consistent with the law of urban expansion. At the same time, due to the newly built reservoirs and other water conservancy projects in this area and the protection policy of Dongting Lake, some cases of cultivated land turning into waterbodies can also be found in the figure. Combining the data and figures for landcover changes from 2005 to 2020, we can conclude that urbanization is the core keyword in the study area. Urbanization has an apparent relationship with social and economic development. The speed of urban expansion is becoming faster and faster, and we should pay attention to the occupation of cultivated land in the process of expansion because cultivated land is directly related to food. Wetlands and waterbodies are directly related to the ecological environment. How to protect wetlands is a significant issue that decision-making departments cannot ignore. As the main features of greening, woodland and grassland also need to be protected and increased as much as possible to achieve sustainable development goals.

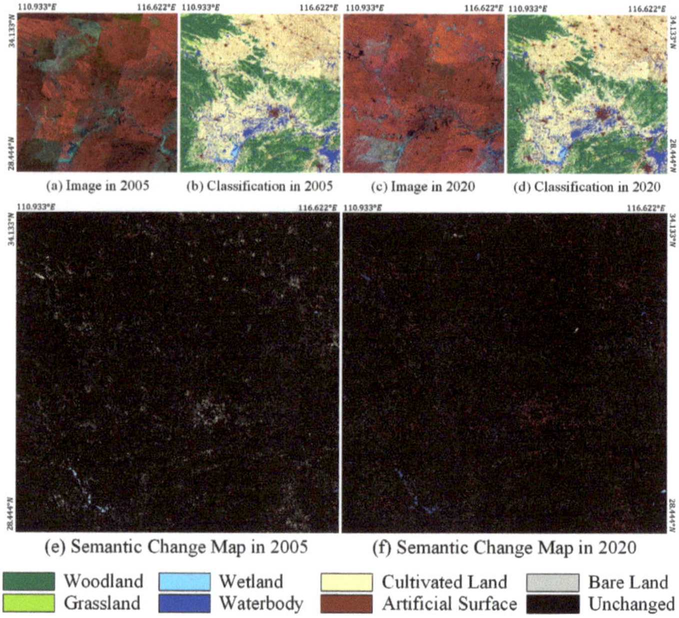

Figure A16. The landcover change map in the past 15 years. (**a**) Image in 2005. (**b**) Classification result in 2005. (**c**) Image in 2020. (**d**) Classification result in 2020. (**e**) Semantic change map in 2005. (**f**) Semantic change map in 2020.

References

1. Ma, L.; Liu, Y.; Zhang, X.; Ye, Y.; Yin, G.; Johnson, B.A. Deep learning in remote sensing applications: A meta-analysis and review. *ISPRS J. Photogramm. Remote Sens.* **2019**, *152*, 166–177. [CrossRef]
2. Hansen, M.C.; Loveland, T.R. A review of large area monitoring of land cover change using Landsat data. *Remote Sens. Environ.* **2012**, *122*, 66–74. [CrossRef]
3. Feddema, J.J.; Oleson, K.W.; Bonan, G.B.; Mearns, L.O.; Buja, L.E.; Meehl, G.A.; Washington, W.M. The importance of land-cover change in simulating future climates. *Science* **2005**, *310*, 1674–1678. [CrossRef] [PubMed]
4. Lepers, E.; Lambin, E.F.; Janetos, A.C.; DeFries, R.; Achard, F.; Ramankutty, N.; Scholes, R.J. A synthesis of information on rapid land-cover change for the period 1981–2000. *BioScience* **2005**, *55*, 115–124. [CrossRef]
5. Evans, T.P.; Manire, A.; De Castro, F.; Brondizio, E.; McCracken, S. A dynamic model of household decision-making and parcel level landcover change in the eastern Amazon. *Ecol. Model.* **2001**, *143*, 95–113. [CrossRef]
6. Rogan, J.; Miller, J.; Stow, D.; Franklin, J.; Levien, L.; Fischer, C. Land-cover change monitoring with classification trees using Landsat TM and ancillary data. *Photogramm. Eng. Remote Sens.* **2003**, *69*, 793–804. [CrossRef]
7. Ma, Y.; Wu, H.; Wang, L.; Huang, B.; Ranjan, R.; Zomaya, A.; Jie, W. Remote sensing big data computing: Challenges and opportunities. *Future Gener. Comput. Syst.* **2015**, *51*, 47–60. [CrossRef]
8. Zhang, B.; Wu, Y.; Zhao, B.; Chanussot, J.; Hong, D.; Yao, J.; Gao, L. Progress and challenges in intelligent remote sensing satellite systems. *IEEE J. Sel. Top. Appl. Earth Obs. Remote Sens.* **2022**, *15*, 1814–1822. [CrossRef]
9. Zhang, B. Remotely sensed big data era and intelligent information extraction. *Geomat. Inf. Sci. Wuhan Univ.* **2018**, *43*, 1861–1871.
10. Zhang, B.; Chen, Z.; Peng, D.; Benediktsson, J.A.; Liu, B.; Zou, L.; Li, J.; Plaza, A. Remotely sensed big data: Evolution in model development for information extraction [point of view]. *Proc. IEEE* **2019**, *107*, 2294–2301. [CrossRef]
11. Bhandari, A.; Kumar, A.; Singh, G. Feature extraction using Normalized Difference Vegetation Index (NDVI): A case study of Jabalpur city. *Procedia Technol.* **2012**, *6*, 612–621. [CrossRef]
12. Xing, D.P. One Method of Urban Land Covers Information Extraction. In Proceedings of the Applied Mechanics and Materials, Hong Kong, China, 17–18 August 2013; Trans Tech Publications: Zurich, Switzerland, 2013; Volume 380, pp. 4011–4014.
13. Xu, H. Modification of normalised difference water index (NDWI) to enhance open water features in remotely sensed imagery. *Int. J. Remote Sens.* **2006**, *27*, 3025–3033. [CrossRef]
14. Wei, W.; Chen, X.; Ma, A. Object-oriented information extraction and application in high-resolution remote sensing image. In Proceedings of the 2005 IEEE International Geoscience and Remote Sensing Symposium, IGARSS'05, Seoul, Korea, 25–29 July 2005; Volume 6, pp. 3803–3806.
15. Lowe, D.G. Distinctive image features from scale-invariant keypoints. *Int. J. Comput. Vis.* **2004**, *60*, 91–110. [CrossRef]

16. Rublee, E.; Rabaud, V.; Konolige, K.; Bradski, G. ORB: An efficient alternative to SIFT or SURF. In Proceedings of the 2011 International Conference on Computer Vision, Washington, DC, USA, 6–13 November 2011; pp. 2564–2571.
17. Pal, M. Ensemble of support vector machines for land cover classification. *Int. J. Remote Sens.* **2008**, *29*, 3043–3049. [CrossRef]
18. Hayes, M.M.; Miller, S.N.; Murphy, M.A. High-resolution landcover classification using Random Forest. *Remote Sens. Lett.* **2014**, *5*, 112–121.
19. Kussul, N.N.; Lavreniuk, N.S.; Shelestov, A.Y.; Yailymov, B.Y.; Butko, I.N. Land cover changes analysis based on deep machine learning technique. *J. Autom. Inf. Sci.* **2016**, *48*, 42–54. [CrossRef]
20. LeCun, Y.; Bengio, Y.; Hinton, G. Deep learning. *Nature* **2015**, *521*, 436–444. [CrossRef]
21. Zhu, X.X.; Tuia, D.; Mou, L.; Xia, G.S.; Zhang, L.; Xu, F.; Fraundorfer, F. Deep learning in remote sensing: A comprehensive review and list of resources. *IEEE Geosci. Remote Sens. Mag.* **2017**, *5*, 8–36. [CrossRef]
22. Lv, Z.; Liu, T.; Benediktsson, J.A.; Falco, N. Land cover change detection techniques: Very-high-resolution optical images: A review. *IEEE Geosci. Remote Sens. Mag.* **2021**, *10*, 44–63. [CrossRef]
23. Lambin, E.F.; Geist, H.J.; Lepers, E. Dynamics of land-use and land-cover change in tropical regions. *Annu. Rev. Environ. Resour.* **2003**, *28*, 205–241. [CrossRef]
24. Bingfang, W.; Quanzhi, Y.; Changzhen, Y.; Zongming, W.; Xinfang, Y.; Ainong, L.; Ronghua, M.; Jinliang, H.; Jingsong, C.; Cun, C.; et al. Land cover changes of China from 2000 to 2010. *Quat. Sci.* **2014**, *34*, 723–731.
25. Verburg, P.H.; Kok, K.; Pontius, R.G.; Veldkamp, A. Modeling land-use and land-cover change. In *Land-Use and Land-Cover Change*; Springer: Berlin/Heidelberg, Germany, 2006; pp. 117–135.
26. Lunetta, R.S.; Knight, J.F.; Ediriwickrema, J.; Lyon, J.G.; Worthy, L.D. Land-cover change detection using multi-temporal MODIS NDVI data. *Remote Sens. Environ.* **2006**, *105*, 142–154. [CrossRef]
27. Sefrin, O.; Riese, F.M.; Keller, S. Deep learning for land cover change detection. *Remote Sens.* **2020**, *13*, 78. [CrossRef]
28. Daudt, R.C.; Le Saux, B.; Boulch, A. Fully convolutional siamese networks for change detection. In Proceedings of the 2018 25th IEEE International Conference on Image Processing (ICIP), Athens, Greece, 7–10 October 2018; pp. 4063–4067.
29. Alcantarilla, P.F.; Stent, S.; Ros, G.; Arroyo, R.; Gherardi, R. Street-view change detection with deconvolutional networks. *Auton. Robot.* **2018**, *42*, 1301–1322. [CrossRef]
30. Zhang, C.; Yue, P.; Tapete, D.; Jiang, L.; Shangguan, B.; Huang, L.; Liu, G. A deeply supervised image fusion network for change detection in high resolution bi-temporal remote sensing images. *ISPRS J. Photogramm. Remote Sens.* **2020**, *166*, 183–200. [CrossRef]
31. Peng, D.; Zhang, Y.; Guan, H. End-to-end change detection for high resolution satellite images using improved UNet++. *Remote Sens.* **2019**, *11*, 1382. [CrossRef]
32. Peng, X.; Zhong, R.; Li, Z.; Li, Q. Optical remote sensing image change detection based on attention mechanism and image difference. *IEEE Trans. Geosci. Remote Sens.* **2020**, *59*, 7296–7307. [CrossRef]
33. Zhou, Z.; Rahman Siddiquee, M.M.; Tajbakhsh, N.; Liang, J. Unet++: A nested u-net architecture for medical image segmentation. In *Deep Learning in Medical Image Analysis and Multimodal Learning for Clinical Decision Support*; Springer: Berlin/Heidelberg, Germany, 2018; pp. 3–11.
34. Zhang, X.; Liu, L.; Chen, X.; Gao, Y.; Xie, S.; Mi, J. GLC_FCS30: Global land-cover product with fine classification system at 30 m using time-series Landsat imagery. *Earth Syst. Sci. Data* **2021**, *13*, 2753–2776. [CrossRef]
35. Ronneberger, O.; Fischer, P.; Brox, T. U-net: Convolutional networks for biomedical image segmentation. In Proceedings of the International Conference on Medical Image Computing and Computer-Assisted Intervention, Munich, Germany, 5–9 October 2015; Springer: Berlin/Heidelberg, Germany, 2015; pp. 234–241.
36. He, K.; Zhang, X.; Ren, S.; Sun, J. Deep residual learning for image recognition. In Proceedings of the IEEE Conference on Computer Vision and Pattern Recognition, Las Vegas, NV, USA, 27–30 June 2016; pp. 770–778.
37. Ioffe, S.; Szegedy, C. Batch normalization: Accelerating deep network training by reducing internal covariate shift. In Proceedings of the International conference on machine learning. PMLR, Lille, France, 7–9 July 2015; pp. 448–456.
38. Paszke, A.; Gross, S.; Massa, F.; Lerer, A.; Bradbury, J.; Chanan, G.; Killeen, T.; Lin, Z.; Gimelshein, N.; Antiga, L.; et al. PyTorch: An imperative style, high-performance deep learning library. In Proceedings of the Advances in Neural Information Processing Systems, Vancouver, BC, Canada, 8–14 December 2019; pp. 8024–8035.
39. Loshchilov, I.; Hutter, F. Decoupled weight decay regularization. *arXiv* **2017**, arXiv:1711.05101.
40. Berman, M.; Triki, A.R.; Blaschko, M.B. The lovász-softmax loss: A tractable surrogate for the optimization of the intersection-over-union measure in neural networks. In Proceedings of the IEEE Conference on Computer Vision and Pattern Recognition, Salt Lake City, UT, USA, 18–23 June 2018; pp. 4413–4421.

sustainability

MDPI

Article

Crop Identification and Analysis in Typical Cultivated Areas of Inner Mongolia with Single-Phase Sentinel-2 Images

Jing Tang [1], Xiaoyong Zhang [1], Zhengchao Chen [2] and Yongqing Bai [2,*]

1 Beijing Key Laboratory of High Dynamic Navigation, Beijing Information Science and Technology University, Beijing 100101, China
2 Airborne Remote Sensing Center, Aerospace Information Research Institute, Chinese Academy of Sciences, Beijing 100094, China
* Correspondence: baiyq@aircas.ac.cn

Abstract: The Hetao Plain and Xing'an League are the major cultivated areas and main grain producing areas in Inner Mongolia, and their crop planting structure significantly affects the grain output and economic development in Northern China. Timely and accurate identification, extraction and analysis of typical crops in Xing'an League and Hetao Plain can provide scientific guidance and decision support for crop planting structure research and food security in ecological barrier areas in Northern China. The pixel samples and the neighborhood information were fused to generate a spectral spatial dataset based on single-phase Sentinel-2 images. Skcnn_Tabnet, a typical crop remote sensing classification model, was built at the pixel scale by adding the channel attention mechanism and the corn, sunflower, and rice in the Hetao Plain were quickly identified and studied. The results of this study suggest that the model exhibits high crop recognition ability, and the overall accuracy of the three crops is 0.9270, which is 0.1121, 0.1004, and 0.0874 higher than the Deeplabv3+, UNet and RF methods, respectively. This study confirms the feasibility of the deep learning model in the application research of large-scale crop classification and mapping and provides a technical reference for achieving the automatic national crop census.

Keywords: crop identification; multispectral remote sensing; Sentinel-2; deep learning; attention mechanism

check for updates

Citation: Tang, J.; Zhang, X.; Chen, Z.; Bai, Y. Crop Identification and Analysis in Typical Cultivated Areas of Inner Mongolia with Single-Phase Sentinel-2 Images. *Sustainability* **2022**, *14*, 12789. https://doi.org/10.3390/su141912789

Academic Editors: Jun Qin and Hou Jiang

Received: 23 July 2022
Accepted: 30 September 2022
Published: 7 October 2022

1. Introduction

Food security lays a solid basis for national security. As the COVID-19 pandemic rages through the whole world, the international situation is severe and complex, and food security is at stake. During China's "two sessions" in 2022, General Secretary Xi Jinping once again mentioned that "food security is the bottom-line task of comprehensively boosting rural revitalization, and it is imperative to keep the red line of 1.8 billion mu of arable land". As a vital granary in Northern China, Inner Mongolia has an area of 6.713 million hectares of arable land [1], and the per capita arable land area is 0.24 hectares ranking first in China [2]. Accordingly, research on the extraction and monitoring methods of grain crops in Inner Mongolia, accurate and timely crop type mapping plays a vital role in crop yield estimation, soil management, and food supply. Furthermore, it is of critical significance to ensure national food security and prevent the tendency of "non-grain" [3].

In the past few decades, remote sensing has gradually become an effective tool for crop identification due to its wide range and strong timeliness. With the enhancement of earth observation ability, much research on crop remote sensing identification have been spawned. Ibrahim et al. [4] used phenological and spectroscopic temporal metrics obtained from Sentinel-2 images for crop type mapping and cropping system mapping with an overall accuracy of 84%. You et al. [5] based on the long sequence Sentinel-2 image of the GEE platform and the Random Forest (RF) algorithm, produced three typical crops in Northeast China for years of classification products. In brief, the existing research methods for crop classification and extraction primarily comprise a hierarchical decision tree based on spectral features [6–8], threshold segmentation method based on time series normalized difference vegetation index (NDVI) [9–11], as well as feature index-based RF method [12–15], etc. The existing research scope is small and concentrated [16,17], and the data source requires multi-temporal images of the entire crop growth period [18,19]. However, continuous multi-temporal images during the crop growth cycle are often missing due to factors (e.g., cloud cover and rainy weather). In addition, data processing operations (e.g., registration and fusion of multi-source and multi-temporal image data) have certain technical thresholds, thus significantly affecting the accuracy of crop classification.

With the rapid development of remote sensing technology and the continuous expansion of application fields, users have increasing requirements for the efficiency and accuracy of crop mapping. Conventional crop identification methods are difficult to solve the data redundancy problem caused by remote sensing of big data. Deep learning has provided a novel idea for crop remote sensing identification for its powerful image feature extraction ability. To be specific, semantic segmentation technology [20] is capable of analyzing the deep semantic information of images and conducting pixel-level supervised classification [21] quickly, which has been favored by many scholars. For instance, Du et al. [22] extracted rice from Arkansas using a semantic segmentation model U-net based on time-series Landsat imagery and the Cropland Data Layer (CDL). Rice could be identified in the heading stage with an overall accuracy of 0.86. Der et al. [23] used drones to obtain high spatial resolution drone images in experimental farms. As well, the SegNet semantic segmentation network was used for crop extraction through the texture gap between different crops. The study achieved an overall classification accuracy of 89.44%. Wang et al. [24] adopted the optimized DeepLabV3+ network to efficiently identify glaciers, lakes, grasslands, and bare land on Sentinel-2 remote sensing images at the source of the Yangtze River, with mAP of 0.639, mIoU of 0.778, and Kappa of 0.825. Since semantic segmentation requires pixel-level sample labels, the production cost is high and the efficiency is difficult to meet the requirements. Thus, reducing the complexity of sample production and using more advanced deep learning methods to achieve rapid and accurate crop extraction is also an urgent problem to be studied.

As an essential ecological barrier in Northern China, Inner Mongolia is vast and sparsely populated. The cultivated land is concentrated in the Hetao Plain in the middle and the Xing'an League in the east, among which sunflower, maize, and rice account for the largest proportions, meeting the needs of crop monitoring based on remote sensing big data. Accordingly, the Hetao Plain and the southwestern part of the Xing'an League were selected as the study area in this study, and single-phase Sentinel-2 images and a small number of samples were employed for automatic identification and analysis of sunflower, maize, and rice based on the optimized Tabnet model. The aim is at solving the difficult problem of capturing optical images in the crop growth cycle, maximizing the timeliness of crop mapping, verifying the applicability of deep learning models in large-scale crop remote sensing identification applications, and providing technical references for the automation of national crop censuses.

2. Materials and Methods

2.1. Study Area

Xing'an League (Figure 1) is located in the transition zone between the Greater Xing'an Mountains and Songnen Plain in the Northeastern part of Inner Mongolia (42°25′–47°65′ N, 119°47′–123°62′ E), with 909,490 hectares of arable land, taking up 13% of the total arable land in the autonomous region [1]. The region exhibits a complex topography, with dense hills in the northwest, flat and thick soil in the southern plain, and sufficient water resources, thus providing convenience for water diversion and irrigation and agricultural machinery operations [25]. The area abounds with rice, maize, and sunflower, and is a vital agricultural production base in Inner Mongolia. Rice is sown in mid-April and harvested at the end of September. Sunflowers are sown in late May, bloom in early July, and harvest in mid-September. Corn is sown in mid-May and matures in mid-to-late September. Two counties in the southwest of Xing'an League (Horqin Right Wing Middle Banner and Tuquan County) were selected as the typical experimental areas to build a crop remote sensing classification model.

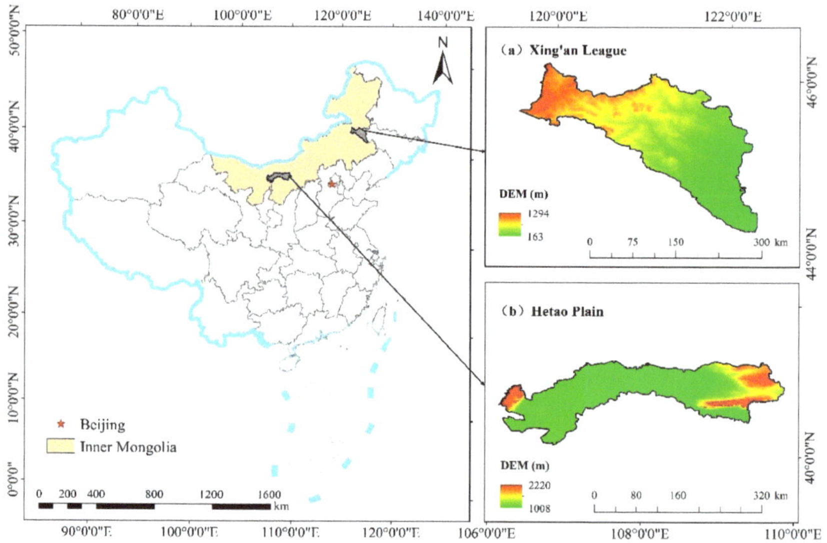

Figure 1. Study area overview map. The location of Inner Mongolia Autonomous Region of China, the study area in Inner Mongolia with two agroecological zones (Xing'an League (**a**) and Hetao Plain (**b**)).

The Hetao Plain, a typical cultivated area in central Inner Mongolia, was selected for model application to verify the applicability of the model. The Hetao Plain is located in the south of Bayannaoer City, Inner Mongolia Autonomous Region of China (40°1′–40°4′ N, 106°1′–109°4′ E), which belongs to a typical continental monsoon climate, with hot and dry summers and cold winters, the annual rainfall is less than 250 mm, while the potential evaporation is 2011 to 2300 mm per year [26]. Although the region has an arid climate, the Yellow River that flows through the region provides valuable water resources for agricultural development. The total arable land area of the region is nearly 733,000 hectares [25], mainly planted with sunflower, maize, and rice. Sunflower and maize have the same phenological period, and they are both sown in May and harvested in September-October. In general, rice is one-season rice, sown in mid-May, and harvested at the end of September.

2.2. Data and Samples

2.2.1. Remote Sensing Data and Processing

This study was primarily based on Sentinel-2 L1C images for crop classification, and the data originated from the European Space Agency (ESA) Copernicus Data Center (https://scihub.copernicus.eu/, accessed on 20 April 2022).

Sentinel-2 comprises two satellites equipped with a Multispectral Imager (MSI) with a revisit period of 5 days and 13 bands (Table 1), including four 10 m resolution bands, six 20 m resolution bands, as well as three 60 m resolution bands. The Sen2Cor (http://step.esa.int/main/third-party-plugins-2/sen2cor/, accessed on 25 April 2022) plugin released by ESA was adopted to analyze the Sentinel-2 L1C raw images for radiometric calibration and atmospheric correction processing since the L1C-level data are not atmospherically corrected. Furthermore, the low-resolution band was resampled to 10 m resolution to acquire the image data for deep learning classification.

Table 1. Detailed information of 13 spectral bands of Sentinel-2.

Bands	Name	Central Wavelength (nm)	Band Width (nm)	Spatial Resolution (m)
1	Coastal aerosol	442.7	21	60
2	Blue	492.4	66	10
3	Green	559.8	36	10
4	Red	664.6	31	10
5	Vegetation red edge	704.1	15	20
6	Vegetation red edge	740.5	15	20
7	Vegetation red edge	782.8	20	20
8	NIR [1]	832.8	106	10
8A	Narrow NIR	864.7	21	20
9	Water vapor	945.1	20	60
10	SWIR [2] Cirrus	1373.5	31	60
11	SWIR	1613.7	91	20
12	SWIR	2202.4	175	20

[1] Near-infrared band. [2] Shortwave-infrared band.

2.2.2. Samples

Real and reliable samples lay a basis for ensuring the accuracy of deep learning model training and classification results. From July 20 to August 30, 2019, a crop planting structure survey was carried out in the Xing'an League research area. A total of 60 corn sampling points, 25 rice sampling points, and 41 sunflower sampling points were acquired using the handheld Global Positioning System (GPS) (GARMIN ETREX 32 X). To avoid the appearance of mixed pixels, the area of the sampling points was greater than 100 m × 100 m. The spatial distribution of sampling points (Figure 3) suggests that sunflowers are largely distributed in the northeastern part of the study area, summer maize is mainly distributed in the southern part of the study area, and rice is distributed in the east along the river.

The optimal classification window was determined from 20 July to 25 August, 2019 in accordance with the phenological characteristics and NDVI index of local sunflower, corn, and rice. Sentinel-2 L1C images with a cloud cover of less than 5% in the study area were selected and downloaded, including five scenes in the Xing'an League study area (T51TVL, T51TVK, T51TUL, T51TUK, T51TUM) and four scenes in the Hetao Plain study area (T48TXK, T48TXL, T49TBF, T49TCF). In view of the problems of uneven distribution and offset of samples collected in the field, the data collected in the field were superimposed and displayed with Sentinel-2 images of the same period in this study. False color synthesis (band8, band11, and band4) of specific band combinations is used to enhance the discrimination between different target features, so as to carry out sample correction. In order to show clearer details, we use small tiles of 256 × 256 pixels for visual analysis of the image. Figure 2 presents the texture and color characteristics of sunflower, maize, and rice on false color images in the Xing'an League study area.

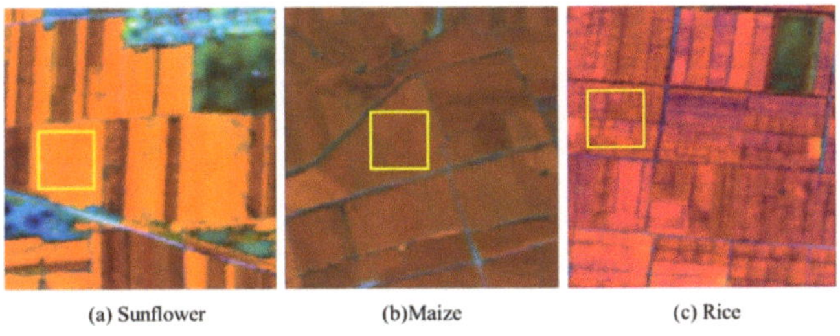

Figure 2. Crop characteristics on false color synthesis Sentinel-2 images (band8, band11, and band4, the stretch type is standard deviations). We use small tiles of 256 × 256 pixels, and the size of the yellow box is 64 × 64.

The detection and classification performance of a deep learning model is largely dependent on the type and quantity of training samples. The richer the types and number of samples, the better the performance of the model and the better the generalization performance will be [27]. In this study, the regions with the same features were visually interpreted, and the Region of Interest (ROI) was delineated to expand the samples based on the texture, color, and other features of existing samples on false color Sentinel-2 images. At the same time, the detailed information on the GF-1 images was used as auxiliary data so that the boundary of the ROI falls within a pure crop field. To avoid the interference of the surrounding complex environment on the crop recognition effect, negative samples (e.g. water bodies and other crops) were added (Figure 3). In addition, manual plotting does not require pixel-level sample labeling, but only the interior of typical crop planting areas and necessary negative samples are labeled with vector polygons. The expanded sample distribution was more uniform, which is beneficial to increase the stability of the model.

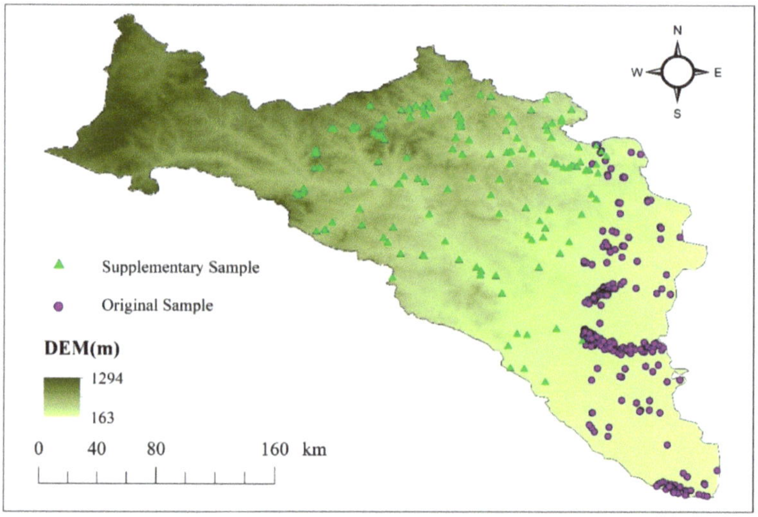

Figure 3. Sample spatial distribution. Original Sample represents samples collected in the field Supplementary Sample represents hand-plotted samples based on visual interpretation of cro features.

The number of ROIs and pixel points of a wide variety of samples is listed in Table 2

Table 2. The number of selected regions of interest (ROI) and number of pixels.

Type	Number of ROIs	Number of Pixels
Maize	471	209,720
Sunflower	326	153,489
Rice	207	130,193
Waters	29	56,079
Other Crops	20	47,701

2.2.3. Auxiliary Data

This study selects Google Earth images (spatial resolution of 1 m) as the direct verification data for the crop identification results in this study. Moreover, the 2019 Inner Mongolia Autonomous Region Statistical Yearbook (http://tj.nmg.gov.cn, accessed on 20 April 2022) was collected, which included data on the sown area and crop yield of a wide variety of crops at the county scale, which was used to indirectly verify the reliability of this study.

2.3. Methods and Models

The main ideas of this research mainly include the following three aspects: (1) The dataset was constructed, including sample data extraction, cleaning, and equalization, and the integration of neighborhood information into the sample; (2) Based on the divided dataset, a classification model was built for three crops of maize, sunflower, and rice; (3) The applicability of the crop extraction model was verified, the model was transferred to the Hetao Plain to identify crops in the same period, the crop distribution map of the Hetao Plain of 2019 was generated pixel by pixel, and Google Earth high-resolution images and statistical yearbook data were adopted to verify crop identification precision.

2.3.1. Sample Data Cleaning and Division

To reduce the labeling cost, the sketched samples are polygon vectors, inconsistent with the pixel-level samples required by the model. Thus, in this study, the ROI and the image corresponded one by one through the sample vector polygon and the geographic coordinates of the image. The ray method [28] was adopted to judge whether the pixel is in the vector polygon; if so, the value of each band of the pixel and the corresponding sample label value were recorded. Since there may be mixed pixels in remote sensing images, quartile box plots (Figure 4) were drawn in this study for three crops (maize, rice, sunflower) and water bodies to ensure the purity of the samples.

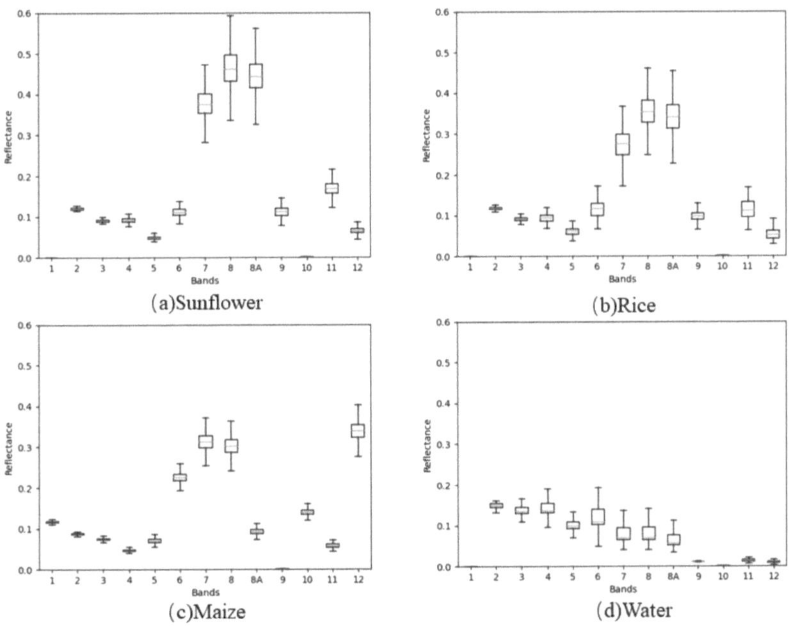

Figure 4. Spectral features of sunflower (**a**), rice (**b**), maize (**c**), and water (**d**). The x-axis represents bands. The y-axis represents the reflectance of each band.

To increase the separability among crops, the reasonable range of spectral values of the respective band of crop samples was obtained, and abnormal samples (As long as one of the IQR values of all bands is out of range, it will be discarded.) beyond this range were deleted to reduce the classification complexity of the feature space. The specific operations are elucidated as follows.

The quartiles of each band of all samples were calculated, including the upper quantile Q_1, the median Q_2, and the lower quantile Q_3. The center points were sorted by the numerical magnitude of each band to obtain the positions of the quartiles:

$$\begin{cases} Q_1 = \frac{n+1}{4} \\ Q_2 = \frac{n+1}{2} \\ Q_3 = \frac{3(n+1)}{4} \end{cases} \tag{1}$$

where n denotes the number of samples. The next step calculates the interquartile range:

$$IQR = Q_3 - Q_1 \tag{2}$$

Subsequently, the reasonable range of each band of the sample is written as:

$$[Q_1 - 1.5IQR \sim Q_3 + 1.5IQR] \tag{3}$$

For model training, sample imbalance will negatively affect the training effect. To obtain the optimal model for crop identification, this study counts the number of samples to obtain the proportion of different crop samples. Proportional replication was performed for samples with a small proportion, and the samples were balanced before training.

To monitor the training situation of the model and verify the accuracy of the model, the sample dataset was randomly divided into a training set and a verification set according to 7:3. The training set was used to train the network, and the validation set was adopted to monitor training and evaluate model performance.

2.3.2. Sample Neighborhood Information Acquisition

Xing'an League is located in the transition zone between the Greater Xing'an Mountains and the Songnen Plain. The cultivated land is fragmented and the fields are scattered. Crop identification faces many interference factors. The existing crop remote sensing recognition algorithms often only employ the grayscale information of pixels without considering the spatial information. Often due to the effect of factors such as noise, partial volume effects, and artifacts, the classification results are inaccurate, and the "salt and pepper phenomenon" occurs.

Existing research suggested that the high correlation between pixels and their neighbors is an essential feature of images [29]. If the neighboring pixels around a pixel are assumed to belong to the same class, the pixel also belongs to this class. Integrating neighborhood information in the classification process can increase the separability between crops for areas with complex crop types and large differences in coverage. Accordingly, in this study, the spectral value and positional relationship of each pixel in the sample vector polygon and its $k \times k$ neighborhood of k^2 pixels in total were saved as text in the order from top to bottom and from left to right. A sample dataset rich in grayscale and spatial information was generated, the anti-noise ability of image segmentation was enhanced, and the effect of crop recognition was effectively improved.

2.3.3. Crop Classification Model Construction

To solve the problem of low accuracy and poor timeliness in large-area crop recognition, this study proposes a crop recognition strategy Skcnn_Tabnet, which uses the tabular network Tabnet as a classifier. By adding the channel attention module to the Tabnet network in the feature extraction stage, the network can pay attention to the spectral features of crops, while placing more stress on the structure and texture of crops. With the use of the soft feature selection mechanism of the Tabnet network, the crop extraction model has a stronger generalization ability and a more reasonable classification effect at the feature classification stage. Moreover, the Convolutional Neural Network (CNN) was used to extract features of different scales, and they were fused with the features extracted by the Tabnet network. The universality of the network was enhanced on remote sensing images of different scales. Compared with conventional machine learning, the complex feature selection module was omitted, resulting in end-to-end training.

Tabnet was originally used to classify tabular data [30]. Based on the performance of decision trees, the network constructed a process with a hyperplane decision manifold similar to decision trees by determining the feature contribution coefficients in the decision-making process. Compared with conventional methods based on Deep Neural Networks (DNN), Tabnet has powerful soft feature selection capabilities in addition to controlling sparsity through sequential attention. For instance, in hyperspectral remote sensing crop classification, Tabnet considers multiple spectral features instead of only focusing on several important red-edge band features. Tabnet's soft feature selection mechanism can avoid complex problems (e.g., "same substance, different spectrum, same spectrum foreign matter") to a certain extent. The basic encoder structure of Tabnet is presented in Figure 5:

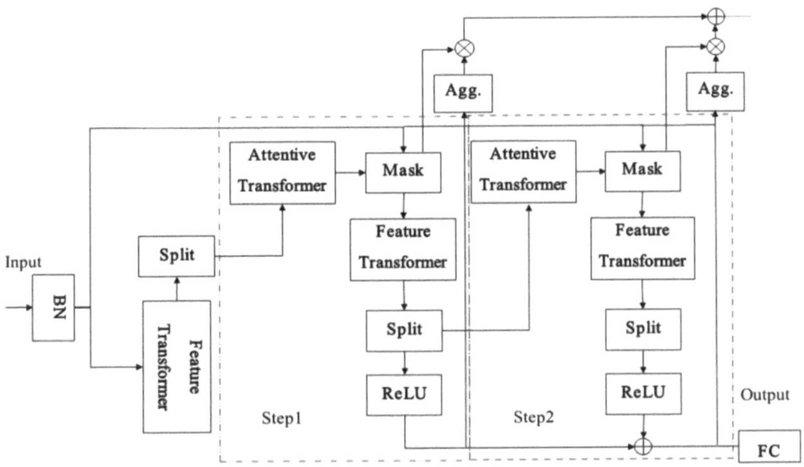

Figure 5. Structure of the Tabnet encoder.BN stands for batch normalization. FC stands for fully connected.

The improvement in this study is mainly to add channel attention to the Attentive transformer module (Figure 6). After the EntMax module, a channel attention module composed of a Maximum Pooling layer (MaxPool), a convolutional layer (Conv2d), and a Fully Connected layer (FC) was added respectively. Where the first convolution is used for channel compression, and the second convolution expanded the channel to input channel C. The sigmoid function was used to map the weights of the C channels between 0 and 1. The channel attention feature map was obtained after multiplying the input feature map with the weights. Lastly, the FC layer generates channel attention feature maps, which were used as input to prior scales to update the abstract features generated by the FC and BN layers within the Attentive transformer. The addition of channel attention reduces the limitations of local perception of convolutional neural networks to a certain extent. Extending single channel to multi-channel optimizes feature extraction and facilitates the improvement of model efficiency and accuracy, which is suitable for hyperspectral remote sensing crop extraction with complex spectral features.

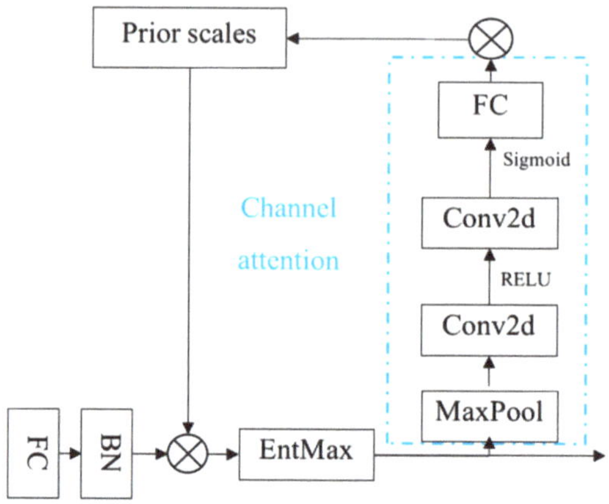

Figure 6. Improved attentive transformer module. Conv2d stands for 2D convolution. EntMax stands for entmax normalization.

2.3.4. Accuracy Evaluation

A total of 30% of the sample data were adopted to examine the classification accuracy of crop types. Four precision evaluation indicators could be obtained: F1 score, overall classification accuracy (OA), precision rate (Precision), and recall rate (Recall). They were adopted to evaluate the precision and compare the classification performance between different models. The specific calculation method is expressed in Equation (4):

$$OA = \frac{\sum_{i=1}^{n} p_{i,i}}{\sum_{j=1}^{n} \sum_{i=1}^{n} p_{i,j}} \tag{4}$$

where $p_{i,i}$ represents the pixel that is classified into the i-th crop and belongs to the i-th crop $p_{i,j}$ denotes the pixel that belongs to the i-th crop and is classified into the j-th crop. OA more effectively represents the overall classification accuracy. By comparing with the sample labels, the total number of correct extractions of crop classification pixels-true positive (TP) total wrong extraction-false positive (FP) and total missing points-false negative (FN), Thus the precision and recall rates of a wide variety of crops are calculated as:

$$Precision = \frac{TP}{TP + FP} \tag{5}$$

$$Recall = \frac{TP}{TP + FN} \tag{6}$$

The F_1 score is an indicator adopted in statistics to measure the accuracy of the classification model. This indicator considers the precision and recall of the classification model simultaneously. It is a harmonious evaluation of the precision and recall. The F_1 score is expressed as follows:

$$F_1 = 2 * \frac{Precision * Recall}{Precision + Recall} \tag{7}$$

3. Experiments and Results

3.1. Neighborhood Size Determination

The neighborhood information of an image has been found as a vital feature for crop recognition. Moreover, the choice of neighborhood size takes on a critical significance to the recognition effect. Excessive neighborhood information will reduce the effect of the central pixel, which may negatively affect the extraction of small fields and boundary points. If the neighborhood information is too small, it cannot be ensured that sufficient features are extracted. During the model building process, the neighborhood information of 3 × 3 pixels, 5 × 5 pixels and 7 × 7 pixels was adopted for the test based on the MLP network, and the model accuracy and the test effect were compared to select the most suitable neighborhood size. Lastly, the optimal neighborhood size was determined as 5 × 5 pixels (Figure 7).

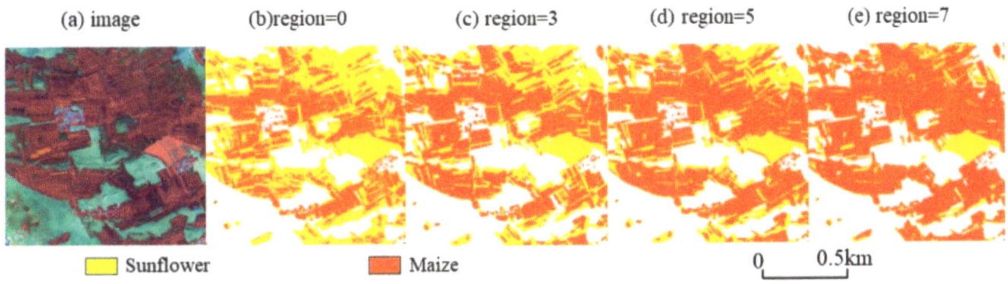

Figure 7. The segmentation effect of different neighborhood sizes based on the MLP network. The region refers to the neighborhood size. The image is 512 × 512 pixels, and is composited with false color of band8, band11, and band4.

The test results showed a considerable number of broken spots before the neighborhood information was added. When the neighborhood size was set to 3 × 3 pixels, the salt and pepper phenomenon was improved, whereas the boundary was still not significant. When the neighborhood size was set to 7 × 7 pixels, numerous sunflowers were mistakenly detected as maize, and the field boundary also showed a corrosion phenomenon. Only when the neighborhood size was 5 × 5 pixels, the sunflowers and maize were accurately distinguished, and the boundary information was effectively extracted.

3.2. Experiments

Three classification algorithms, including the common crop identification algorithm, RF, and two mainstream deep learning segmentation algorithms (UNet and Deeplabv3+), were selected in this study based on the same Sentinel-2 image data in the Xing'an League area to more comprehensively evaluate the performance of Skcnn_Tabnet. Model training was conducted, and the corresponding crop extraction results were predicted. RF is a classification method based on multi-decision tree voting proposed by Breiman [31]. Chen et al. [32] proposed the Deeplabv3+ method, which is a hybrid architecture based on a backbone network and codec, preserving the resolution of feature maps using atrous convolution and extracting features at different scales based on ASPP (atrous spatial pyramid pooling) module. The UNet method was proposed by Ronneberger et al. [33]. UNet is capable of retaining the features of the respective level in the encoder, up-sampling the feature map of the same size as the original image level by level in the decoder, and fusing it with the low-level features of the corresponding level in the encoder.

The software and hardware environment, parameter configuration, loss function, and optimization mechanism of the four network models in this study are consistent. The setting of the respective optimal parameter underwent several parameter adjustments and trials and had errors to ensure the reliability of the experiment. Lastly, the learning rates of the three deep learning models were determined based on the WarmUp strategy and the adaptive learning rate strategy. The initial learning rate was 1e-4 at the WarmUp stage, which was increased to 1e-3 after 10 epochs. At the adaptation stage, when the accuracy of the validation set no longer was increased for 10 consecutive epochs, the learning rate was multiplied by a factor of 0.3. The maximum training epoch was 300 epochs. The loss function was the sum of cross entropy and Lovasz Loss, and the optimizer employed Adam. The key parameter number of estimators for RF was set to 300 with a max depth of 25. To make the accuracy more objective, we randomly trained each model ten times. We computed the average accuracy of each model as the metric of the final accuracy comparison. We also presented the performance variation range (Absolute deviation) with ±.

The best overall accuracy and single-class accuracy of the extraction results corresponding to the four network models were calculated based on pixels in accordance with the accuracy evaluation method proposed in Section 2.3.4 (Table 3).

Table 3. Accuracy comparison of classification results of four different methods.

Method	Accuracy Category	Maize	Sunflower	Rice	Waters	Others	Average
Deeplabv3+	IOU	0.7258 (±0.028)	0.6092 (±0.037)	0.7254 (±0.030)	0.9141 (±0.021)	0.4172 (±0.044)	0.6783
	F1 score	0.8462 (±0.039)	0.7541 (±0.026)	0.8368 (±0.029)	0.9543 (±0.035)	0.5834 (±0.026)	0.7949
	Overall accuracy	0.8149 (±0.031)					
UNet	IOU	0.7650 (±0.043)	0.6476 (±0.046)	0.7461 (±0.027)	0.9226 (±0.033)	0.4477 (±0.038)	0.7058
	F1 score	0.8663 (± 0.029)	0.7827 (± 0.032)	0.8534 (± 0.050)	0.9586 (± 0.027)	0.6176 (±0.033)	0.8157
	Overall accuracy	0.8266 (± 0.038)					
RF	F1 score	0.7684 (±0.051)	0.6798 (±0.060)	0.7503 (±0.026)	0.9244 (±0.031)	0.6706 (±0.042)	0.7587
	Overall accuracy	0.8396 (±0.043)					
Skcnn_Tabnet	IOU	**0.9063** (±0.026)	**0.8432** (±0.027)	**0.8738** (±0.037)	**0.9822** (±0.036)	**0.6951** (±0.029)	**0.8601**
	F1 score	**0.9428** (±0.034)	**0.9103** (±0.029)	**0.9289** (±0.026)	**0.9878** (±0.031)	**0.7562** (±0.028)	**0.9052**
	Overall accuracy	**0.9270** (±0.026)					

Deeplabv3+ optimizes the segmentation effect of objects of different scales by introducing ASPP convolution. However, the overall accuracy is low due to the low classification accuracy of sunflower and other categories. UNet retains more detailed information by fusing context information. Both the single-class accuracy and the overall accuracy of crop recognition have been increased to a certain extent. The accuracy scores of the classification results of Skcnn_Tabnet suggest that the soft feature selection mechanism and channel attention of the Skcnn_Tabnet network can significantly increase the accuracy of crop remote sensing classification results. The overall accuracy of these classification results reaches 0.9270, which is 0.1121, 0.1004, and 0.0874 higher than Deeplabv3+, UNet, and RF methods, respectively. For the single class, the average IoU and F1 Scores of Skcnn_Tabnet for five types of ground objects are 0.1818 and 0.1103 higher than Deeplabv3+, and 0.1543 and 0.0895 higher than UNet. The above analyses reveal that the Skcnn_Tabnet network is highly promising in the field of crop remote sensing classification.

Four sets of local images in Xing'an League are selected in this study from the perspectives of multi-type mixed distribution, field size, and complex terrain to further evaluate and analyze the detailed characteristics of crop remote sensing classification results. The local results of the four network models in crop remote sensing classification (Figure 8) were compared and studied based on the standard false-color sentinel-2 images and referring to auxiliary data (e.g., Google Earth images). In order to show clearer details, we use small tiles of 512 × 512 pixels for visual analysis of the results. The first group of constituencies has a variety of crop types (e.g., sunflower, maize, and rice), and maize is primarily distributed in contiguous patches. The Skcnn_Tabnet method outperforms the other three methods to extract small plots of rice and sunflower mixed in the maize planting area. Deeplabv3+ and UNet misclassify rice as sunflower, whereas Skcnn_Tabnet is capable of accurately identifying rice. The second group of constituencies is relatively neat and has clear boundaries, mainly rice. The other three methods exhibit different degrees of corrosion in extracting field boundaries, and the field roads are wrongly divided into rice. In addition, the extraction effect of the detailed features of the field boundaries is significantly lower than that of Skcnn_Tabnet. The third group of constituencies is mountainous areas exhibiting complex topographies, of which a small amount of cultivated land and small water bodies are distributed in the valleys. The classification results showed that Deeplabv3+ misclassified numerous mountain shadows into sunflowers and maize, and the small water bodies between the valleys were not extracted. UNet and RF methods misclassified small water bodies as rice. The fourth group of constituencies is dominated by strip-shaped fields, in which some river water bodies and rice along the banks are also included. Except for the Skcnn_Tabnet classification results closest to the original images, the other three methods have significant errors in extracting field and water boundaries. The other three methods all misclassified a small number of unplanted or harvested fields as maize, and the details (e.g., the inner ridge of the field) are not as finely indicated as the Skcnn_Tabnet method.

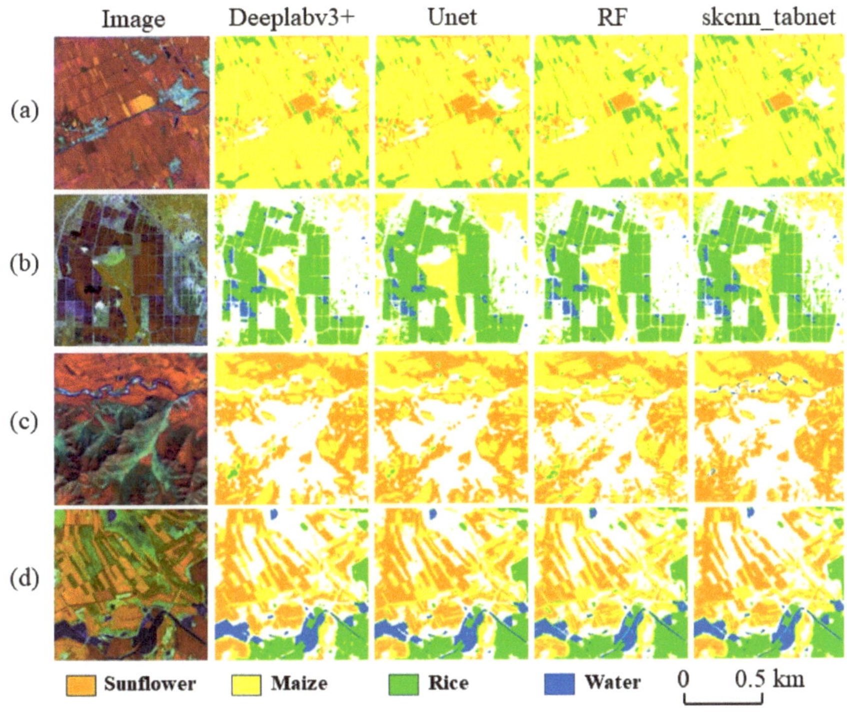

Figure 8. Some examples of the results on the Sentinel-2 data. Comparison between our skcnn_tabnet and other methods. The image is 512 × 512 pixels, and is composited with false color of band8, band11, and band4. (**a**) Multi-Type mixed distribution. (**b**) Neatly distributed area of fields. (**c**) Mountainous areas. (**d**) Strip distribution area of fields.

In general, the cultivated land in the target area is complex, with sunflowers and corn staggered, and numerous small fields exist in the form of broken spots. The other three methods cannot effectively extract small fields. Notably, sunflowers in many corn fields were misclassified by the UNet model. Moreover, Skcnn_Tabnet is capable of extracting small fragmented fields. The reason for this finding is the addition of channel attention to the network, making the network more sensitive to the feature differences between corn and sunflower. Thus, the accuracy of crop remote sensing classification results is increased. In addition, the decoders in Deeplabv3+ and UNet networks lose boundary detail information during the upsampling process. As a result, the extraction results were gradually over-smoothed, and the tiny roads in some fields were corroded or misclassified as sunflowers. Skcnn_Tabnet is capable of extracting slender roads and ridges due to the soft feature selection mechanism and multi-scale feature extraction of the Tabnet network. The adaptive receptive field of the model is achieved, thus effectively increasing the overall classification accuracy.

3.3. Accuracy Verification

Five main grain-producing areas and counties (Dengkou County, Hangjinhou Banner Linhe District, Wuyuan County, and Wulateqian Banner) in the Hetao Plain were selected for the crop extraction experiments in the same phenological period and verify the effect, so as to verify the application ability of Skcnn_Tabnet in large-scale space. The distribution and area of sunflower, maize, and rice in the Hetao area in 2019 were examined and compared with the spatial distribution of the crop statistical area. The extraction results are presented in Figure 9.

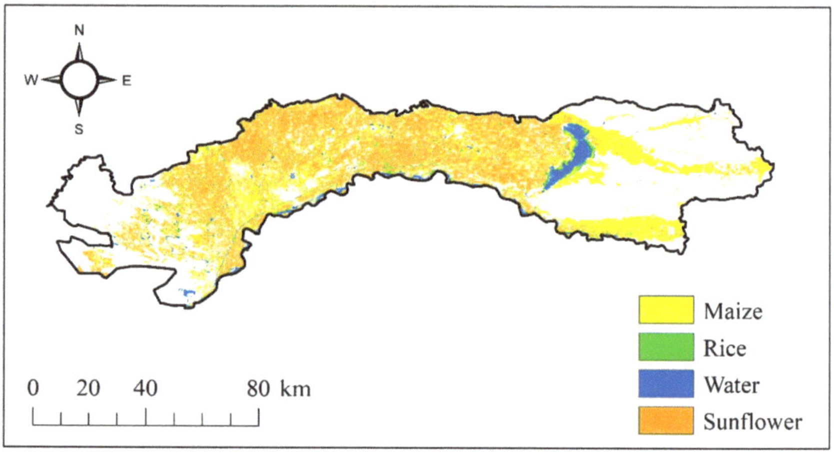

Figure 9. Distribution map of crop types in the Hetao Plain in 2019.

In general, sunflowers are planted in a large area, spread over the entire study area, primarily in connected plots, and some are cross-planted with corn. The corn planting areas are concentrated largely in the northern part of the Hetao Plain, the central part of the Linhe District, and the coast of Wuliangsuhai Lake. The rice planting area is small and relatively scattered in the Yellow River and its tributaries, lakes, and other water-rich basins (e.g., Shuanghe Town in Linhe District, Dengkou County, Fuxing Town in Wuyuan County, as well as other counties and cities). The planting areas of the three crops in the Hetao area were obtained as 1734.76 km^2 for corn, 2743.38 km^2 for sunflower, and 118.53 km^2 for rice by calculating the pixel points of each crop in ArcGIS. To further verify the extraction accuracy, the data were found in this study (e.g., the 2019 Inner Mongolia Statistical Yearbook and the 2019 Bayannaoer City Statistical Yearbook (http://tjj.bynr.gov.cn, accessed on 15 June 2022)), thus indirectly verifying the validity of this study. The comparison result suggests that the regional proportions of rice and corn planting areas and statistical areas extracted by the Skcnn_Tabnet model are nearly the same. The sunflower area is 386.52 km^2 more than the statistical area, and the relative error is slightly larger.

The survey suggests that Northeast China has implemented a policy of adjusting the planting area and structure of crops over the past few years, thus encouraging different crop rotation and interplanting patterns [34]. There are a considerable number of sunflower-soybean intercropping and intercropping patterns in Linhe District, Wuyuan County, and Wulateqian Banner. Considerable sunflowers may be misclassified as soybeans since the phenological and spectral characteristics of sunflowers and soybeans in the Hetao area are highly similar, thus reducing the accuracy of remote sensing classification.

4. Discussion

This study was based on single-phase Sentinel-2 images and a small number of samples. The optimal crop identification model was transferred to the Hetao Plain to identify crops in the same period. The crop distribution map of the Hetao Plain in 2019 was generated pixel by pixel. In addition, the statistical yearbook data verification suggests that the overall verification accuracy of the crop identification model in the Hetao area has reached 85%. In this paper, the single-phase Sentinel-2 image was used instead of the long-time series images, which provided a method reference for crop recognition, especially under long-term rainy weather in Southern China. For example, the flood disaster in Henan Province in 2020 caused a large area of crop disasters, and the compensation work of affected farmers often needs to be combined with remote sensing data statistics. However, the long-time rainy weather made it difficult to obtain the available long-time series remote sensing images. In this case, the advantages of the proposed method were reflected, which only needed remote sensing data of a single-phase to realize crop recognition. In terms of crop growth analysis, we often judged crop growth according to changes in NDVI data. However, it did not distinguish which crops were growing. This study can identify the crop species and grasp the growth situation of various crops. In agricultural insurance claims, this method can assist agricultural insurance companies to ensure the rationality and fairness of insurance claims by providing objective and real crop growth situations and area data.

Compared to the method that generates "training samples" based on historical information [35], our method uses the current year sample and its extended samples to ensure that the trained crop extraction model is more accurate. Due to the differences in inter-annual environment, inter-annual spectral curves of the same crop can be inconsistent. Applying the sample data of the classified years to this link can avoid the influence of the differences in the spectral curves. Compared with methods that only use spectral curves [36], our method considers both spectral information and neighborhood information, which can increase the discrimination of crops with similar spectral curves. We compared this paper with the research of You's team [5], which produced three typical crop classification products in Northeast China based on GEE platform long-sequence Sentinel-2 images. In this study, the same recognition effect can be achieved without long sequence images, thus increasing crop recognition efficiency. Moreover, the effect of cloudy and rainy weather on the research was eliminated. During the production of the Dong Crop Map, 22,171 samples were used in Northeast China for model training and testing in 2019. Its classification process is highly complex and comprises a feature selection process, RF classifier training for the respective agroclimatic region, and then the identification of the farmland and the identification of the crops within the field. Although the overall validation accuracy of this study is slightly lower than the former (87%), 126 samples collected in the wild were only used, and one model was only trained to achieve multi-classification, which makes the classification process more concise, thus confirming the feasibility of the method proposed in this study for large-scale crop identification applications. This study provides a technical reference for achieving the automatic national crop census and the sustainable development of cultivated land resources.

5. Conclusions

Based on single-phase Sentinel-2 images and a small number of samples, this study applies the improved Skcnn_Tabnet network to crop classification mapping for the first time, and compared the crop mapping results of three different network models. The results show that: (1) The Skcnn_Tabnet method after adding channel attention has the optimal effect in the application of crop classification and extraction in the cultivated land area of Inner Mongolia. In this study, RF, DeepLabv3+, and Unet all have F1 less than 85%, whereas Skcnn_Tabnet's F1 score and ACC accuracy are higher than 90%. (2) Crop recognition based on single-phase Sentinel-2 images confirmed that adding 5×5 pixels neighborhood information based on the spectral information can significantly increase the overall crop classification effect. (3) A small amount of training data was employed in this study for large-area crop recognition, verifying the spatial scalability and robustness of the Skcnn_Tabnet model. The result suggests that the county-level spatial scale exhibits high applicability in the Hetao area. The crop planting area obtained by the model is well consistent with the statistical data, which reveals that the classification method proposed in this study can meet the requirements of refined crop extraction in large areas. The research results achieved in this study can provide scientific, technical, and data support for the pattern of cultivated land resources and the optimization of agricultural structure in the floodplain.

Author Contributions: J.T. wrote the manuscript, designed the methodology, and conducted experiments; Z.C. and X.Z. supervised the study and reviewed the manuscript; Y.B. pre-processed the remote sensing images and municipal statistical data; J.T. made the datasets. All authors have read and agree to the published version of the manuscript.

Funding: This work is supported by China high-resolution earth observation system (Grant No. 03-Y30F03-9001-20/22) and the National Natural Science Foundation of China (Grant No. 42071407

Institutional Review Board Statement: Not applicable.

Informed Consent Statement: Not applicable.

Data Availability Statement: Not applicable.

Acknowledgments: The authors thank the editors and anonymous reviewers for their valuable comments, which greatly improved the quality of the paper.

Conflicts of Interest: The authors declare no conflict of interest.

References

1. Yu, B.; Shang, S. Multi-year Mapping of Maize and Sunflower in Hetao Irrigation District of China with High Spatial and Temporal Resolution Vegetation Index Series. *Remote Sens.* **2017**, *9*, 855. [CrossRef]
2. Li, Z.; Bagan, H.; Yamagata, Y. Analysis of Spatiotemporal Land Cover Changes in Inner Mongolia Using Self-organizing Map Neural Network and Grid Cells Method. *Sci. Total Environ.* **2018**, *636*, 1180–1191. [CrossRef] [PubMed]
3. Yang, Q.; Zhang, D. The Influence of Agricultural Industrial Policy on Non-grain Production of Cultivated Land: A Case Study of the "one Village, One Product" Strategy Implemented in Guanzhong Plain of China. *Land Use Policy* **2021**, *108*, 105579. [CrossRef
4. Ibrahim, E.S.; Rufin, P.; Nill, L.; Kamali, B.; Nendel, C.; Hostert, P. Mapping Crop Types and Cropping Systems in Nigeria with Sentinel-2 Imagery. *Remote Sens.* **2021**, *13*, 3523. [CrossRef]
5. You, N.; Dong, J.; Huang, J.; Du, G.; Zhang, G.; He, Y.; Yang, T.; Di, Y.; Xiao, X. The 10-m Crop Type Maps in Northeast China During 2017–2019. *Sci. Data* **2021**, *8*, 1–11. [CrossRef] [PubMed]
6. Tian, H.; Huang, N.; Niu, Z.; Qin, Y.C.; Pei, J.; Wang, J. Mapping Winter Crops in China with Multi-source Satellite Imagery and Phenology-based Algorithm. *Remote Sens.* **2019**, *11*, 820. [CrossRef]
7. Li, X.; Sun, C.; Meng, H.; Ma, X.; Huang, G.; Xu, X. A Novel Efficient Method for Land Cover Classification in Fragmented Agricultural Landscapes Using Sentinel Satellite Imagery. *Remote Sens.* **2022**, *14*, 2045. [CrossRef]
8. He, Z.; Li, S.; Wang, Y.; Dai, L.; Lin, S. Monitoring Rice Phenology Based on Backscattering Characteristics of Multi-temporal Radarsat-2 Datasets. *Remote Sens.* **2018**, *10*, 340. [CrossRef]
9. Jiang, L.; Shang, S.; Yang, Y.; Guan, H. Mapping Interannual Variability of Maize Cover in a Large Irrigation District Using a Vegetation Index–phenological Index Classifier. *Comput. Electron. Agric.* **2016**, *123*, 351–361. [CrossRef]
10. Ming, Z.; Qing, B.Z.; Zhong, X.C.; Jia, L.; Yong, Z.; Chongfa, C. Crop Discrimination in Northern China with Double Cropping Systems Using Fourier Analysis of Time-series Modis Data. *Int. J. Appl. Earth Obs. Geoinf.* **2008**, *10*, 476–485. [CrossRef]
11. Johnson, M.D.; Hsieh, W.W.; Cannon, A.J.; Davidson, A.; Bédard, F. Crop Yield Forecasting on the Canadian Prairies by Remotely Sensed Vegetation Indices and Machine Learning Methods. *Agric. For. Meteorol.* **2016**, *218*, 74–84. [CrossRef]
12. Abubakar, G.A.; Wang, K.; Shahtahamssebi, A.; Xue, X.; Belete, M.; Gudo, A.J.A.; Shuka, K.A.M.; Gan, M. Mapping Maize Fields by Using Multi-temporal Sentinel-1a and Sentinel-2a Images in Makarfi, Northern Nigeria, Africa. *Sustainability* **2020**, *12*, 2539. [CrossRef]
13. You, N.; Dong, J. Examining Earliest Identifiable Timing of Crops Using All Available Sentinel 1/2 Imagery and Google Earth Engine. *ISPRS J. Photogramm. Remote Sens.* **2020**, *161*, 109–123. [CrossRef]
14. Wang, S.; Azzari, G.; Lobell, D.B. Crop Type Mapping Without Field-level Labels: Random Forest Transfer and Unsupervised Clustering Techniques. *Remote Sens. Environ.* **2019**, *222*, 303–317. [CrossRef]
15. Htitiou, A.; Boudhar, A.; Chehbouni, A.; Benabdelouahab, T. National-scale Cropland Mapping Based on Phenological Metrics, Environmental Covariates, and Machine Learning on Google Earth Engine. *Remote Sens.* **2021**, *13*, 4378. [CrossRef]
16. Wang, X.; Zhang, J.; Xun, L.; Wang, J.; Wu, Z.; Henchiri, M.; Zhang, S.; Zhang, S.; Bai, Y.; Yang, S.; et al. Evaluating the Effectiveness of Machine Learning and Deep Learning Models Combined Time-series Satellite Data for Multiple Crop Types Classification Over a Large-scale Region. *Remote Sens.* **2022**, *14*, 2341. [CrossRef]

17. Shelestov, A.; Lavreniuk, M.; Kussul, N. Large Scale Crop Classification Using Google Earth Engine Platform. In Proceedings of the 2017 IEEE International Geoscience and Remote Sensing Symposium (IGARSS), Worth, TX, USA, 23–28 July 2017; pp. 3696–3699.
18. Zhang, C.; Di, L.; Lin, L.; Li, H.; Guo, L.; Yang, Z.; Yu, E.G.; Di, Y.; Yang, A. Towards Automation of In-season Crop Type Mapping Using Spatiotemporal Crop Information and Remote Sensing Data. *Agric. Syst.* **2022**, *201*, 103462. [CrossRef]
19. Ofori-ampofo, S.; Pelletier, C.; Lang, S. Crop Type Mapping from Optical and Radar Time Series Using Attention-based Deep Learning. *Remote Sens.* **2021**, *13*, 4668. [CrossRef]
20. Mo, Y.; Wu, Y.; Yang, X.; Liu, F.; Liao, Y. Review the State-of-the-art Technologies of Semantic Segmentation Based on Deep Learning. *Neurocomputing* **2022**, *493*, 626–646. [CrossRef]
21. Chen, Y.; Song, X.; Wang, S. Impacts of Spatial Heterogeneity on Crop Area Mapping in Canada Using Modis Data. *J. Photogramm. Remote Sens.* **2016**, *119*, 451–461. [CrossRef]
22. Du, M.; Huang, J.; Wei, P.; Yang, L.; Chai, D.; Peng, D.; Sha, J.; Sun, W.; Huang, R. Dynamic Mapping of Paddy Rice Using Multi-temporal Landsat Data Based on a Deep Semantic Segmentation Model. *Agronomy* **2022**, *12*, 1583. [CrossRef]
23. Der, Y.M.; Tseng, H.H.; Hsu, Y.C. Real-time Crop Classification Using Edge Computing and Deep Learning. In Proceedings of the 2020 IEEE 17th Annual Consumer Communications & Networking Conference (CCNC), Las Vegas, NV, USA, 10–13 January 2020; pp. 1–4.
24. Wang, C.; Zhang, R.; Chang, L. A Study on the Dynamic Effects and Ecological Stress of Eco-environment in the Headwaters of the Yangtze River Based on Improved Deeplabv3+ Network. *Remote Sens.* **2022**, *14*, 2225. [CrossRef]
25. Yang, L.-T.; Zhao, J.-F.; Jiang, X.-P.; Wang, S.; Li, L.-H.; Xie, H.-F. Effects of Climate Change on the Climatic Production Potential of Potatoes in Inner Mongolia, China. *Sustainability* **2022**, *14*, 7836. [CrossRef]
26. Zhang, X.; Guo, P.; Zhang, F.; Liu, X.; Yue, Q.; Wang, Y. Optimal Irrigation Water Allocation in Hetao Irrigation District Considering Decision Makers' Preference under Uncertainties. *Agric. Water Manag.* **2021**, *246*, 106670. [CrossRef]
27. Shen, H.; Lin, L.; Li, J.; Yuan, Q.; Zhao, L. A Residual Convolutional Neural Network for Polarimetric Sar Image Super-resolution. *ISPRS J. Photogramm. Remote Sens.* **2020**, *161*, 90–108. [CrossRef]
28. Haines, E. Point in Polygon Strategies. *Graph. Gems* **1994**, *4*, 24–46.
29. Zhang, L.; Zhang, Q.; Du, B.; Huang, X.; Tang, Y.Y.; Tao, D. Simultaneous Spectral-spatial Feature Selection and Extraction for Hyperspectral Images. *IEEE Trans. Cybern.* **2016**, *48*, 16–28. [CrossRef] [PubMed]
30. Arik, S.Ö.; Pfister, T. Tabnet: Attentive Interpretable Tabular Learning. In Proceedings of the AAAI Conference on Artificial Intelligence, Vancouver, BC, Canada, 2–9 February 2021; pp. 6679–6687.
31. Breiman, L. Random Forests. *Mach. Learn.* **2001**, *45*, 5–32. [CrossRef]
32. Chen, L.; Zhu, Y.; Papandreou, G. Encoder-decoder with Atrous Separable Convolution for Semantic Image Segmentation. In Proceedings of the European Conference on Computer Vision (ECCV), Munich, Germany, 8–14 September 2018; pp. 801–818.
33. Ronneberger, O.; Fischer, P.; Brox, T. U-net: Convolutional Networks for Biomedical Image Segmentation. In Proceedings of the Medical Image Computing and Computer-Assisted Intervention–MICCAI 2015, Munich, Germany, 5–9 October 2015.
34. Yang, L.; Wang, L.; Huang, J.; Mansaray, L.R.; Mijiti, R. Monitoring Policy-driven Crop Area Adjustments in Northeast China Using Landsat-8 Imagery. *Int. J. Appl. Earth Obs. Geoinf.* **2019**, *82*, 101892. [CrossRef]
35. Hao, P.; Wang, L.; Zhan, Y.; Wang, C.; Niu, Z.; Wu, M. Crop Classification Using Crop Knowledge of the Previous-year: Case Study in Southwest Kansas, USA. *Eur. J. Remote Sens.* **2016**, *49*, 1061–1077. [CrossRef]
36. Siesto, G.; Fernández-Sellers, M.; Lozano-Tello, A. Crop Classification of Satellite Imagery Using Synthetic Multitemporal and Multispectral Images in Convolutional Neural Networks. *Remote Sens.* **2021**, *13*, 3378. [CrossRef]

MDPI AG
Grosspeteranlage 5
4052 Basel
Switzerland
Tel.: +41 61 683 77 34

Sustainability Editorial Office
E-mail: sustainability@mdpi.com
www.mdpi.com/journal/sustainability

www.ingramcontent.com/pod-product-compliance
Lightning Source LLC
Chambersburg PA
CBHW041400210926
51850CB00013B/3143
9783725826100